Basic Design · Design Basics

基础设计·设计基础

[瑞士] 赫伯特 · 克莱默 顾大庆 吴佳维 著 | Herbert Kramel Gu Daqing Wu Jiawei

中国建筑工业出版社 | China Architecture & Building Press

前言一

顾大庆

Preface I

Gu Daqing

关于建筑设计的入门课程，大家普遍接受的一个认识就是在正式进入建筑设计之前，我们先要掌握与做建筑设计相关的一些基本的概念、知识和技能。据此，一般的设计基础课程的结构安排都是这样的：先是一系列有针对性的专项训练，最后以一个建筑设计作业作为结束。但是，什么才是建筑设计的基础呢？不同的时代有不同的解答。就我们国内的建筑教育而言，1980年代前基础训练都是围绕表现技法来展开的，如字体练习、作图练习、渲染练习等。而1980年代后的基础训练则转向设计形式语言的训练，典型的是三大构成（平面构成、立体构成和色彩构成）抽象练习。到了1990年代以后，设计基础训练的内涵不断扩展，训练的内容更加多样化。空间抽象练习、1:1的实体搭建、装置设计等开始流行。但是，无论怎么变化，课程的基本结构却依然保持不变，这表明人们对建筑设计基础课程应该如何教已经形成了一个根深蒂固的观念，即没有必要的基本观念、知识和技能方面的准备，学生是不能够直接做建筑设计的。

我在1987年去苏黎世联邦理工学院之前从来没有想到过建筑设计基础课程可以直接从建筑设计开始。那时候我已经做了两年的老师，还开始在东南大学筹划了一个新的设计基础课程，但是当时的思考还是局限于单项的练习，特别是抽象的空间练习。所以，当开始接触到克莱默教授的基础课程时，对我来说是看到了一个全新的、未知的可能性。 哇，原来建筑设计基础课程还可以这样来教！

在克莱默教授的基础课程教案中，一年级新生设计课的第一天就要开始做设计，还不是设计一个建筑物，而是设计一个建筑群的总体布局！学生用一堆大小两种尺寸的蓝色卡纸折成的“U”形立方体模件在一块A2的基地上设计一个划艇俱乐部的总体布局。这个作业要在设计课当日完成。在学生对建筑学完全无知，没有给予充分信息以及缺乏必要的模型制作和作图技能训练的情况下，他们又是如何做到的呢？对于已经是老师的我来说，在与瑞士的一年级新生做着同样的练习时，想的却是如何教和如何学的问题。这个完全新鲜的体验使我开始对以往形成的某些关于建筑设计基础教学的既定观念产生质疑。比如，没有模型制作和作图的技能其实并不重要，学生可以在完成设计作业的过程中学习如何做模型和画图。没有充分的场地设计的知识也没有什么关系，学生完全可以凭借他们的生活经验来完成这个看似不可能的任务。

这个一天的场地设计练习有一个重要的功能，即为一个学年的设计训练奠定一个大框架。1987/88学年教案的大题目是苏黎世湖边的划艇俱乐部，有4个设计课题（阶

One common understanding of the introductory course in architectural design is that before embarking on formal architectural design, we must first master some basic concepts, knowledge and skills related to architectural design. Based on this assumption, the structural arrangement of most design foundation courses is as follows: first, a series of exercises on singular issues and skills, followed by a concluding architectural design project. However, what is the basis of architectural design? Different times have occasioned different answers. As far as China's architectural education is concerned, before the 1980s, basic training had been conducted around representational techniques such as font writing, drawing exercises, and rendering exercises. The foundation course after the 1980s had turned to the training of formal languages, typically the abstract exercises of Sanda Goucheng (three abstract courses on form-making: 2D composition, three-dimensional composition, and color composition). After the 1990s, the connotation of basic training for design has been continuously expanded and the training content has become more diversified. Space abstraction exercises, 1:1 scale construction, and installation design have become popular. However, regardless the changes, the basic structure of the curriculum has remained unchanged, which shows that people have long adhered to an entrenched concept of how basic architectural design courses should be taught: students cannot directly deal with architectural design without a substantial foundation of basic concepts, knowledge and skills.

Before I went to ETH Zurich in 1987, I had never thought that the fundamentals of architectural design could begin directly with architectural design. At that time, I had already worked as a teacher for two years and started a new design foundation course at Southeast University. However, my interest at that time was still limited to individual exercises, especially abstract exercises on space. So, when I joined Prof. Kramel's introductory courses, it opened up a brand new and unknown possibility for me. I thought to myself, "Wow, the architectural design foundation course could also be taught in this way!"

In Prof. Kramel's basic course, the freshmen were confronted with a design problem in their first studio day, not for the design of a building, but that of the overall layout of a building complex! Students were given the task of making a layout plan for a rowing club with a set of blue cardboard U-shaped cubic modules of two dimensions. The task had to be completed before the end of the studio session. How could the students do it when they had been basically ignorant of architecture, lacking adequate knowledge as well as the necessary skills in model-making and drafting? For me, already a teacher at the time, my main concern as I was doing the same exercise as a freshman, was how to teach and how to learn. This completely fresh experience led me to begin to question some of the established ideas about the teaching of architectural foundation courses. For example, I began to realize that model-making skills and drawing skills are not really the prerequisites for design. Students can learn how to make models and draw drawings as they work on design problems. A lack of basic knowledge on site planning is also not a critical issue as students can draw on their life experience to complete this seemingly impossible task.

This first-day design exercise on site layout had an important function, which was to establish a framework for the design training of the whole academic year. The major assignment of the 1987-88 year program was to design a lakeside rowing club in Zurich, which com-

段）：入口、宿舍、瞭望塔和划艇会所。这些建筑设计课题的选择和编排都有特定的训练目的，每个设计过程细化到每次课的具体任务，这些任务要解决特定的设计问题，要用到特定的设计知识，要使用特定的设计媒介。因为每个设计的步骤都有清晰的规定，难易程度适合大多数学生的能力水平，故而是可以完成的。一个接一个的练习做下来，最后学生就能够完成一个复杂的设计任务。而且，每一个设计任务又通过场地而相互关联。前一个完成的设计可能对下一个设计产生影响，最初看似很幼稚的场地设计随着每个设计项目的推进而不断改进。

可以想见这样的一个设计基础教学教案是需要非常精巧的设计的。那么，克莱默教授是如何来构思这套教学体系的，其理论和方法学基础是什么，教学的实际操作是如何实施的，教学的演变过程又是如何？这些问题可以在本书中找到答案。本书的目的就是全面梳理克莱默教授在苏黎世联邦理工学院的这套设计基础教学体系。本书的产生有两个动因，2012年我获得了香港研究资助局的经费资助，研究设计教学法，其中的一个子课题是关于基础课程的结构问题，克莱默教授的教案是一个重要的案例研究对象。此外，吴佳维博士的研究课题是东南大学和苏黎世联邦理工学院的交流作为当代教学法移植的一个典型案例。在这两个因素的推动下，吴佳维获得香港中文大学研究生院的资助在苏黎世逗留了近一年，通过深入访谈，文献调查以及资料收集就这个专题做了深入的调研。现在我们终于能够完成此书。

本书的题目《基础设计 · 设计基础》是克莱默教授经常挂在嘴边上说的一个词组，体现了他对设计基础教学的精辟概括。“基础设计”在西方建筑教育的语境中是有着特定含义的，它指的就是包豪斯的设计基础课程，霍斯利于1959年从美国回苏黎世联邦理工学院执教，也将一年级的设计基础课程更名为“基础设计”。其本意在包豪斯是设计的先修课程，目的是帮助学生选择专业。而在霍斯利的课程中最鲜明的特色就是那些抽象的空间练习。于是，“基础设计”的一般理解是“作为设计的基础”，而不是设计本身。克莱默教授对此的理解有所不同，“基础设计”是指简单、低复合度的设计。也就是说这个课程所针对的始终是作为一个整体的建筑设计。相应地，课程所涉及的“设计基础”也必须基于这个整体的概念。

本书由四章所组成。第一章“原理”是克莱默教授对其教育思想的梳理，包含三个部分，“设计原理”阐述了他对建筑设计和建筑教育的一些基本观点，“结构化的基本建筑设计”解释了他的课程讨论建筑设计问题的三个层次，“三要素”则是指明了他

prised four projects (phases): entrance, dormitory complex, watchtower, and rowing clubhouse. The selection and arrangement of these building projects had specific teaching objectives. Each project was broken down into several stages, each of which focused on solving a specific design problem, using specific design knowledge as well as specific design media. Students could complete these tasks because each design stage had been clearly defined and the degree of difficulty carefully calculated to suit most students' ability level. One step after another, students could eventually complete a complex design task. Moreover, all four design projects were related to each other through the site. The previous completed design may have had an impact on the next design. The seemingly naïve site design in the beginning continued to improve after successive design projects throughout the whole year.

It can be imagined that such an architectural foundation course needs a very delicate design. So, how did Prof. Kramel come up with this teaching system? What was its theoretical and methodological basis? How was the program implemented? How did the teaching evolve? Answers to all these questions can be found in this book. The purpose of this book is to fully grasp Prof. Kramel's design of the foundation program at ETH. There are two motivations for the production of this book. In 2012, I received a grant from the Research Grants Council of the Hong Kong Special Administrative Region to study design pedagogy. One of the sub-topics was on the structure of the basic curriculum. Prof. Kramel's foundation course was selected for the case study. In addition, Dr. Wu Jiawei discussed the exchange program between SEU and ETH as an exemplary case of contemporary teaching transplantation in her Ph.D dissertation. Driven by these two factors, Wu Jiawei spent nearly a year in Zurich with a grant from the Graduate School of the Chinese University of Hong Kong and conducted in-depth investigations on this topic with interviews, literature surveys, and data collection. Now, we are able to finally compile the materials into the current book.

The title of this book, "Basic Design·Design Basics," is a phrase that Prof. Kramel often has on his lips, embodying his ingenious interpretation of architectural fundamentals. "Basic Design" has a specific meaning in the context of Western architectural education. Originally, it refers to the Bauhaus Vorkurs. When Prof. Bernhard Hoesli returned from the United States to ETH in 1959, he restructured the foundation course with the new title "Basic Design." At the Bauhaus, the purpose of Vorkurs was to prepare students to choose their most suitable subject in the upper year. The most distinctive feature of Hoesli's course is those abstract exercises on space. Thus, the general understanding of "Basic Design" is "the basis for design" but not design itself. For Prof. Kramel, "Basic Design" refers to architectural design at the level of low-complexity. In other words, this course has always emphasized architectural design as a whole. Correspondingly, the "Design Basics" involved in the course must also be based on this overall concept.

This book consists of four chapters. The first chapter "Principles" is about Prof. Kramel's educational thoughts. It consists of three parts. "Principles of Design" presents some of his basic views on architectural design and architectural education. In "Structured Basic Architectural Design," he explains the three levels of architectural design problems dealt with in his course. In "The Three Elements," he articulates his definition of architectural design. The second chapter "Pedagogy" is about Prof. Kramel's teaching methods. It consists of two parts. In "Pedagogic Model," he discusses his views on the

对建筑设计问题的定义。第二章“教学”是克莱默教授对其教学法的梳理，包含两个部分，“教学模型”讨论了他的教学实践所处的专业和对教育环境的系统性看法，“教学过程”则是他对基础课程教学整体教案设计的一些原则性思考。第三章“教案”通过具体的实例来呈现具体的教案，“课程结构”是一个典型的教案的基本架构和内容，“演化”是对历年教案的梳理以呈现一个教学模型在一个长时期的演进过程，“实例”通过三个作业来呈现教学的实际操作。第四章“回响”则是对克莱默教授的教学法传到中国后的一个简单的描述，包含四个部分，“南京交流”简述了东南大学和苏黎世联邦理工学院的交流历史，“对话”收集了几位交流参与者对克莱默教授教学的个人体会，而“影响”则呈现了东南大学、南京大学和香港中文大学的不同教案，从中可以看到与克莱默教授的渊源关系。“解读”是一篇关于克莱默教授的教学思想渊源、教学法及教案设计和实施的系统解读，作为本书的结束。

本书的研究过程并非如预料的一帆风顺。由于克莱默教授已经退休多年，资料散失严重，也给本书的内容带来些许遗憾。比如，实际的教学是以模型来推动的，可惜的是这个特点由于素材不足而不能凸显出来。此外，书中呈现的几个教学的作业案例都来自于东南大学的老师，这也是缺乏学生作业案例完整资料的缘故，不过无意中为本书增添了“中国”特色。

我是1987年10月去苏黎世联邦理工学院，与克莱默教授相识，从此而展开了一段长达三十多年的友谊。东南大学与苏黎世联邦理工学院的交流从1986年开始至2011年结束，在这期间33位东大的青年老师中有20位接受过克莱默教授的指导。在三十年前，一位远在瑞士的教授愿意为另一个遥远的国度培养年轻设计教师而付出大量心血，现在看来实在是非常有远见之举。记得当年克莱默教授要在ETH建筑系办一个关于两校交流的展览，我们设想一个有中国元素的标题，最后借用了中国成语“立竿见影”，以一个很业余的翻译“Big Stick Big Shadow”来强调这个交流计划对中国建筑教育的影响，可是那个时候谁也不会预料到这根交流之“竿”在三十年后会在中国投下如此深远的影子。

2018年6月9日

professional and educational contexts of architectural education in which his teaching practice is rooted. "The Teaching Process" is a summary of his pedagogic thoughts on the design of the foundation course. Chapter 3 "Program" presents his course at three levels. First, "Program Structure" explains the exercise sequences and contents of a typical program. Second, "Evolution" is a summary of different versions of the program over a long period of time, showing the evolution of a pedagogic model. Third, "Cases" illustrates the implementation of the program with students' works. The fourth chapter "Resonance" is about the impact of Prof. Kramel's pedagogical approach on China's architectural education. It consists of four parts. "Nanjing Exchange" outlines the history of the exchange program between SEU and ETH. "Dialogues" is a collection of participants' personal reflections on their experiences of Prof. Kramel's teaching. "Impacts" is a collection of foundation programs at SEU, Nanjing University, and the Chinese University of Hong Kong, all of which are related to Prof. Kramel's teaching to different extents. "Interpretation", as the conclusion of the book, is a summary of Prof. Kramel's thoughts on architectural education, his design pedagogy, and his program and implementation.

The research process behind this book has not been as smooth as expected. Since Prof. Kramel has retired for many years, resulting in a severe loss of information, which has brought some regrets to the book in terms of content. For example, the actual teaching was driven by physical models. Unfortunately, this feature cannot be highlighted due to the lack of photographic materials. In addition, several examples of teaching assignments presented in the book came from SEU teachers due to the lack of complete information on students' work cases. However, it has inadvertently imparted a "Chinese" characteristic to the book.

I met Professor Kramel in October 1987 as a visiting scholar to ETH. From then on, we have built up a friendship that has lasted more than 30 years. The exchange program between SEU and ETH began in 1986 and ended in 2011. In this long period, nearly 20 of 33 junior teachers from SEU had been taught by Prof. Kramel at ETH. Thirty years ago, a professor from Switzerland was willing to devote a lot of effort to training young design teachers from a distant country. In retrospect, it has proved to be a very far-sighted move. Once, Prof. Kramel wanted to make an exhibition on the exchange program between the two schools. We tried to give it a nice title with a Chinese characteristic. Eventually, we borrowed the Chinese idiom "立竿见影" with an amateurish translation "Big Stick Big Shadow" to emphasize the immediate impact of the exchange program on Chinese architectural education. However, at that time, no one could foresee that the "Big Stick" of this exchange would come to cast such a profound shadow over China after 30 years.

June 9th, 2018

前言二

赫伯特·克莱默

Preface II

Herbert Kramel

关于如何开启学习建筑学的旅程提出了多个问题。

第一年的学习任务是什么？答案因学校的文化背景和历史而异。

第一年是否应该以激发学生创造力为目标？

是否意味着将学生从过往的人生阶段中解脱出来？

艺术学院教学的主导思想往往是把建筑视为艺术。引领学生进入未知领域全赖教师对条件、作业或者练习的构想。此种方法的评价一向褒贬不一，如今大多在英语国家奉行。

另一思路则是将建筑视为一个行业，有明确的惯例、需求及技术。由此入门的话，教师的职责是通过课程及一系列举措，向学生引介其未来的职业。这一方向并不探索学生的潜能，而重传授及指引。这种方法必然包括分辨对错，教学与考试紧密关联。

第三种思路视建筑为一门学问。对学问的求索、尝试逐步获得洞察、知识、方法及经验。这种思路下，知识主体所定义的学习过程意味着逐层递增的复杂性，而非堆砌互不关联的知识包。

或许，还有第四种理解，将第一年视为往后研修生活的开端。那么信息搜集与研究就是首要的教学方式。学习的过程将会非常个人化，相应的教学方式以指导为主。“如何”将取代“什么”成为首要问题，借此磨砺思维，将研究的过程体系化。

将以上种种思路并置便产生了问题。哪种思路最好？什么时期运用哪种思路？孤立地看，对方式的选取仅仅取决于个人偏好。一旦我们保持观察距离，便能发现每个思路都有自身的合理性。这又回到开头的设问——时间、地点、学校的历史、教程的性质以及文化环境共同决定了教学的执行。显而易见的是，选择某一条道路必定会远离其他的道路，因为每条路都通向不同的终点。

包豪斯最先提出了设计入门课程的概念，即“基础课程”（Vorkurs）。第二次世界大战期间及其后，这一概念在美国继续发展。它以“基础设计”的形式广布于众多院校的课程体系中。作为当时得州大学的年轻教员，本纳德·霍斯利在战后的建筑教育进程中发挥作用。成为ETH教授之后，他发展了苏黎世模式的“基础课程”（Grundkurs）。在海因茨·罗纳教授和汉斯·艾斯教授[1]的协助下，霍斯利尝试将他的美国经验与瑞士的具体情形结合起来。作为他的继任者，我肩负了组织新的基础设计课程的责任，并且将建筑设计和建筑技术结合。这一教程从1985/86学年开始，到

[1] 罗纳负责的是构造课，艾斯负责的课程名称为“制图与透视”（Zeichnen, Perspektive），实际授课内容主要是平面视觉设计。

[1] Ronner was responsible for the construction course; Ess was responsible for the course name "Drawing and Perspective" (Zeichnen, Perspektive), yet the actual teaching content was mainly graphic visual design.

The beginning of a course of studies in architecture raises a number of questions.
What is the purpose of the first year of learning in architecture? Depending on the cultural environment as well as the history of the school, the answer will differ.
Is the goal of the first year to open up the creative potential of a student?
Is it meant to free the student from his/her former phases of life?
Architecture as an art is the line of thought that often prevails when taught at an art academy. It is then often up to the teacher to invent conditions/assignments/exercises that provoke the student to enter a yet unknown territory. As usual, there are some arguments for and against this approach— an approach that, in our days, is often taken in English-speaking countries.
Quite a different line of thought is based on the notion of architecture as a profession with clearly defined conventions, demands, and techniques. In this form of introduction, it is the responsibility of the teacher to develop a course and a course of action that introduces the student to his/her future profession. This direction does not explore potentialities but rather teaches and instructs. Distinguishing right from wrong is an essential part of this approach; teaching and examinations are closely related.
The third line of thought looks upon architecture as a discipline. In this case, it involves exploratory learning, the step-by-step acquisition of insight, knowledge, methodologies as well as experiences. In this line of teaching, there is a body of knowledge that defines the field for exploration, and teaching incorporates layers of complexity instead of additive packages of knowledge.
Perhaps, there is a fourth line of thought that takes the first year as an introduction to the rigorous work at the university level. Data collection and research would be the predominant methods of teaching and learning. The learning process then would be highly individualistic, and teaching would take the form of tutoring. Not the what but the how would be in the foreground; the shaping of the mind and the systematization of research processes would be emphasized.
Putting several approaches next to each other raises questions. Which line is the best, and when should one pursue what? Taken individually, it is the personal preference that excludes one in favour of the other. Yet, from a distance, one could also see that there is validity in each line presented. Again, it is the time, the place, the history of the school, the nature of its program, and the cultural environment that determine the course of action. It is clear that going down one road excludes the others. Each road leads to a different destination.
It was at the Bauhaus that the idea of a new introductory course for design, the "Vorkurs", was created. During World War II and in the following years, this idea was further developed in the United States. As "Basic Design", it was instituted in many schools and became an integral part of the architecture curricula. As a young teacher in Texas, Bernhard Hoesli took part in this development. After being elected a professor at ETH, he developed the Zurich model of his "Grundkurs". With the assistance of Heinz Ronner and Hans Ess,[1] Hoesli attempted to apply his American experiences with special consideration of the specific Swiss circumstances. As his successor, I was assigned the responsibility to organize a new Basic Design program, integrating architectural design and design technology. This new program was executed for the first time in the academic year

1995/96学年，不间断地进行了11年。这本书记录了这一历程。

这本小册子的标题是“基础设计 · 设计基础”。它的背后是三名学者关于建筑基础教育的交流。这是一个幸运的组合，30岁的吴博士，55岁的顾教授和80岁的克莱默教授。如果说25年正好是一代人的跨度的话，这便是三代人的相遇。这种层面的交流尤为可贵。我们试图总结和梳理这11年的“基础设计”。

这一过程体现了我们所选择的思路。每一种思路或者模式，都无疑给出了某种“操作”的可能性。一种可能性是以“纯粹”的方式执行一个模式，结果是得益于这个模式特性的同时，以排斥其他模式为代价。另一种可能性是将不同模式进行结合。虽然并不是所有模式都可以相融，却可以用模块的方式将不同方向统一起来，适应具体学校、课程的需求与目标。

对于我们，则是向学生强调建筑为一个行业及一门学科，因此视建筑为艺术或者科学推理的门径将退居次位。这个前提的优势和代价也是显而易见的。重要的是通过选择特定的课程模式，理解可供选择的范围及其包含的“代价”与“收益”。

我们的工作可以总结为：

· 理性的设计过程；
· 高度结构化；
· 教与学的过程；
· 关联概念和建造的重要性；
· 用“无名建筑”检验我们的成果。

这本书是吴佳维博士、顾大庆教授和克莱默教授互动的成果，其背后是长达30年跨度的研究所得。而顾教授是这一切的核心人物，是他使这一切成为可能！

2016年11月

1985/86 and was continuously carried out until 1995/96, as documented in the following pages.
The title of this rather short documentation is "Basic Design · Design Basics". For the preparation of this work, three professionals in the area of Basic Design met for an exchange about educational concerns. It is a most fortunate configuration that a young professional, Dr. Wu, aged 30, met Professor Gu, aged 55, and me (Professor Kramel), aged 80. Between the three of us, we thus represented two generations of concerns. This exchange opportunity across generations is indeed quite outstanding. This documentation is our attempt to sum up the work of 11 years of Basic Design.
Through this documentation one can reflect upon the road taken. Looking at each of these approaches, or models, it becomes apparent that they offer "operational" opportunities. One could execute the model in a "pure" mode, gaining from its properties and paying the price of excluding everything else. The other way to go is by combining different models. Not all models can be combined. However, it is possible to incorporate modules, like parts, of each direction into a more unified form, as befitting the needs and goals of the school and its course.
In our specific case, the emphasis is on the introduction to architecture as a profession and as a discipline. Consequently, the notion of architecture as an art and as a framework for the development of scientific reasoning takes a secondary position. With it, it should be apparent what can be gained and what is the price to be paid. Again, it is important that in taking a specific course of action, one understands the options available as well as the 'costs and benefits' involved.
Our work could be summarized as:
— A rational process of design;
— Highly structured;
— A process of teaching and learning;
— The importance of relating concept with construction;
— The verification of our work is the "anonymous architecture".
This documentation is the result of an interaction between Dr. Jiawei Wu, Prof. Dr. Daqing Gu, and Prof. Herbert Kramel. Behind it stands 30 years of research. The key person, behind it all, is Prof. Gu. He made it possible!

November 2016

目录 | Contents

15 **1 原理 | Principles**

16 1 设计原理 | Design Principles
18 关于现代建筑 | On Modern Architecture
22 关于无名建筑 | On Anonymous Architecture
26 关于基本建筑教育 | On Basic Architectural Education
30 关于设计与建造 | On Design and Construction

34 2 结构化的基本建筑设计 | Structured Basic Architectural Design
38 模块一：房间 | Module 1: The Room
46 模块二：住宅 | Module 2: The House
52 模块三：集群 | Module 3: The Village

60 3 三要素 | The Three Determinants
62 空间/功能 | Space / Function
64 材料/建造 | Material / Construction
66 场地/场所 | Site / Place

69 **2 教学 | Pedagogy**

70 1 教学模型 | The Educational Model
72 教学环境 | The Environment
78 教师，学生 | The Teacher, the Students
80 基础课程 | The Basic Courses

82 2 教学过程 | The Teaching Process
84 要点 | Issues
88 过程 | Process
92 方法 | Means
98 教案设计清单 | Checklist for Program Design

101 **3 教案 | Programs**

102 1 教案结构 | Program Structure
106 第一阶 | Phase I
112 第二阶 | Phase II
118 第三阶 | Phase III
124 第四阶 | Phase IV

130 2 演化 | Evolution
134 历年教案汇总 | Summary of Teaching Plans
136 教案选例 | Examples

144 3 实例 | Cases
146 1987/88 划艇俱乐部 | Rowing Club
164 1991/92 游客中心 | Tourist Centre
182 1998/99 旅馆 | Hotel Complex

201 **4 回响 | Resonance**

202 1 南京交流 | Nanjing Exchange

212 2 对话 | Dialogues
212 单踊 | Shan Yong
214 顾大庆 | Gu Daqing
218 丁沃沃 | Ding Wowo
220 赵辰 | Zhao Chen
222 张雷 | Zhang Lei
224 龚恺 | Gong Kai
226 吉国华 | Ji Guohua
228 张彤 | Zhang Tong
230 鲍莉 | Bao Li

234 3 影响 | Impacts
244 东南大学 1990/91 一年级 | SEU Year 1, 1990/91
246 东南大学 1997/98 二年级 | SEU Year 2, 1997/98
248 东南大学 1999/2000 三年级 | SEU Year 3, 1999/2000
250 南京大学 2003/04，2004/05 硕一 | NJU MArch 1, 2003/04, 2004/05
252 香港中文大学 1996/97 一年级 | The CUHK Year 1, 1996/97
254 香港中文大学 2016/17 一年级 | The CUHK Year 1, 2016/17

256 4 解读 | Interpretation

276 **附录 | Appendix**

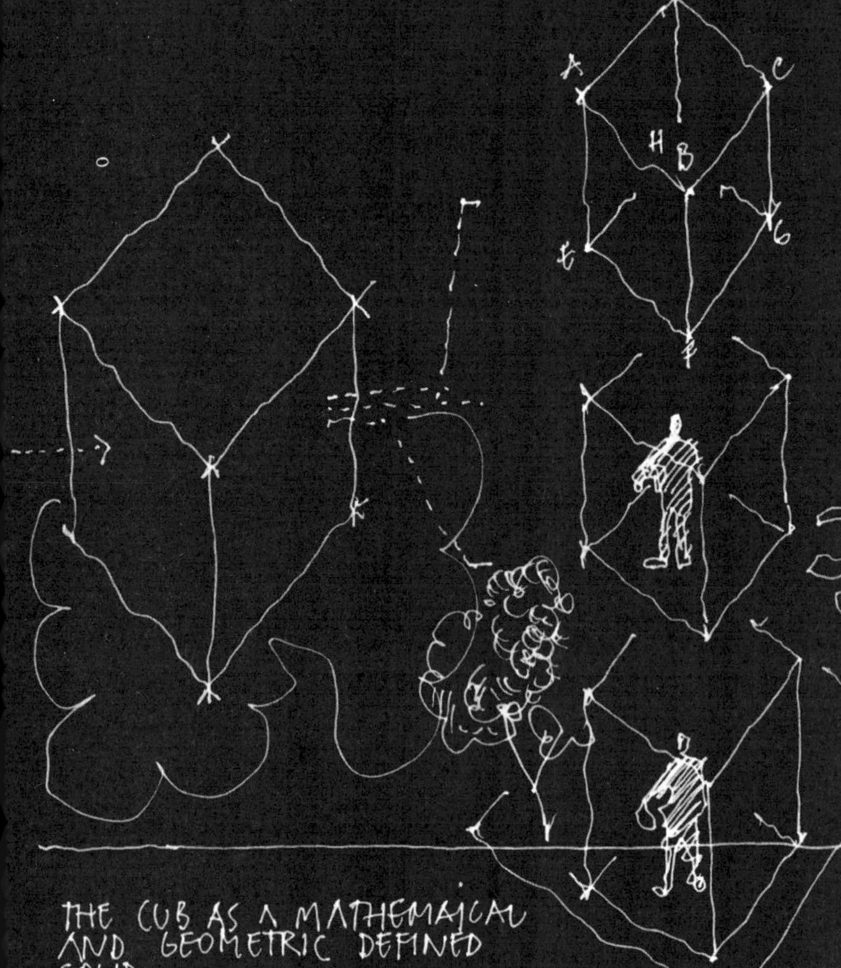

THE GEOMETRIC DIMENSION

THE HUMAN DIMENSION

THE ENVIRONMENTAL DIMENSION
(THE BUILT AND THE NATURAL ENVIRONMEN

THE CUB AS A MATHEMAICAL AND GEOMETRIC DEFINED SOLID.

THE CUBE CAN BE CONSIDERED AN 'IDEA'
IT CAN BE SEEN AS AN IDEA OF A CUBE IN THE PLATONIC SENCE.

AND IT CAN BE CONSIDERED AS A 'CONCEPT' IN DESIGN TERMS.

- THE CUBE AS UNIT
- MEN
- LAND (SITE)
- TREE.
- SUN. + ...

THE CUBE AS A GEOMETI DEFINED SOLID -
- CAN SERVE AS A BASIC ARCHITECTURAL UNT (B.A.U) IN OUR RESEARCH.

赫伯特 · 克莱默手稿 | Herbert Kramel's original manuscript

1 原理 | Principles

1 设计原理 | Design Principles

关于现代建筑 | On Modern Architecture
关于无名建筑 | On Anonymous Architecture
关于基本建筑教育 | On Basic Architectural Education
关于设计与建造 | On Design and Construction

2 结构化的基本建筑设计 | Structured Basic Architectural Design

模块一：房间 | Module 1: The Room
模块二：住宅 | Module 2: The House
模块三：集群 | Module 3: The Village

3 三要素 | The Three Determinants

空间/功能 | Space / Function
材料/建造 | Material / Construction
场地/场所 | Site / Place

1 设计原理 | Design Principles

建筑作为一门学科及一种职业，具有常量和变量。常量是建筑的基本原则，被每一代人的不同文化所诠释。理论与两者相关，提供了操作的框架。理论基础必须成为整体教育过程及具体教案中的重要问题，因为它建立了历史和知识的连续性。教育的内在结构完整性需通过其面对常量和变量时的一致态度来获得。

“基础设计”的基石是现代运动的建筑案例及宣言。这一历史进程是以有意识的建造，多个关键人物以及将形式理解为社会议题的映射为标志的。这决定了对建筑形式及其产生过程的理解。而后者尤其决定了教与学的过程。建筑形式被视为功能/空间、建造/材料和场地/场所之间互动的结果，直截了当地阐明了理论基础对此教案的影响。

由文化决定的变量对建筑原理的重新阐释形成了“要义”系统。指引现代运动发展的“要义”，专注于建筑形式的根本概念，剥除了世纪之交流行的陈旧传统和装饰，通过环境设计实现社会变革。这反映在空间概念的转变及其对形式的影响。

在这个意义上，当今苏黎世联邦理工学院建筑学院对建筑的理解可以说是现代运动的直接延伸。

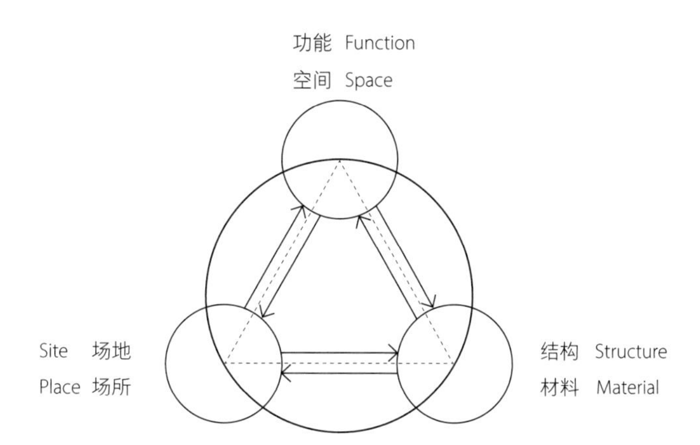

复杂建筑系统图解 | Diagram of a complex architectural system

Architecture as a discipline and, relatedly, as a profession, operates within a field of constants and variables. The constants form the fundamental principles of architecture that are then reinterpreted by culturally determined variables in each generation. Theory, being related to both, thus provides a framework in which operational decisions can take place. The theoretical base must be an important question in the educational process in general, as well as in the program specifically, as it establishes a historic and intellectual continuum. Integrity within the educational structure can only be achieved through a consistent stance towards the constants and the variables.

In the case of the Basic Design, the architecture and manifestos of the Modern Movement serve as a base. Historically, the movement is marked by conscious construction, by some key figures, and by an understanding of form as a response to social issues. This determines the understanding of architectural form as well as the process of its creation. It is specifically the latter that structures the teaching/learning process. Architectural form, seen as the result of the interaction between function/space, construction/material, and site/place, clarifies the way the theoretical base influences the making of the program in a very direct way.

The reinterpretation of architectural principles by culturally determined variables forms the system of vital ideas. The vital ideas, which have guided the development of the Modern Movement, focus on the fundamental notion of architectural form, clear out obsolete traditions and decorations that were prevalent before the turn of the century, and emphasize a notion of social change through environmental design. This is manifested in a changed concept of space and its impact on form.

In this sense the understanding of architecture at ETH today is arguably a direct extension of the Modern Movement.

关于现代建筑

密斯·凡·德罗将现代建筑定义为借时代之手段，以时代之技术、材料，解决时代之问题[1]。我们或许还记得，现代建筑曾追求客观、国际化及科学。我们还必须意识到“时代”的“要义”已经改变，我们几乎时刻身处大众社会之中。

建筑教育应呼应这些境况。基本建筑设计必须遵循这一事实。现代运动希冀呈献甚至定义一个更好的“新世界”，然而所谓的“社会契约”并未实现。人类处境业已急剧地，并将持续地改变。但我们若暂且忽略当前形势，便能够重新审视在20世纪发展出来的概念。它们仍有价值，甚至得到了转化。回顾风格派1920年的宣言十分有趣，它定义了对于彼时甚至现今的未来建筑。

当今建筑问题的关键为何？是技术的抑或经济的？或仅仅是形式的？从瑞士建造业来看，有实力的建筑师及建造商能够胜任各种复杂度的项目。解决技术、科技及经济问题的知识与能力与日俱增。显然，建筑的当务之急不单在科技及经济层面。然而，当代的建筑形式却体现了塑成环境方法上的根本性错误。

最严重的问题恰恰出在我们构想并组织赋形过程的方式中。我们从根本上迷失了方向。目标的缺失助长了企业家和建筑师干扰性的个人表达倾向。建筑师原有的基于集体利益的判断力被个人价值观所挟持。一种文化中间立场的缺失、过度的功能主义以及委托人在形式问题上的无能都加剧了这一危机。然而，不能简单地归咎于此，建筑师及建筑院校难辞其咎。二者的失职导致了一系列行为模式，造成了文化实体及生存环境的瓦解。

讽刺的是，我们观察发现当下建筑行业发展与现代运动的目标大相径庭。当时的建筑师试图为20世纪的人们创造一个文化与社会的人性世界。为实现这个目标他们运用了科学上的发现。为建筑打下理性基础的尝试奠定了新运动自我呈现及行动的基调。为投身未来，现代运动背离了历史。拒绝建筑史典型的英雄崇拜及对“大师”的膜拜，格罗皮乌斯构想的集体合作模式取代个人灵感占了上风。新的技术和科技运用在依据功能排布的城市房屋中以满足人性关怀。城市与住区被重新定义，并与地形有机结合。建筑师在现代运动提供的愿景与规划中找到自身定位。

1930年后，复杂的世界没有变得简单，恰恰相反，关键的进程正在加速。问题接踵而至，愈演愈烈。全球经济、政治、社会网络的复杂性将看似毫不相干的，彼此牵涉其中。按照现代运动所标榜的范畴及信念，其方法应顺应当代建筑的新形势，进行扩充并建立新的联系。

[1] 密斯·凡·德·罗撰于短文《办公建筑》(Bürohaus) 前的四行诗。原文为“建造的艺术是定格在空间中的我们时代的意志。鲜活。变化。崭新。/ 不是昨日，不是明天，只有今天是可塑的。只有这种建造活动能够赋形。/应任务之本质、借时代之手段以赋形。/ 这是我们的任务。”《G: 基本造型材料》(G Material zur elementaren Gestaltung), 1923年第一期，第四页。

[1] "The art of building is the will of our time captured in space. Living. Changing. New. /Not yesterday, not tomorrow, only today can be formed. Only this practice of building gives form. /Create the form from the nature of the task with the means of our time. /That is our task." Mies van der Rohe, Ludwig. "Bürohaus". *G: Material zur elementaren Gestaltung*, no. 1(1923):4. English translation by Detlef Mertins, Michael William Jennings.

On Modern Architecture

Mies van der Rohe once defined modern architecture as solving the problems of today by the means of today with the methods and materials of today [1]. And we might remember that the modern architecture has aspired to be objective, international, and scientific. We also have to realize that the "vital ideas" of the "time" have changed and we constantly find ourselves within a mass society.

Architectural education should respond to these conditions. Basic architecture design has to accede to this fact. The modern movement has wanted to contribute or even to define a "new world" for a better world. Well, needless to say, the "social contract" has not been realized. We can see that the human situation has drastically changed and is changing ever more. Irrespective of the present situation, it seems meaningful to consider the concepts developed in the 20th century. They are still valuable, even transformational. It is interesting to look back to the De Stijl manifesto of 1920, which has defined architecture of the future at that time and even today.

What are the important problems in architecture today? Are they technical problems or problems of economics? Or, is it merely in the area of architectural form that questions about contemporary architecture emerge? If we consider the construction industry in Switzerland, we see that competent architects and contractors are in a position to create buildings of any desired complexity. The knowledge and ability to solve technical, technological, and economical problems are available and continue to grow at a stunning speed. It is clear that the urgent problems that confront architecture today are not to be found in the areas of technology and economics alone. However, today's formal architectural embodiments offer us examples of fundamental faults in our contemporary approach to the shaping of our environment.

It is in the very way in which we conceive and organize the form-giving process that our greatest problems lie. We suffer from a fundamental lack of direction. Aimlessness has permitted a disruptive inclination towards self-representation on the part of the entrepreneur and the architect. Individual values impair our ability to formulate a criticism based on collective interests. The lack of a cultural middle ground, overt functionalism, and the incompetence of the client in formal issues all deepen the crisis. However, it is too simple to find fault with these alone. Architects and architecture schools are themselves partially responsible and guilty. Moreover, their offenses have resulted in patterns of behaviour that facilitate the dismemberment of first, our cultural substance and second, our living environment.

Ironically, one observes in the present condition of architectural involvement, a sharp contrast with the goals of the architects of the Modem Movement whose desire was to create a cultural and socially humane world for the people of the twentieth century. Towards this aim the discoveries of science were applied. The attempt to create a rational basis for architecture had set the conditions by which the new movement sought to establish itself and carry out its mission. In order to be able to follow a path to the future, the Modem Movement turned away from history. The hero worship that had characterized architectural history, as well as its idolization of the "great masters", was rejected. In its place was Gropius's conception of work as a collaboration of many, rather than the inspiration of one assumed wider authority. New techniques were used and new technologies applied in order to meet the demand for humanistic housing set in functionally ordered cities. The city and the settlement were redefined and were brought into a coherent relation with the

事实却并非如此。“大师”陨落之后，先锋派的代言人们被置于新的历史讲坛。这些精英们通往圣坛之路伴随着狂热的崇拜，当代媒体为其扫清了成为今日之星，乃至巨星的前路。历史再一次复苏了，在后现代时期沦为形式的自选商场。

“英雄的历史”、明星的出现、建筑史转而记录个别特例并成为以商业为导向的出版物——这一切深深地烙印在建筑及建筑教育上。学生从一开始便被教育笃信一种自我认识及表现的模式。在这种竞争模式中，只有最好的学生才获得关注，而余下的所有人只能在及格或平均线上沉浮。人们早已忘记演奏大师和乐手之间的区别；早已忘记大师诞生于普通乐手所造就、维系的音乐文化中。教育应以培养后者为目标，而大师是可遇而不可求的。

大师的形象主导了大多数建筑教育。可是学生主体并不能达到这种教育计划的目标。能力与勤奋已经远远不够。精度与手作不被重视。我们心系建筑教育，以上问题可以，也必须在我们的教程中有所应对。我们不采取居高临下的姿态，而是将我们的教程视为一种建筑教育的新途径。

最后，我想回到先前的问题，症结不在建筑或房屋的技术或经济层面，我们所面临的是学校所折射出的一种混乱。因此，这一代及下一代都必须将为建筑奠定理性基础奉为宗旨，将科学与艺术再次在理性原则的基础上结合起来。我们仍旧希冀为每个人塑造出更美好的世界。

landscape. Thus, the Modem Movement had a vision and a program through which individual architects could define their roles.

Since the 1930s, the complex world has not become any less complex. Quite the opposite, the pace of vital processes has accelerated. Problems appear in more rapid succession and in ever more critical forms. The complexity of our global economic, political, and social networks implicates us in situations with which we would seem to have no apparent relation. Considering the scope and intensity of belief that the Modem Movement represents, one would have thought that its methods could have been expanded and brought into relationship with the new circumstances that confront contemporary architecture.

However, this has not been the case. After the downfall of the "great masters," the representatives of the Avant-garde were placed on the podium that history offered. The cult that accompanied the passage to prominence of these personalities has, with the help of contemporary media, paved the way for the stars and superstars of the present. History has repeated itself and consequently, has been degraded into a formal supermarket in the period of the post-Modern. This development of the "history of heroes", of the appearance of stars, of the transformation of architectural history into a documentation of exceptional situations and a product of commercially oriented publishing concerns, has left deep marks on architecture and architectural education. Very early on, students are taught to believe in a model of expression characterized by self-recognition and self-representation. This is a competitive model in which only the best counts, leaving all the rest to sink or swim around the passing or average line. One forgets all too readily the differences between the musician and the master. One forgets that the musician is essential for creating and sustaining the music culture that nurtures the master. The goal of education is to produce the musician. The master is but an extraordinary occurrence.

The image of the master dominates most architectural education. The majority of students cannot meet the standards of such an educational program. Competence and diligence in carrying out an exercise are regarded as insufficient. Precision and craftsmanship are given limited value. Because we are also intensively involved in architectural education, these questions can and should be addressed in our own program. We do not claim to work from a position of superior knowledge. Rather, our program should be understood as a pursuit for a new way in architectural education.

In closing, I refer back to the introductory observation. We have no real technical or economic problems in architecture or building. We are confronted, however, by a chaos that is reflected in our schools. It must be the goal of this, as well as the next generation, to create a rational basis for our architecture that is founded on rational principles, bringing science and art together once again. We still wish to furnish a better world for everybody.

关于无名建筑

通过对建筑出版物的研究，我们发现意大利正统建筑与非正统建筑（大写的建筑学和小写的建筑学）之间的区别。我们同样可以在文丘里的著作中找到这种差别，即主要建筑和普通建筑。然而这种分化并非人所共知，因此楼宇常常等同于建筑。

20世纪60年代，希腊建筑规划学家C·A·道萨迪亚斯在建造总体研究中发现，建筑总量中可能只有20%可以说是受到建筑师的影响，其中仅有2%的建造物是完全在建筑师的掌控之下，而构成总量基础的其余80%的建造则被其称为"自然建筑"，是按居住者自身的构想或传统形式产生[1]。其中一些"没有建筑师的建筑[2]"可被看作是"无名建筑"。

我们把建筑定义为"建造之艺"，"建筑大师"的称谓便来自于这一思路。与此相反的立场则定义"建筑作为艺术"，建筑师因而被称为艺术家。新的明星建筑师便是这一思路的产物。其结果是，所有的人工环境都是"被"建成的。楼宇仅因复杂程度及建筑师的设计能力而异。

第一次世界大战后的建筑响应了政治上对住宅的需求。建筑必须回应社会和社会主义的路线，因此要发展一种完全脱离于战前历史主义的住房新形式。在这种战后建设的新视点所催生的图景中，重新发现"穷人的住宅"是逻辑的必然。20世纪50年代开展了对无名建筑现象的研究和出版。从那时起该领域就与对各种形式的人类居所及环境的兴趣一同发展起来。希比尔·莫霍利-纳吉所著《无名建筑的自然风采》介绍了各种建成环境。这些建筑在功能上呼应了环境的逻辑，建造方法与当地材料相符，方式最优且道德正确[3]。

威森霍夫住区展是德意志制造联盟展示其目标的必然行动。当住区按照密斯的总平面规划以及对住宅设计要求建成平屋顶、白色外观之时，被当时的保守派讽刺为"阿拉伯村"[4]。

许多社会责任在其形成的村落或催生的群体中不但印证了寻求生存的过程，还有原型、形制日臻完美的进程。从前，获得罗马大奖的建筑师在前往罗马之前会先抵达伊维萨岛、米科诺斯或北非[5]。文丘里很久以后申辩道"主街挺不错的。[6]"他还曾游历拉斯维加斯。所有这一切意味着，无名建筑在多个意义上验证了新的现代运动。

与此同时，新的无名建筑成为全球无产者群众运动的框架。

[1] C·A·道萨迪亚斯，《转型中的建筑学》（伦敦，1963年），第75页。
[2] 伯纳德·鲁道夫斯基，《没有建筑师的建筑：简明非正统建筑导论》（伦敦，1964年）。
[3] 西比尔·莫霍利－纳吉，《无名建筑中的自然风采》（纽约，1957年）。
[4] 弗兰茨·舒尔兹，爱德华·温德霍斯特，《密斯·凡·德·罗》（芝加哥，2012年），第96-97页。
[5] 伊维萨岛（西班牙）和米科诺斯（希腊）都是地中海的岛屿。它们在第一次世界大战后成为欧美建筑师探访乡土建筑的热点。
[6] 罗伯特·文丘里，《建筑的复杂性与矛盾性》（纽约，1966年），第10页。
[1] Doxiadis, Constantinos A. *Architecture in Transition*. London: Hutchinson, 1963: 75.
[2] Rudofsky, Bernard. *Architecture without Architects: A Short Introduction to Non-Pedigreed Architecture*. London: Academy, 1964.
[3] Moholy-Nagy, Sibyl. *Native Genius in Anonymous Architecture*. New York: Horizon Press, 1957.
[4] Schulze, Franz, Edward Windhorst. *Mies van der Rohe*. Chicago: The University of Chicago Press, 2012: 96-97.
[5] Ibiza (Spain) and Mykonos (Greek) are islands in the Mediterranean. They became popular places for visiting vernacular buildings among European and American architects after World War I.
[6] Venturi, Robert. *Complexity and Contradiction in Architecture*. New York: The Museum of Modern Art, 1966:10.

On Anonymous Architecture

Through the study of publications on architecture, in Italy, we found the distinction between major architecture and the minor one, meaning "architettura maggiori e minori". We can also find the same distinction in the publications of Venturi, namely between major architecture and architecture of the ordinary. This differentiation is not common knowledge. Often, any building is simply architecture. In the sixties, the Greek architect Constantinos Apostolou Doxiadis observed in his study of overall constructions that out of the total number of constructions, only 20% of which may have been influenced by architects, and only perhaps 2% were created under the complete control of architects. The other 80% making up the base were called natural architecture. They were conceived by the inhabitants or generated by traditional forms.[1] Some could be seen as "anonymous architecture" in the sense that they were "architecture without architect [2]".

We have defined architecture as the "art of building". The idea of the "master builder" derives from this position. The opposite position defines "building as art"; the architect is thus identified as the artist. The new star architect is a logical culmination of this thinking. The consequence of this line of thought is that all parts of the man-made environment are considered built. Buildings only differ according to the levels of complexity and the architects' competence.

The architecture after WWI followed the political needs for housing. The buildings had to respond to the social and socialist lines. Therefore, a new form of housing totally removed from the pre-war historicism had to be developed. This new perspective of the post-war building inadvertently led to the "rediscovery" of the "housing for the poor". It was in the 1950s that the phenomenon of anonymous architecture entered the fields of investigation and publication. Since then, the fields have grown with the rising interest in the many diverse forms of human habitation and the environment. Sibyl Moholy-Nagy's book, Native Genius in Anonymous Architecture, introduces an array of built environments. These buildings respond functionally to the environment. Given the congruence of native materials and building methods, these buildings are constructed in the most optimal and morally sound way [3].

It was logical to build a settlement, the Weissenhofsiedlung, as a demonstration of the goals of the Werkbund. When the settlement was completed according to Mies's layout and houses designed to abide by his requirement of flat roofs and white exteriors, it was derided by conservatives as an "Arab village" at the time [4].

Much social responsibility in forming villages or in generating groupings is not only examples of a process of survival but also of the optimization of types and prototypes. Previously, architects, prior to arriving in Rome for the Rome Prize in Architecture, first travelled to Ibiza, Mykonos or North Africa. Venturi later recalled, "The main streets were quite alright. [5]" He has also travelled to Las Vegas before. All of this signifies that anonymous architecture serves in many ways as verification for the new "Modern Movement".

At the same time, new anonymous architecture has become the framework for mass movements of the desperate all over the globe.

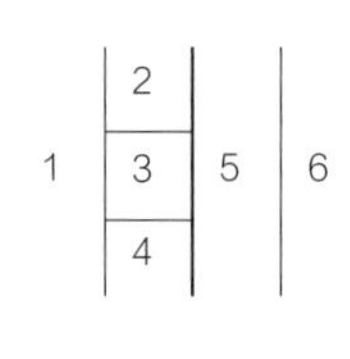

1
2
3
4
5
6

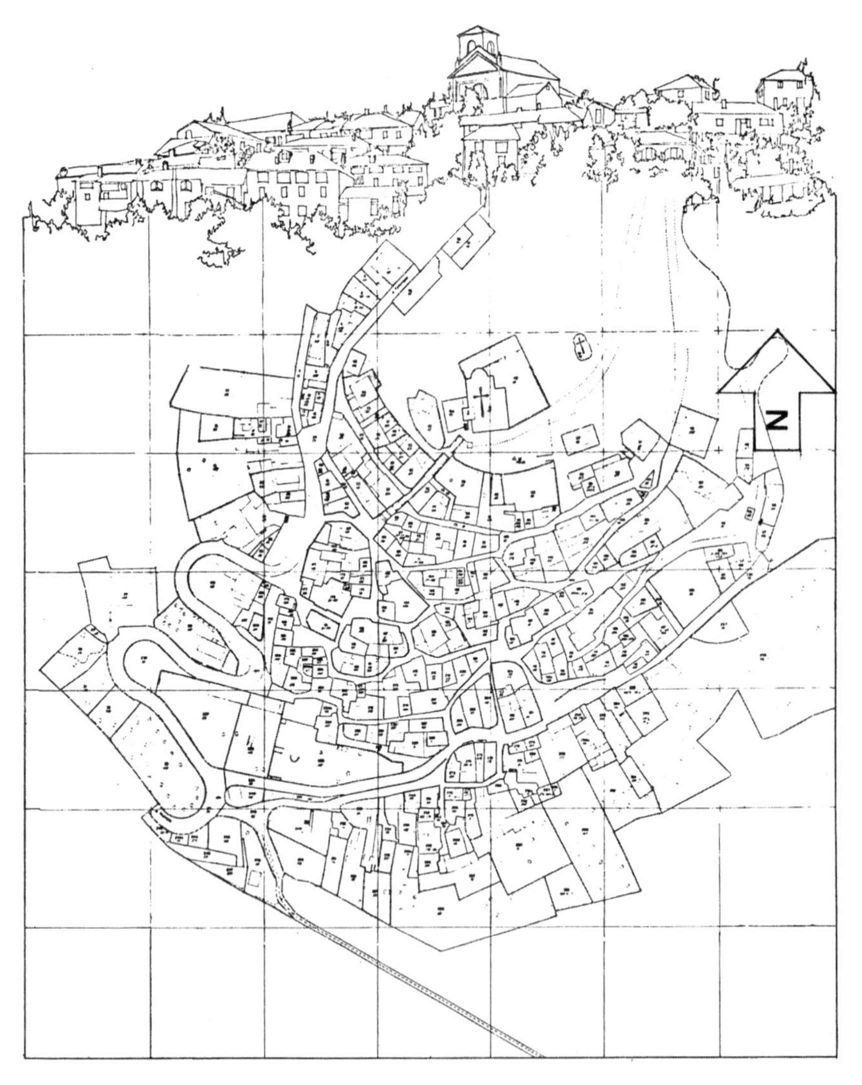

N

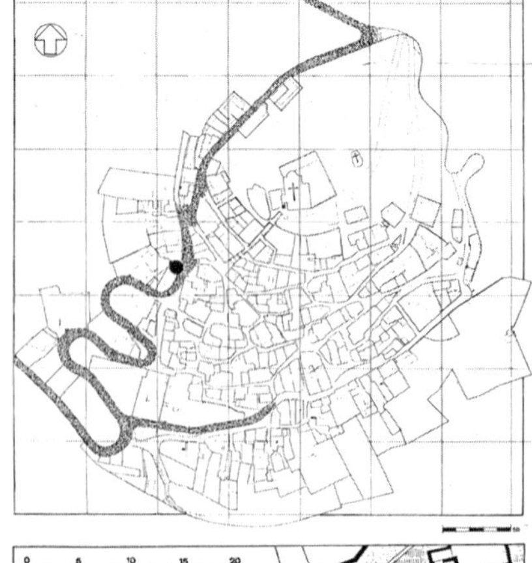

0 5 10 15 20

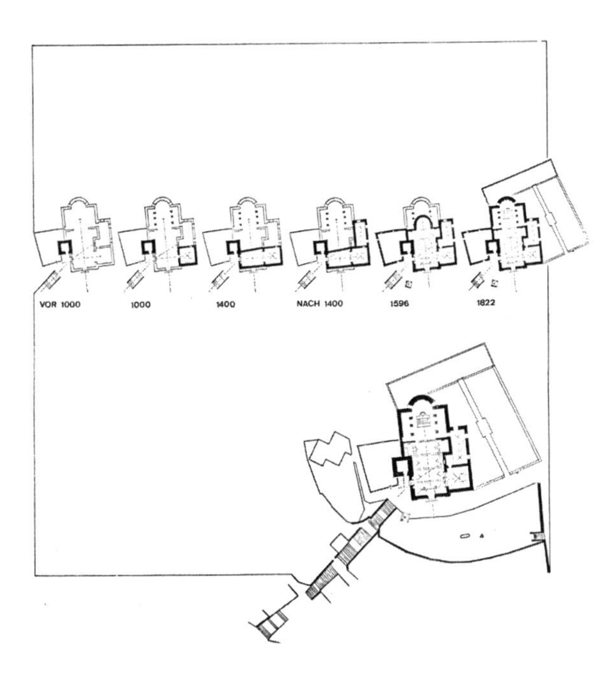

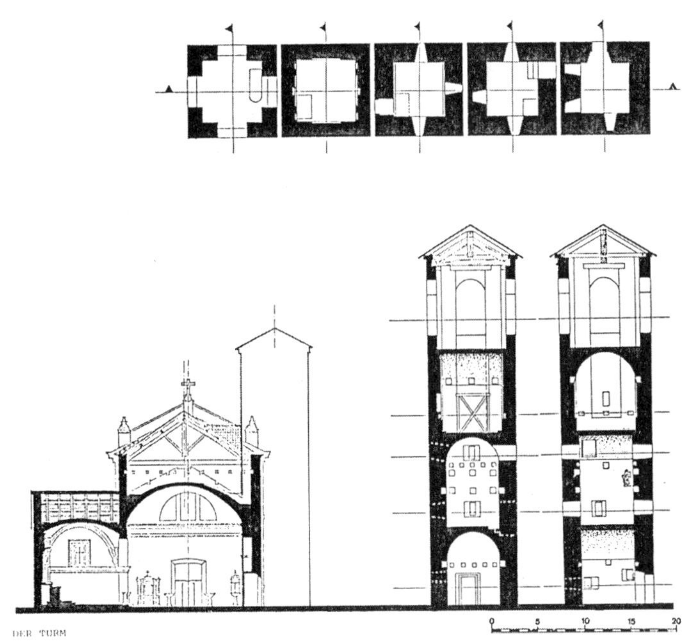

克莱默教席于1974年研讨周组织的布列诺村调研部分成果：
1 村落的平面与形态记录 2 建筑分布图 3 新建道路系统 4 建筑首层开口与外部空间 5 村内教堂历史演化分析 6 教堂与钟塔剖面及钟塔各层平面

Documentation of the suvey of Breno organized by Chair Herber Kramel in the 1974 Seminar Week. 1. Planar and perspective record of the village 2. Building volumes 3. The new circulation system 4. Exterior space and opennings in the first floor plan 5. The analysis of the evolution of the church in the village 6. Plans and sections of the tower and section of the church

从20世纪70年代中期到90年代，克莱默教席在瑞士南部提契诺的多个村庄（如马尔坎托内、阿兰诺、皮亚诺）进行了对无名建筑的研究。研究与学院的研讨周和选修课结合。1974、1975年在布列诺的田野工作记录了村落建筑的体量和屋顶平面、开放与私密空间、图—底关系、交通组织、当地建造技术及房屋类型，后续的分析推演了主要建筑物的历史演变。

The projects on anonymous building led by Kramel from the mid-seventies to the nineties were investigations of Ticino villages in the southern part of Switzerland, such as Malcantone, Aranno, and Madonna del Piano. The research projects were organized in conjunction with the seminar weeks and electives at the department. In 1974 and 1975, fieldworks were carried out in the seminar weeks in Breno. The students recorded the volumes and roof plans, open and private spaces, figure-ground relationship, circulation, building technology, and housing types. The subsequent analysis revealed the historical formation of major buildings.

关于基本建筑教育

霍斯利（左）与克莱默 | Hoesli (left) and Kramel

每个学校都必须明确其出发点。

与大多数学校的简单意图不同，巴黎美术学院则有明晰的课程架构。1919年包豪斯成立，“基础课程”的确立顺理成章。艺术与工业的融合体现了工业化背景下的特定发展阶段。大众需要大规模的工业化生产，设计的目标是更多、更快、更廉价，最好能兼具优质。

历史主义建筑面向学院风格的大师们。现代运动带来了翻天覆地的变化。其核心是概念与建造、设计和发展。设计面向建成环境的总和。格罗皮乌斯认为，一切皆建筑[1]。

“基础设计”是先于专业导向而迈出的第一步，但它必须整合艺术和工业。教师须身兼艺术家与生产工匠，就如我们在包豪斯所见。

“基础设计”在包豪斯从1919年发展至1933年，学校因政治原因关停。纵观历史便会发现，流亡者们使得下一步的发展主要发生在美国。密斯 · 凡 · 德罗和路德维希 · 希尔伯塞尔默在芝加哥的伊利诺伊理工继续教学。格罗皮乌斯去了哈佛。纳吉成为芝加哥的新包豪斯的院长。阿尔伯斯创办了黑山学院。还有的回到瑞士等地。这些思想在第二次世界大战后散播开来并融合到不同学派的系统中。马克斯 · 比尔参与成立的乌尔姆设计学校对于德国和欧洲都是一个新的开始。

20世纪50年代有了新的进展。其中之一是德克萨斯大学奥斯汀分校建筑学院的课程建构，形成了一个后来被称为“得州骑警”的团体。目前已有不少出版物再现了这段历史并肯定了“得州骑警”的贡献。

从我个人角度看，历史学家柯林 · 罗、建筑师本纳德 · 霍斯利、约翰 · 海杜克及沃纳 · 塞利格曼和画家罗伯特 · 斯拉茨基在其中作出了重大贡献。他们虽借鉴了包豪斯，但更重要的是引入了勒 · 柯布西耶的新建筑、立体派绘画以及现代运动的建筑。虽然我对他们在工业化方面的应对毫无印象，但他们对现代运动的研究和教学极其关键。后来这个团体再次分散到不同的大学：霍斯利回到苏黎世联邦理工学院（ETH）；罗在康奈尔大学自成一派；海杜克和斯拉茨基执教于库柏联盟；塞里格曼后来执掌雪城大学建筑学院。

霍斯利在ETH任教25载，成就了瑞士年轻一代的建筑师。我把他的贡献概括为：将现代建筑带到了ETH。霍斯利以柯布西耶、赖特及密斯的建筑作品为基础建立自己的教学，把建筑设计作为一个操作过程来传授，引入了一套建筑语言。他将教学结构

[1] 沃尔特 · 格罗皮乌斯，《整体建筑学的范畴》（纽约, 1970 年）。

[1] Gropius, Walter. *Scope of Total Architecture*, New York: Collier Books,1970.

On Basic Architectural Education

Every "school" of architecture has to clearly define its point of initiation.

Most schools are rather simple in their intention, while schools like the Ecole de Beaux-Arts have a clearly structured curriculum. When the Bauhaus was initiated in 1919, the establishment of the "Vorkurs" was a logical development. The integration of art and industry epitomized a specific developmental phase in the climate of industrialization. Mass society demanded industrial mass production. The goal was to produce more, faster, and cheaper, sometimes also better.

The masters of the academic styles were oriented towards historicist architecture. The modern movement brought about drastic changes. It was then essential to talk about concept and construction, design and development. Design was oriented towards the totality of the built environment. Gropius considered everything architecture [1].

The Basic Design was the first step that preceded professional orientation. But it had to be organized as an integration of art and industry. The teacher then had to be both the artist and the master of production, as we can see in the Bauhaus.

The Basic Design was implemented within the Bauhaus from 1919 to 1933 when, for political reasons, it was closed down. By taking a look at the overall development, one discovers that it was the exiles who brought the next stage of development over to the US. Mies van der Rohe and Ludwig Hilberseimer developed their work at the IIT in Chicago. Gropius went to Harvard. Nagy later became the director of the New Bauhaus in Chicago. Albers went to the Black Mountain. Some returned to Switzerland and to other places. This dissemination of ideas started after the War and became integrated into the "systems" of various schools. The Ulm School of Design founded by Max Bill was, to Germany and Europe, a brand new beginning.

In the 1950s, new developments took place. One of these was the development of the program at Texas. A group formed, which later became known as the "Texas Rangers". Their history and their work were well published.

From my own perspective, the historian, Colin Rowe, the architects Bernhard Hoesli, John Hejduk, and Werner Seligmann, and the painter, Robert Slutzky contributed crucially to this development. While they did reference the Bauhaus, it is of greater significance that they were involved in the new architecture of Le Corbusier, the Cubist painting, and the architecture of the Modern Movement. As far as industrialization is concerned I cannot remember any line of development in their work. Yet their research and teaching of the Modern Movement are quite important. And again, the group later dispersed to various universities. Hoesli went to ETH Zurich. Rowe established a school of his own at Cornell. Hejduk and Slutzky taught at the Cooper Union. Seligmann later became Dean and Professor of Architecture at Syracuse University.

From 1959, Hoesli had worked at ETH for 25 years, fostering a generation of young Swiss architects. In an attempt to sum up Hoesli's work I would say that he introduced the Modern Movement to ETH. He built his teaching upon the work of L.C., Frank L. Wright, and Mies van der Rohe. He taught architectural design as an operational process and introduced an architectural language. He structured his teaching and became, for many young architects, the exemplary teacher. Also, beginning with Hoesli, and together with Ronner, Chair of Construction back then, we can trace a development of the notion that recognize problem types, instead of building types in architecture, as the base for design. Design Basics and Basic Design are thus rendered complimentary.

克莱默基础设计教学成果集《作为教案的教学》封面，1985 | The cover of Kramel's book on basic design teaching *Die Lehre als Programm*, 1985

化，并成为众多年轻建筑师心中的教师典范。

也正是从霍斯利开始，与当时构造教席的罗纳合作发展出一个以问题类型代替建筑类型的设计思路。因此设计基础和基础设计呈互补关系。

由于种种原因，从1984年开始我同时负责建筑设计和建筑构造的教学。我将这两门课程及计算机辅助设计入门整合成一种基础教学的新形式——"结构化的基本建筑设计"。同时，为适应庞大的学生数量（约300人），高度的结构化必不可少。

这一教程主要有两个想法：其一是一年级建筑设计教育模式的当代发展，其二是保有延续性。新教程并非从零开始，霍斯利25年的教学依然具有现实意义，是我们思想的源泉，而非仿效的对象：

· 教学基于一系列独立的步骤，它们可以被单独审视、理解及验证；
· 将学习的过程解读为一种工作及设计过程；
· 最小的教学单元是"练习"，而非"项目"；
· 基于建筑设计中的典型问题，或称问题类型；
· 通过教案的设置向学生传授知识并产生体验，而不全依赖教师；
· 基于整体的观念（即强调部分与整体间的互动）；
· 以特奥·凡·杜斯伯格于1924年发表的宣言《走向塑形的建筑》为起始点[2]；
· 在自身文化的范畴内设定建筑参照，因此瑞士建筑，尤其是现代运动时期的例子，是重点参照。我们相信自身的传统是理解国际发展的前提；
· 建筑史不是英雄史，而是发生在更宏大的系统背景之下的现象与进程的历史；
· 此外，建筑是建造之艺，是设计过程与建造过程之间有意义的关联；
·"基础设计"呈现了现今一种教育的可能性，但并不排斥其他建筑教育和建筑设计的途径。

[2] 乌里奇 · 康拉德斯，《20 世纪建筑的程序与宣言》（剑桥，1971 年），第 78-80 页。

[2] Conrads, Ulrich. *Programs and manifestoes on 20th-century architecture.* Cambridge, Mass.: MIT Press. 1971: 78-80.

For a number of reasons I became the professor responsible for architectural design and construction in 1984. I combined these two areas with an introduction into CAD to form a new kind of basic design called "structured basic architectural design". Meanwhile, there was a need to develop a high degree of structure to adjust to the large group of students (almost 300).

The major concerns in the development of the program have been two-fold. Our foremost concern was the development of a contemporary educational model for the first year in architectural design. At the same time, continuity was an important issue. The new program did not start from scratch. The work of Bernhard Hoesli, started 25 years ago, was still relevant, and served as a source of inspiration instead of as a prototype for exact imitation.

— It is based on discrete steps of teaching and learning that are examined, comprehended and verified separately.

— It interprets the process of learning as a working and design process.

— The smallest teaching unit takes the form of the "exercise" and not the "project".

— It is based upon the idea of typical problems or problem types in architectural design.

— It transmits knowledge and generates experiences via the teaching program and not predominantly via the teacher.

— It is based on a holistic concept (thus emphasizing the interplay between the parts and the whole).

— It considers Theo van Doesburg's manifesto "Towards a Plastic Architecture" in 1924 the starting point [2].

— It establishes its architectural references within the boundaries set by its own culture. Thus, Swiss architecture, especially examples from within the Modern Movement, serves as points of reference. It is our belief that international developments must be understood through one's own traditions.

— Architectural history is not seen as the history of heroes but rather the history of phenomena and processes that take place in the context of a larger system.

— Furthermore, architecture is understood as the art of building, allowing a meaningful interrelationship between the process of design and the process of construction.

— Basic Design is presented as one possible option existing today that does not exclude the possibility of other approaches to architectural education and architectural design.

关于设计与建造

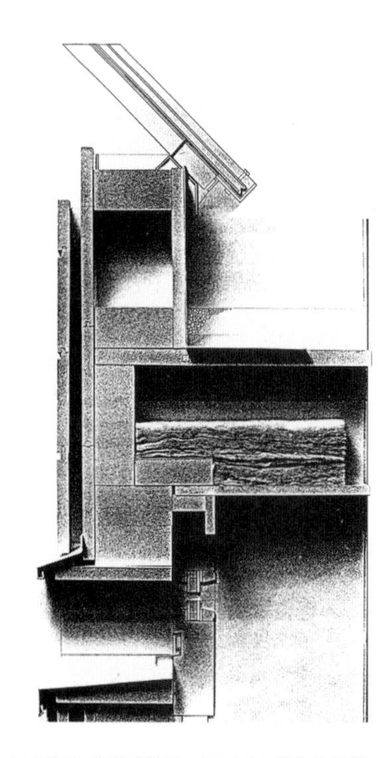

1:1 细部构造拼贴模型，1983/84 学年构造课
| The 1:1 collage model showing construction details, Construction course, 1983/84

从20世纪60年代初开始，ETH的基础设计课程包含三门主课：建筑设计，建筑构造和视觉设计。每门课既各自独立，又相互依存。建筑设计与构造两门课在教学的前期就已紧密关联。

一年级构造课的目标包括：1.建立学校的设计课程和技术领域的联系，以减少学生与职业间的差距；2.注重传授当下（尤其是瑞士境内）建筑实践所需的技术知识，进一步引介建造技术、生产技术及建材市场；3.使学生具备面向未来的适应力，能够应对快速的变化及环境转变。

为达成这些目标，构造课需构建一个知识体系，或者说一套工具。这包括将建筑理解为一套由子系统、组件和元素构成的体系。此外还要介绍源自基本建造方法和材料的建造原理。

我们坚持一个前提：每一个特定的建筑历史阶段都承载着自身的教育模式，并持续从中引出主题和范例。这一模式建立了技术和设计之间的关系。这一假设可以在现代主义运动，尤其是功能主义和理性主义的主题中得到印证，并且富有启发性。ETH在20世纪60、70年代的建筑设计与构造课程正是按照这个方向发展的。我们的设计始于“概念”。从概念到项目的转化，设计过程首先考量一个固有系统愈渐清晰并（通过绘制剖面图等）进行技术“证明”。因此，技术课程要处理结构系统（以框架结构为主）及外围护结构的技术问题。

乡土建筑同样提供了许多例证。传统及乡土建筑的逻辑、结构与简洁的形式为我们的教学提供了诸多范本。

20世纪60年代，结构与围护在技术层面的解决方法似乎是无限的。一切皆有可能。20世纪70年代发生了一些变化。“无限”被“增长的极限”所替代，人们逐渐意识到科技的局限。生物学法则——“形式只能在限制中演化”——得到重视。这一对于限制的意识影响了我们之后对于建筑与技术的设想。设计过程的重点是要建立对问题内在逻辑的理解。问题的结构，而不是解法的结构，才是工作的主体。我们发现，通过优化程式及问题描述能够加深对一个系统特性的理解。

我们相信形式与空间的深化也应遵循这一理解。如此，建筑的形式表达将不会与它的内涵相冲突。由此，建造的现实便不会是优秀设计的绊脚石，而是它的发生器。建筑重新成为建造的艺术。

On Design and Construction

Since the early 1960s, the Basic Design courses at ETH had consisted of teaching in three areas: architectural design, construction, and visual design. Each of these three courses was in part autonomous and in part coordinated with one another. Architectural design and construction were in coordination with one another at an early point in the educational process.

The objectives of the 1st-year construction course are: 1. to establish a link and bridge the gap between design courses and technical fields; 2. to focus on the teaching of technical knowledge presently required in the architectural practice (in Switzerland), and furthermore, to introduce the student to building technology, production techniques, and the building market; 3. to prepare the student for future conditions, rapid change, and environmental transformation. To fulfil these goals, a body of knowledge — a set of tools — had to be developed. These included the understanding of the building as a system consisting of subsystems, components, and elements. Furthermore, principles of construction derived from basic construction methods and materials had to be introduced.

We held fast to the hypothesis that each distinct phase of architectural history inherently possesses its own educational model from which it draws its sustaining themes and examples. It is through this model that the relationship of technology and design is established. The hypothesis, when applied to the Modern Movement, especially regarding the themes of functionalism and rationalism, was verifiable and productive. Our courses in architecture and design technology in the 1960s and 1970s had been clearly oriented in this direction. Our point of departure in design was the "concept". The transformation of the concept into a project, the clarification of an inherent system and its technological 'Verification" (through the use of drawings such as the section) had been the predominant concerns in the design process. As a consequence, the technology courses had to deal with the structural system (mostly skeletal) and the technological problems of the building envelope.

The same was true for indigenous architecture, which offered many examples supporting our themes. Because of their logic, structure, and simplicity, traditional and rural buildings have provided us with working models in our teaching.

At that time, technological solutions to structure and envelope seemed unlimited. Anything and everything appeared to be possible. In the late 1970s, some changes took place. The notion of "infinitude" was replaced by the 'limits of growth". A consciousness of the limits of technological means slowly emerged. The biological law stating that form evolves only through limitations gained traction. This emergent awareness of limitations informed our subsequent conceptual thinking in architecture and design technology. In the design process, it was essential to develop an understanding of the inherent logic of a problem. The structure of problems and not the structure of solutions was the subject of our work. We found that through the improved formulation and description of problems, insights into the properties of a system could be gained.

We believe that the further development of form and space should be based upon this understanding. The formal expression of a building cannot be in conflict with its substance. In this sense, the reality of the construction process is not a roadblock to good design but its generator. Architecture thus becomes, once again, the art of building.

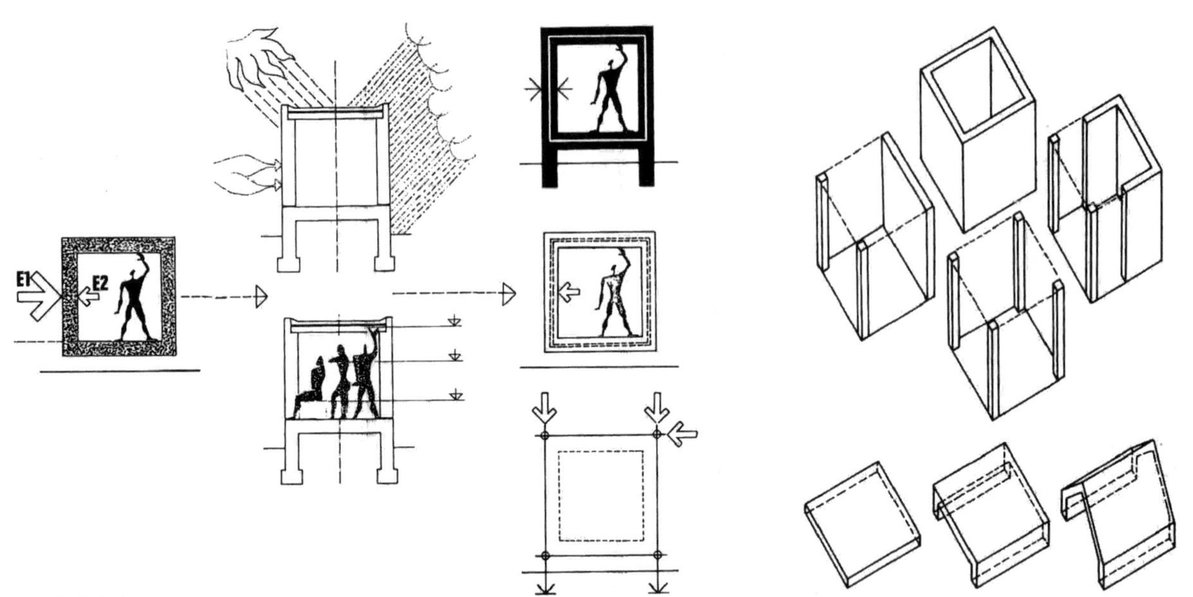

从原型到细部 | From prototype to details

思维模型首先将结构和建造的系列相关问题有序地整合在一起。考虑人的精神、物质及心理需求，将其转化为构造上的问题，便是建筑围护、空间围合及建筑结构。限定用方形体量进行操作使得划分空间与空间的限定元素与多种构造关联变得可能。

A thinking model first brings a variety of problems about structure and construction into an orderly context. The mental, physical, and physiological needs of humans are taken into consideration. In terms of construction, this leads to a categorization of the issues into the problems of building envelope, the conditions of the enclosed space, and the issues of building structure. The restriction to cube-shaped volumes makes it possible to correlate space and space-defining elements with various constructions.

在此，我们处理 1 或 2 层的建筑并运用“砖及杆件”的建造技术。给定的符合模数的场地及模块；对页：设计过程，构造课，1981/82 学年

Here, we deal with buildings of one or two floors and we talk about the "brick and stick" technology. Given site plan and elements with modulus; facing page: design process, Construction course, 1981/82

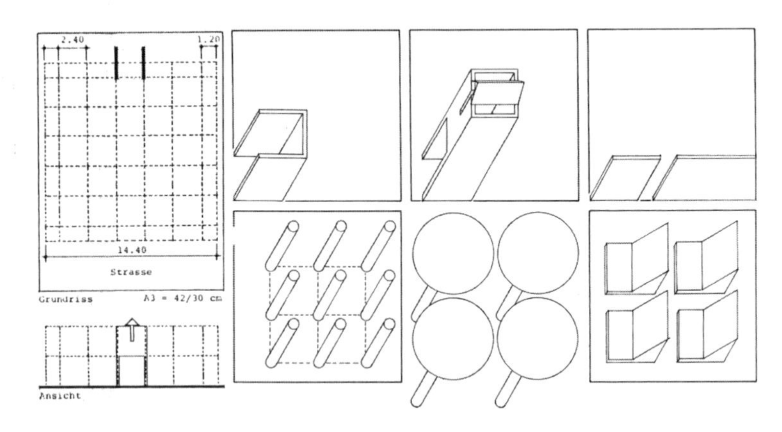

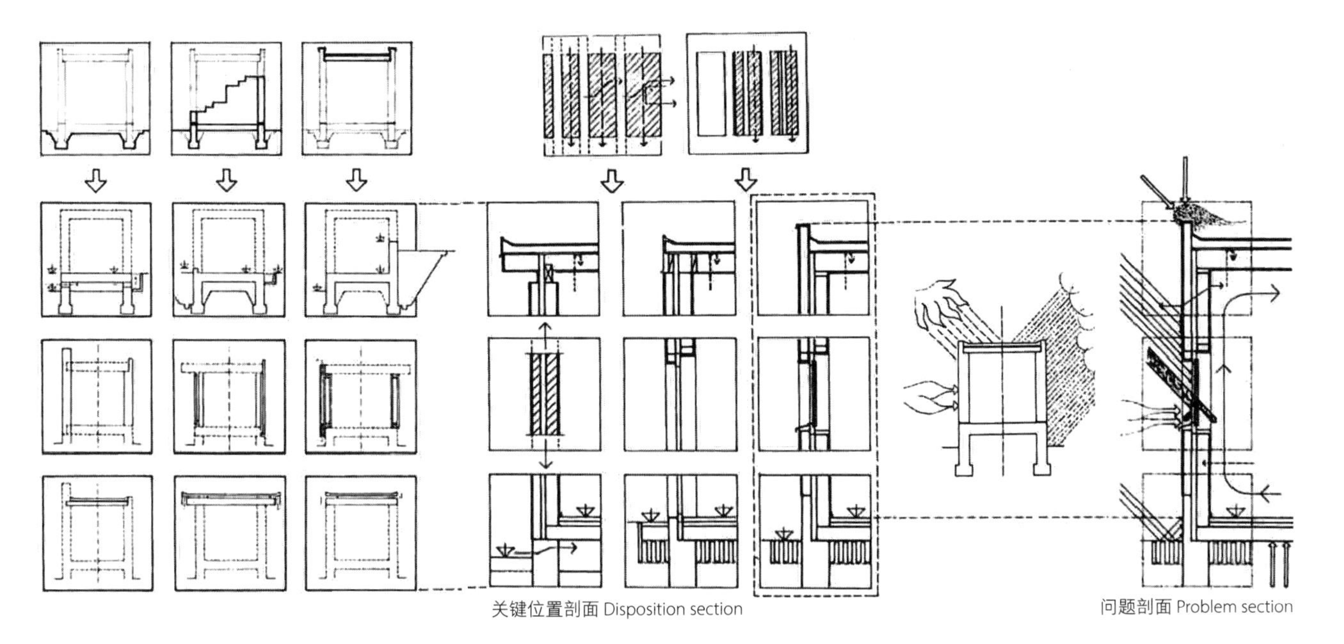

关键位置剖面 Disposition section

问题剖面 Problem section

接着，思维模型把结构概念和设计过程相结合。对于构件和建造过程的比较可以得到多种类型的构造策略，同样可以涉及材料和生产方式。学生在设计过程中一有疑问，则应画“关键位置剖面”来推敲。之后再以各种极端气候情形进行叠加验证，最后在“问题剖面”中具体化。

Secondly, the model combines structural concepts and design processes. A comparison of components and construction processes results in an array of solution principles, which can involve material and production methods. During the design process, students are asked to draw “disposition sections” to make an assumption when in doubt. Then, students are required to re-examine their assumptions by “superimposing” extreme climate cases. Problems of the respective assumptions should become visible in a “problem-section”.

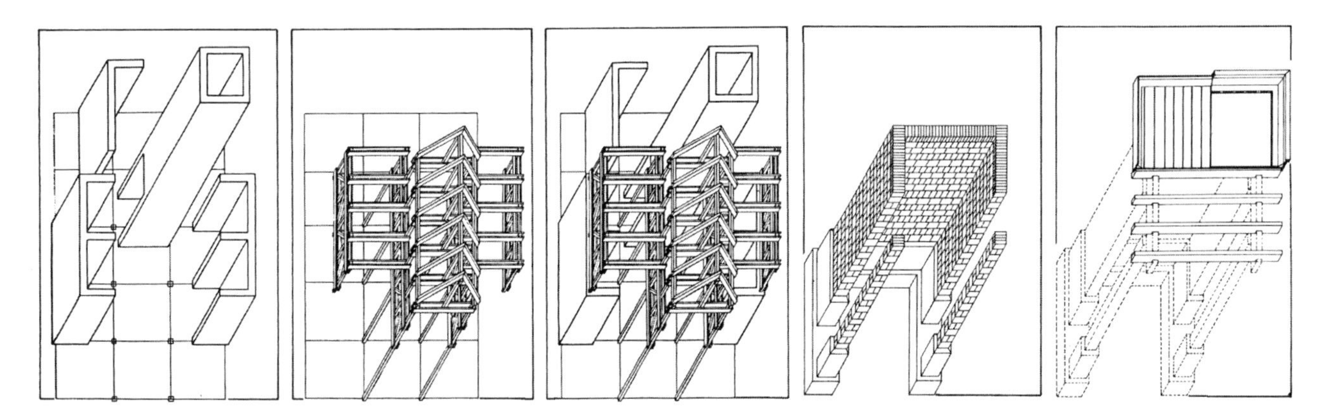

2 结构化的基本建筑设计 | Structured Basic Architectural Design

“基础设计”所涉很广，故需将讨论限定在“建筑教育”，更具体为“结构化的基础建筑设计”；要进一步将“建筑”界定为“建造之艺术”（奥托·瓦格纳）。凭借此定义，“概念”和“建造”必须关联起来以便从操作层面进行设计。这最终导向了概念——构思的行为与建造的行为一一对应。“形式”概念是“构思”的关键。形式是生命的基础，构思无法脱离形式。这是我们存在、认知以及实现概念的基础。

我们生活在自然环境中并不断努力改善自己的生存环境。我们持续地创造人工环境，重塑自身的栖息地。人工环境与建造的进程息息相关。如果将“建造之艺”理解为建筑，那么建筑师的职业生涯便与人类的存在休戚相关。

建造意味着转化，而形式只能在限制中演进。转变的原因是试图在不同层面上改善人类的生存、安全及社会交往。

我们将基本建筑设计称为“结构化”的教程。其中包含的理念是，在首要地处理建筑物体的同时，也处理它的“构件”。在另一尺度，“城市”将介入设计。

进入设计范畴，具体指进入上述“基于结构的建筑设计”，我们必须明白，初学者在各方面都毫无头绪。他没有工作的范畴，没有工作的语汇，没有操作的工具，没有工作的目标。正因为这样的“零点状态”，作为施教者，我们必须设定“基础设计”的初始条件。

我们的工作框架有3个模块，每个模块代表一个学习阶段。每个模块包含一组限制条件，一种建筑设计“语言”的开端和一套“媒介和方法”。这种框架构成工作复杂

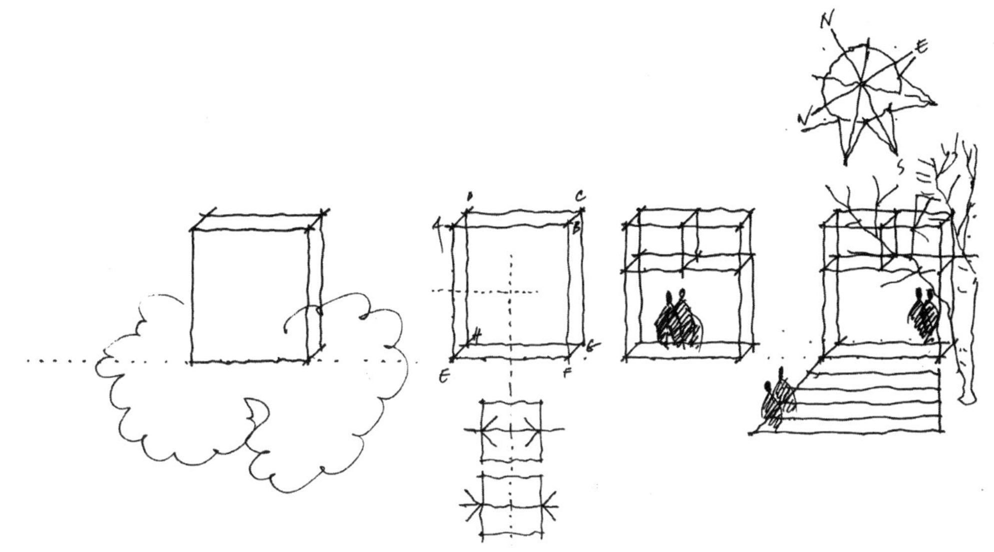

立方体：基本建筑单元 | The "cube": basic architectural unit or BAU

While Basic Design covers a very large field, it is necessary to limit our discussion to "architectural education" and more specifically, to "structured basic architectural design". It is furthermore necessary to define "architecture" as the "art of building" (Otto Wagner). With this definition, it is necessary to relate "conception" to "construction" in order to develop the architectural design in operative terms. This leads eventually to the question of conception, the act of conceiving, vis-à-vis the act of construction. In order to "conceive", the concept of "form" is essential. Form is essential to existence. We cannot conceive without form. It is the foundation of our existence, our cognition, and our concept of reality.

We live as part of the natural environment and continuously strive to improve our existence. We are creating a man-made environment. We are continuously transforming our habitat. The man-made environment is something that is essentially related to the process of building. If the "act of building" is understood as architecture, we may then define our professional life existentially.

To build means to transform. And form evolves only through limitations. And the reason to change, to transform is an attempt to improve human existence in terms of survival, security, and social interaction.

We defined basic architectural design as a "structured" program. Part of this structure is the understanding that while we are dealing, first of all, with the architectural object, we are also dealing with the "components" of this object. At a different scale, we may conceive the "urban" level as an integral part of the design.

If we enter the field of design, and in our specific case, "structured basic architectural design", we have to understand that the "student" is ignorant in all aspects. They do not have a field of actions, no limitations to work within. They do not have a vocabulary, a language to work with. They do not have instruments to operate. And they have no goals to work towards. Given this rather-zero-condition, we, the teacher, have to define the conditions for the beginning of "Basic Design".

The framework of our work consists of three modules. Each module represents a phase of learning. Each module represents a set of limi-

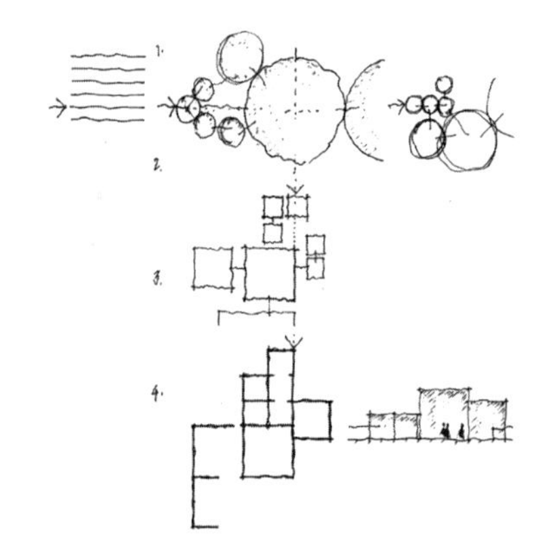

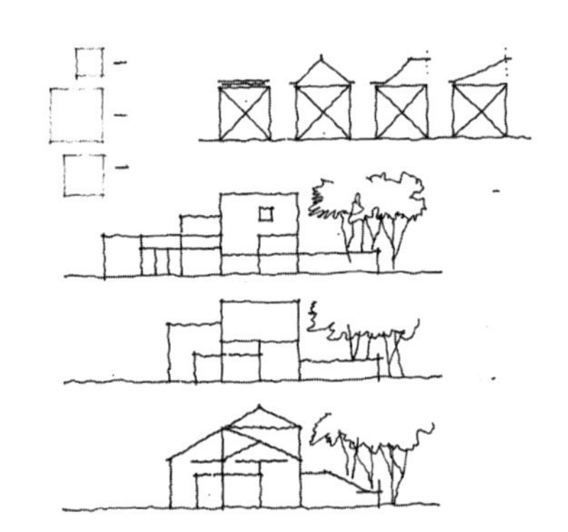

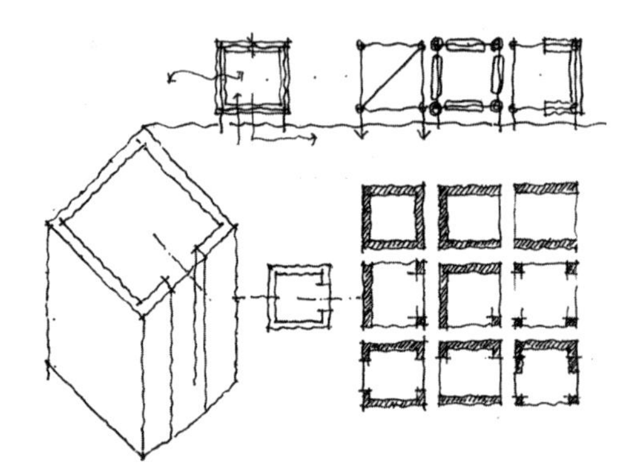

程度的递增。3个模块及它们的目标可以用简单的术语定义如下：首先是“建筑物体”；进而是“居所”；最后，形成“集群”。

模块可在彼此之上展开并在方法学上关联。这是一种叠加的工作方式，它们的共同点是“立方体”。3个模块分别是：

1. 物体 - 房间 - 艺术家工作室
2. 居所 - 房子 - 艺术家的住宅
3. 集群 - 村 - 艺术家村

此外，还需要一个操作的“场”以及操作工具。为此我们设定一块场地，以6mx6m的格子划分。作为操作工具的是边长为3m的立方体。立方体作为中性的几何实体并无好坏之分。它是一个“基本建筑单元”(Basic Architectural Unit, BAU)，代表了一种最小的空间度量，成为学生在设计初始阶段着手或“把玩”的元素。各步骤产生的模型和图纸是对核心问题的回答，应作为设计与学习过程的“日志”详细记录。如此想来，“设计”意味着提出下一个问题。

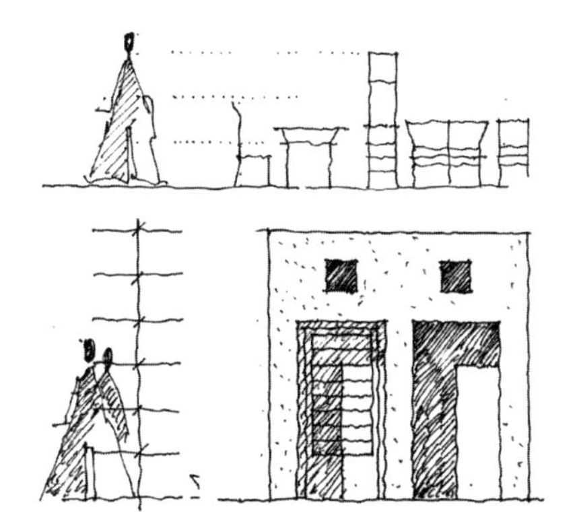

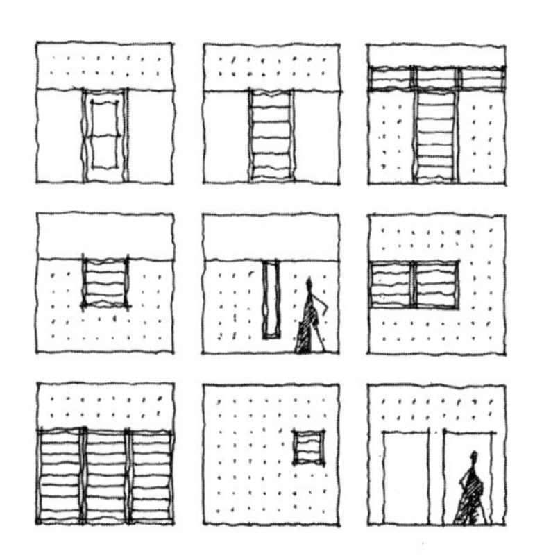

从泡泡图转化为方块图，进而转化为图解式的平面图；从图解转化为空间，建筑空间及体量；内与外的关系，外围护与家具。| The bubble diagram is transformed into a block diagram and then a diagrammatic floor plan; the diagrammatic plan is transformed into a space plan, then architectural space and volumes; the relationship between exterior and interior, the envelope and furniture.

tations, the beginning of a "language" of architectural design and a set of "media and methods". It represents a heightening of complexity in the work. These modules and their goals are defined in simple terms. First, we want to develop an "architectural object". Secondly, we want to develop an "architectural habitat". Finally, we are going to develop an "architectural compound".

Each of these modules should expand upon each other. They are interrelated in methodology. The work is additive in nature. The common denominator is – "the cube". The program progresses in "modules" are:

1. Object – room – artist's studio
2. Habitat – house – artist's home
3. Compound – artist's village

We also have to define a "field" of action, and some instrument for action. For the field, we defined a piece of land and divided it into grids of 6mx6m. As the "instrument for action," we developed a cube of 3mx3mx3m. The cube as a neutral geometric solid is neither good nor bad. It simply is. The dimension of 3mx3mx3m is meant to be a "basic architectural element" (BAU) and represents a minimal form of space. It furnishes the student with an element for working or "playing" in the initial phase. The work in steps should represent questions that have to be answered with models and drawings. Each step should be documented. This document, a "log book", should be understood as the design and learning process. In this sense, "design" means also to ask the next question.

模块一：房间

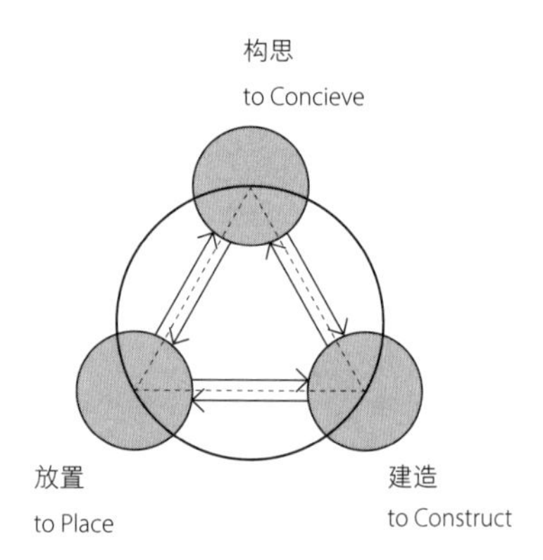

“构思-建造-放置” | To conceive, to construct and to place.

“基础设计”开始于教程的“编排”。我们不但希望能够“着手设计”，还能够“批判性地”进行评估。

本教程的第一阶要形成一个简单的建筑物体，雕塑家的工作间。引入空间与形式、条件与功能的关系。首先引入一些启动设计过程的前提条件。给出一块地、树木和一个已存在的旧塔，设定建造材料。在“结构化的基本建筑设计”教程中比其他条件更重要的是引入“基本建筑元素”（BAE）。“基本建筑元素”是“催化剂”，也是共同的基本点。进一步将其量化为：1个边长6m的立方体，4个边长3m的立方体和1个边长4.5m的立方体。从而将“基本建筑元素”转化为“基本建筑单元”。这6个单元代表的使用需求包括画廊空间、辅助空间和储物空间，它们是设计的启动器。

前文介绍了“复杂建筑系统图解”（见第4页）。它将建筑设计看作一个包含3个子系统——构思、建造、放置——的集合。我们把建筑设计看作一个过程，“构思”和“建造”是相互关联的，因此本质是一个理性的过程。

我们将会发现建筑空间，意识到“建筑语言”的开端；关联空间形态和功能；操作空间定义元素，辨认空间类型；把功能和建筑开口相关联；讨论了内/外空间的关系；处理围护结构和立面。

我们必须面对媒介，表达的尺度以及多种形式的信息、平面、剖面等的角色问题。关注“对象的表达”问题，定位轴测图和透视图的角色，制作作为最复杂的表现形式的实体模型。设计过程、反馈以及通过计算机作为可视化工具进行建模和表现也是工作的组成部分。

Module 1: The Room

The "program" of Basic Design is a beginning. We aspire to not only "act" and to design but also to understand how to "critically" evaluate our work.

In the first phase of our "structural basic architectural design" program, we develop a simple architectural object, the workshop of a sculptor. It introduces architectural space and form in relation to "givens" and functions. The first step introduces some givens to kick-start the design process. A piece of land is introduced together with some high trees and an old tower. Materials for the building are at hand. But more than other requirements, we want to introduce into the "structured basic architectural design" program, a "Basic Architectural Element" (BAE). This BAE is, first of all, a "catalytic agent" and a common denominator in our work. We add as further quantity requirements, one cube of 6mx6mx6m, four cubes of 3mx3mx3m, and one cube of 4.5mx4.5mx4.5m. In this way, BAE turns into BAU. The six units represent functional requirements, gallery space, supporting spaces, and a storage space. They are starters in our work.

Earlier in this documentation, we have introduced the "diagram of a complex architectural system". It relates architecture design as a system to three sub-systems: to conceive, to construct, and to place. We want to define architectural design as a process in which "to conceive" and "to construct" are interrelated. As such, it is an essential rational process.

We will achieve some gains. We discover architectural space. We recognize the beginning of an "architectural language". We relate space, form, and functions. We operate with space-defining elements and identify types of space. We relate functions and openings. We discuss the inside/outside relationship of spaces. We deal with the envelope and the façade.

We have to deal with the role of media, of scales of representation and various forms of information, the plan, the section. The "expression of the object" is a concern. The roles of isometrics and perspectives are discussed. The physical model as the most complex form of representation is also a concern. The design process, feedbacks, and the computer as a visualization tool for modelling and representation have been part of our work.

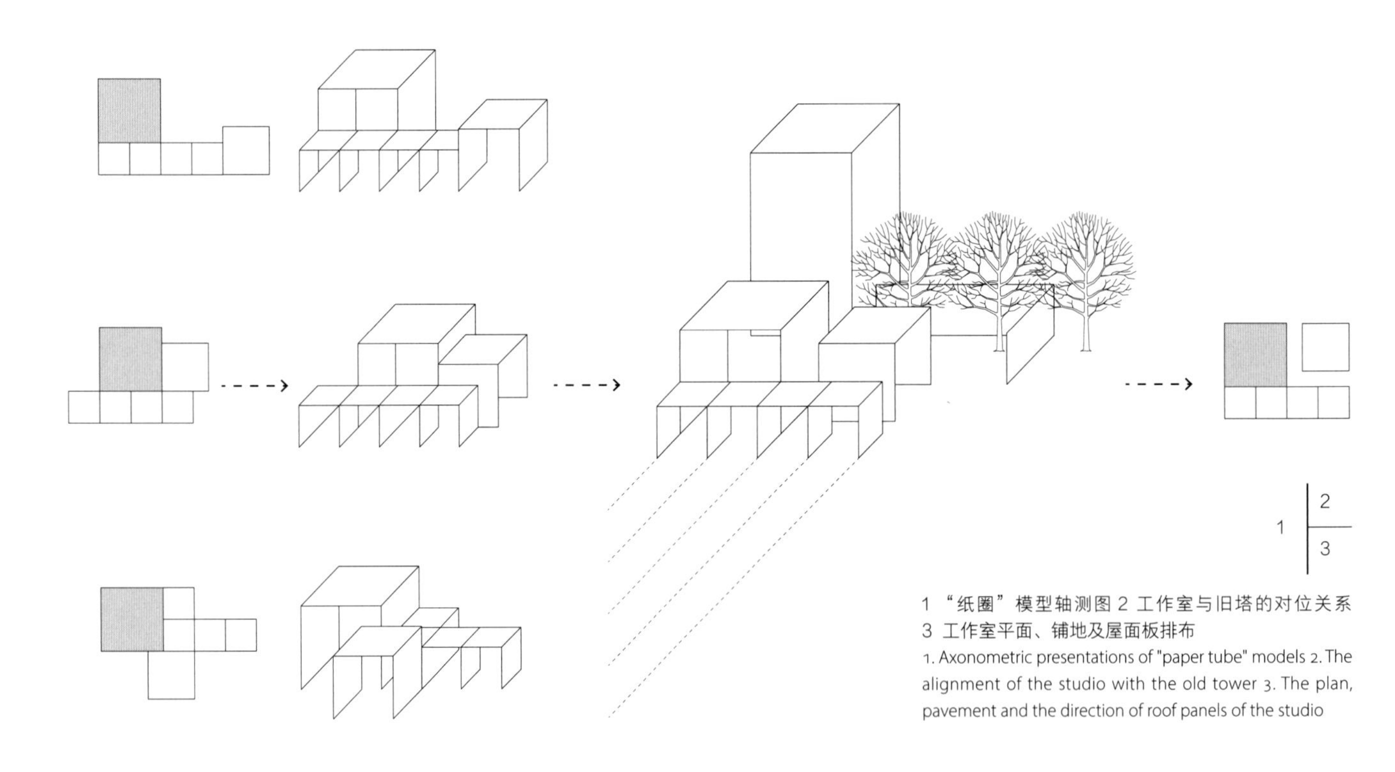

1 “纸圈”模型轴测图 2 工作室与旧塔的对位关系
3 工作室平面、铺地及屋面板排布

1. Axonometric presentations of "paper tube" models 2. The alignment of the studio with the old tower 3. The plan, pavement and the direction of roof panels of the studio

第一轮试验：由6个立方体构成的工作室。开始尝试3种不同的组合方式。将第二个组合进一步发展。

The given of this exercise: the “studio” formed by six cubes. Three kinds of combinations are made at the beginning. Scheme 2 has been further explored.

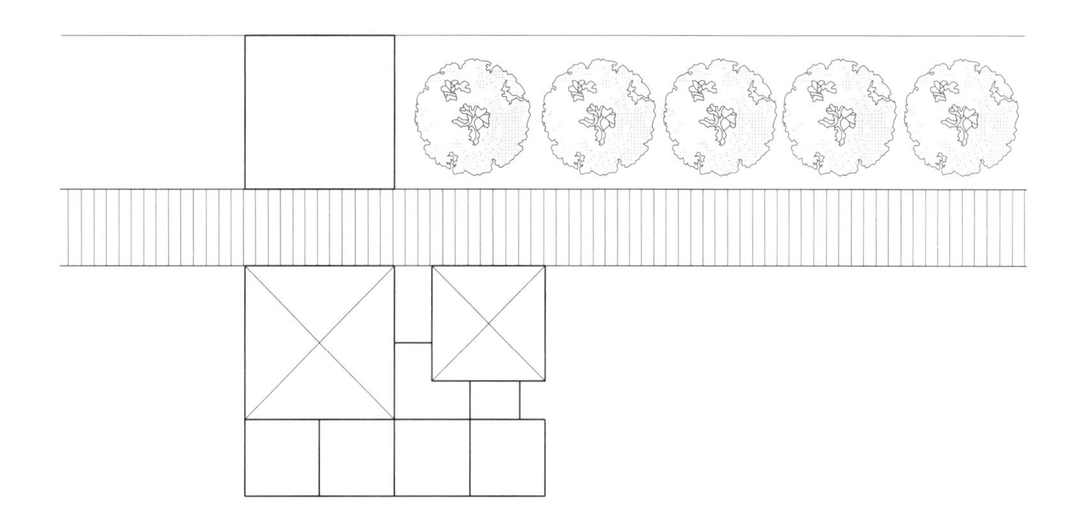

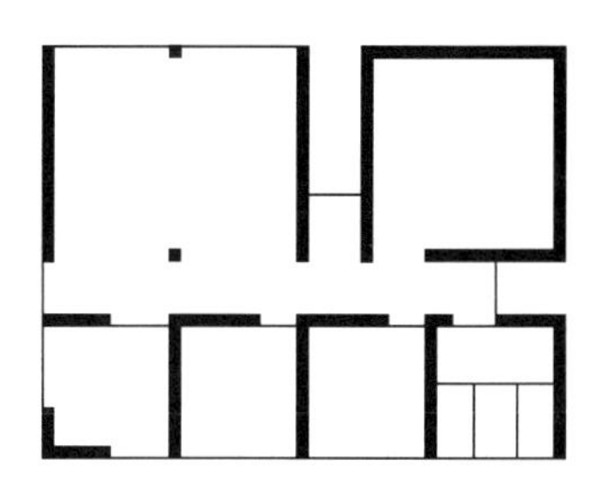

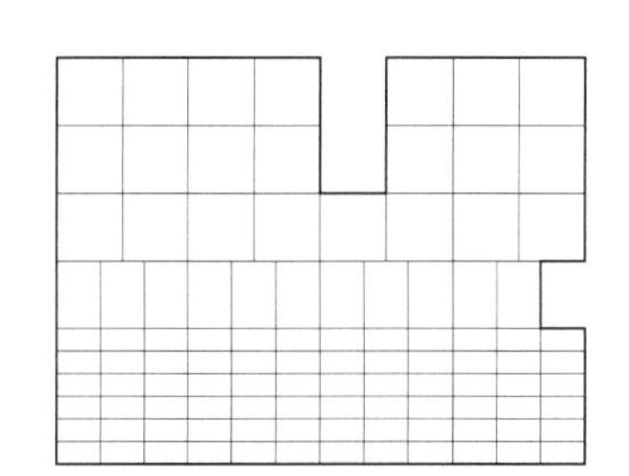

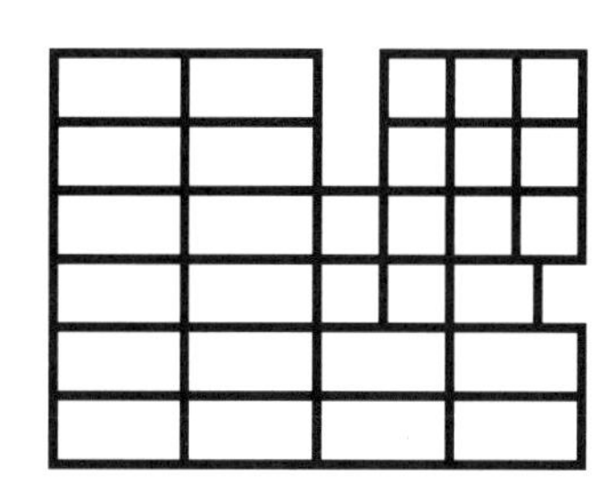

逐步演化出的平面以及空间之间的联系；铺地以几何形式呼应平面关系；屋面板排布呼应建筑体量。

尝试了几种不同的体量定义方式。概念似乎很简单。假若我们仅着眼于体量或外观，那么确实如此。假若我们着眼于内部空间的组织、功能的联系、开口呼应内外的布局，最后考虑由体量产生的空间，那么复杂性就产生了。

The floor plan has evolved and the relation and interaction of spaces have developed. The floor is geometrically developed in an attempt to support the plan. The direction of roof panels is conceived to support the volume.

Several attempts to define the volume have been made. The present concept seems to be very simple. It appears so if we look only at the volumes or the outside. However, complexity can be discerned if we look at the internal organization of spaces, the relation of functions, the inside-outside dispositions of openings and finally, the spaces generated by the volumes.

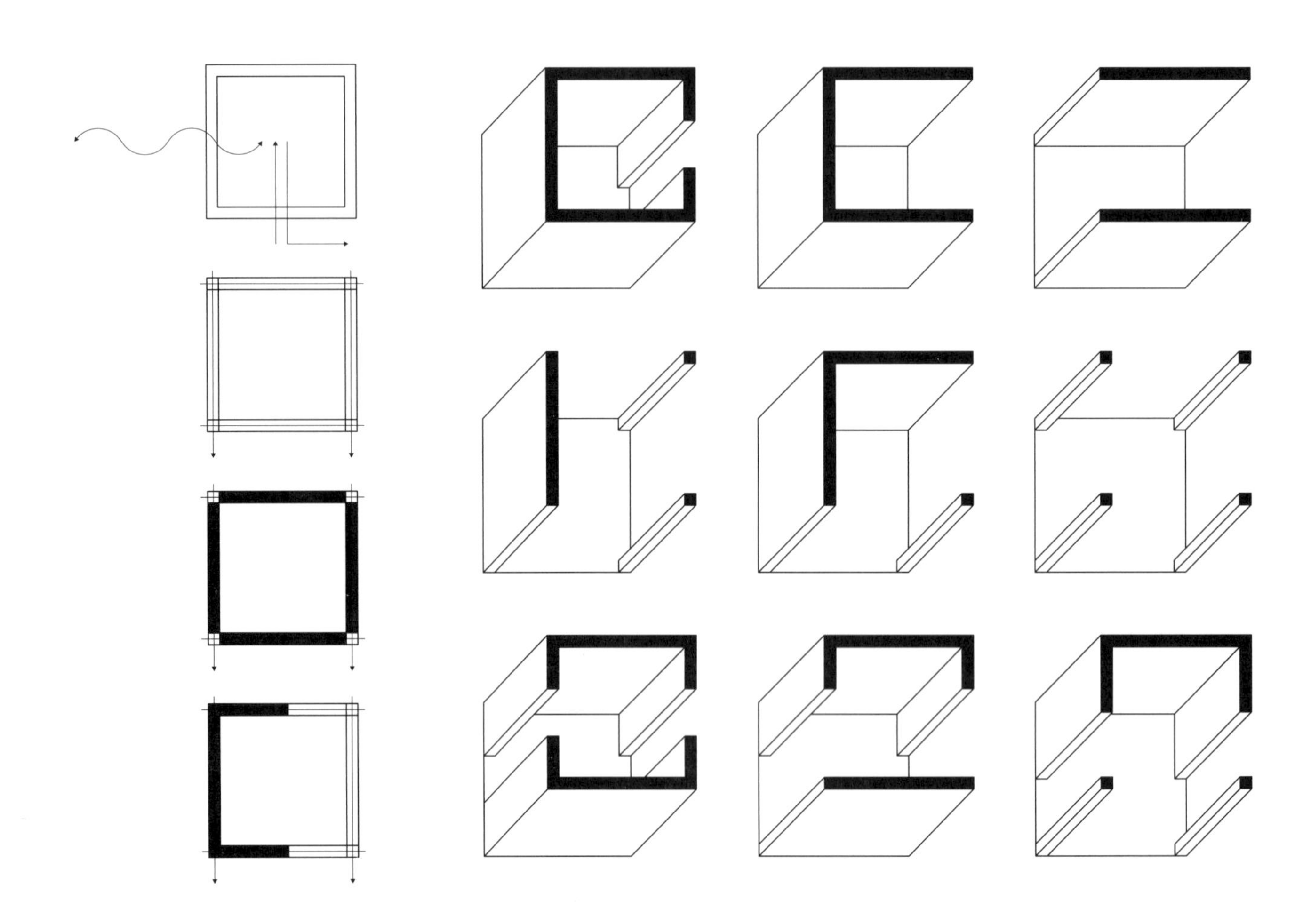

深化第二个设计，引入开口的问题。平面元素定义了使用功能、与家具相关联的空间界定和空间使用。

Scheme 2 has been further developed. Openings have been introduced. Elements of the floor plan have stipulated the functional use, furniture-related space definition, and space use.

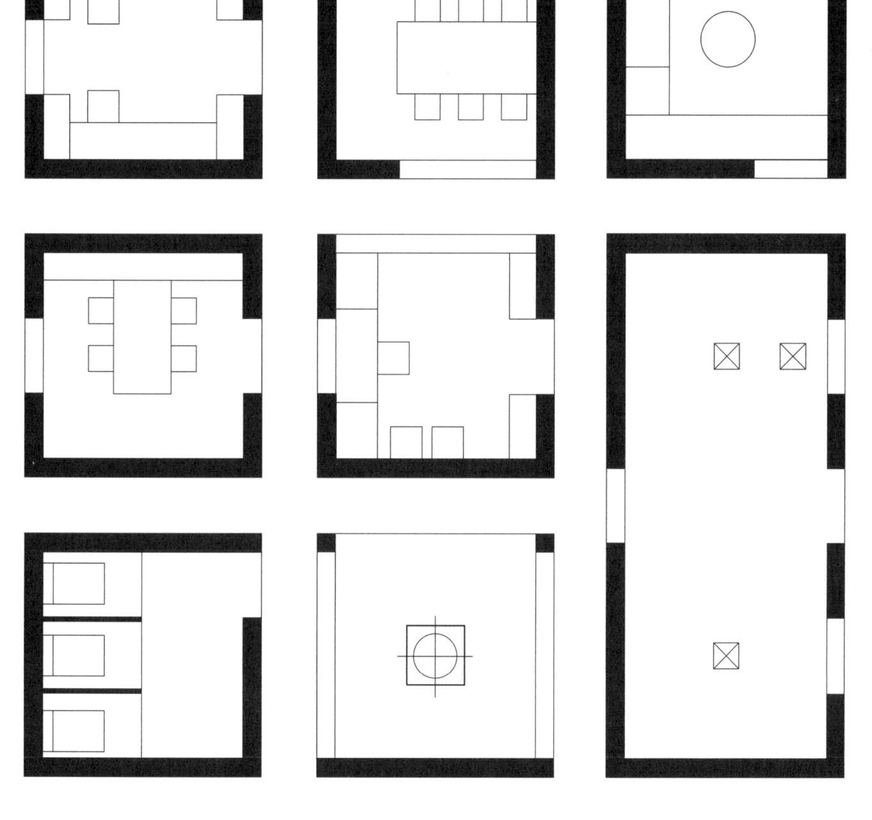

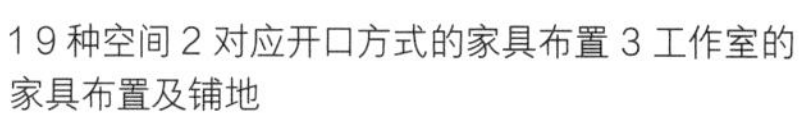

1 9 种空间 2 对应开口方式的家具布置 3 工作室的家具布置及铺地

1. Nine types of space 2. Furniture position in accordance to the opennings 3. Plan of the studio with furniture and pavement

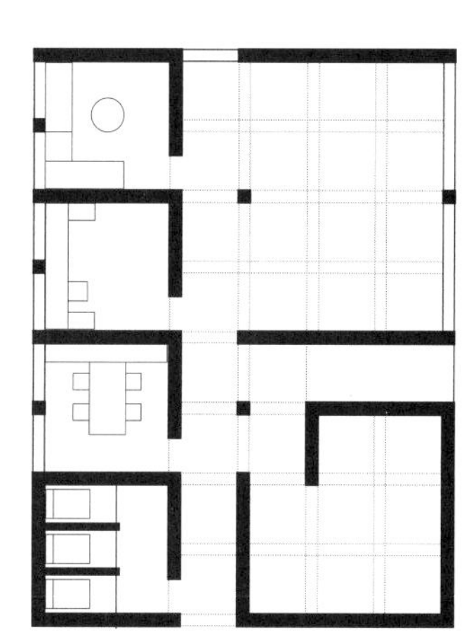

此处我们处理

· 形式
· 内外关系
· 空间
· 9种空间类型
· 空间定义元素
· 门窗洞
· 空间组织与组织的形式

We are dealing with

· Form
· Inside-outside relationship
· Space
· 9 types of space
· Space-defining elements
· Openings
· Organization of space and organized form

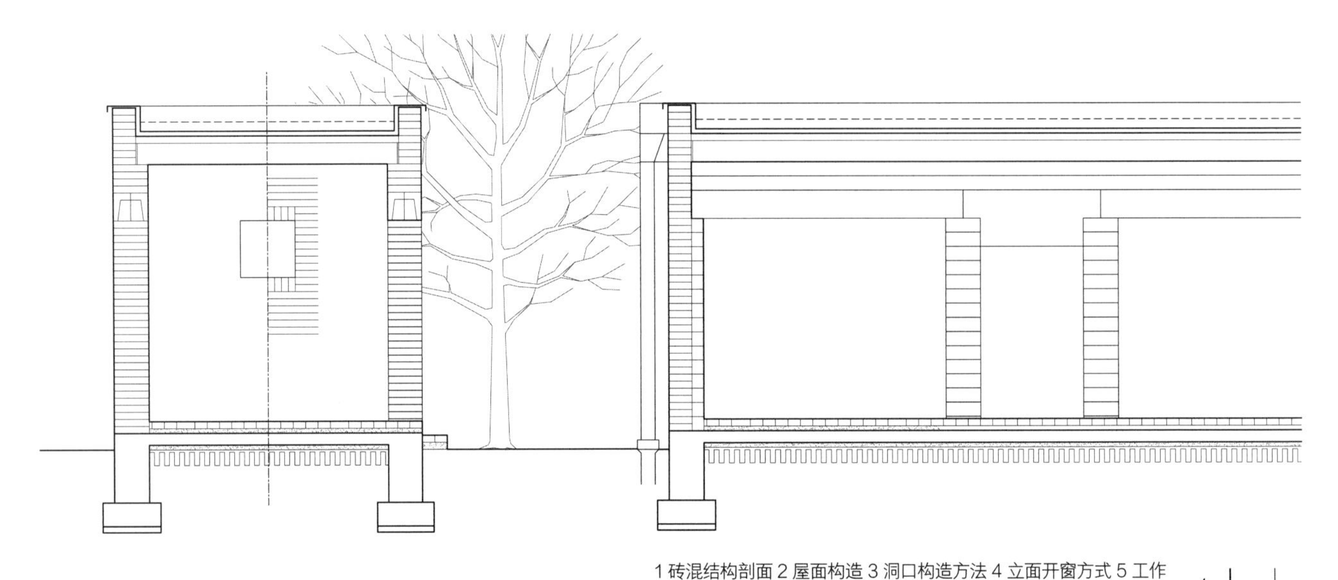

1 砖混结构剖面 2 屋面构造 3 洞口构造方法 4 立面开窗方式 5 工作室和场地的关系并加入绿植

1. Sections of a brick structure 2. Roof construction 3. Opennings in masonry 4. Types of fenestration 5. The relationship between the studio and the site with plantation added

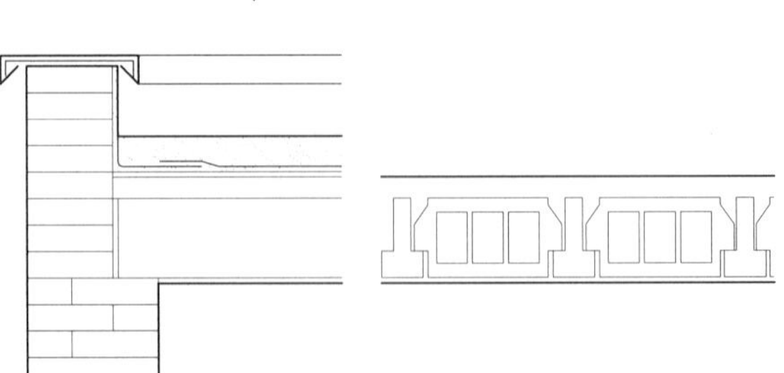

建造：结构系统的深化基于选材；基本建筑单元建立骨骼框架；构件与围护结构同时深化。

构件与细部：引入建造的逻辑。砖块的尺寸使地面、墙体、屋顶彼此关联起来。

The construction: the development of the volume's structural system is based upon the materials selected. The BAU constitutes the skeletal structure. The components, together with the envelope, have been developed.

The components, the details: the logic of building is introduced. The brick dimensions interrelate with the floor, the walls, and the ceiling.

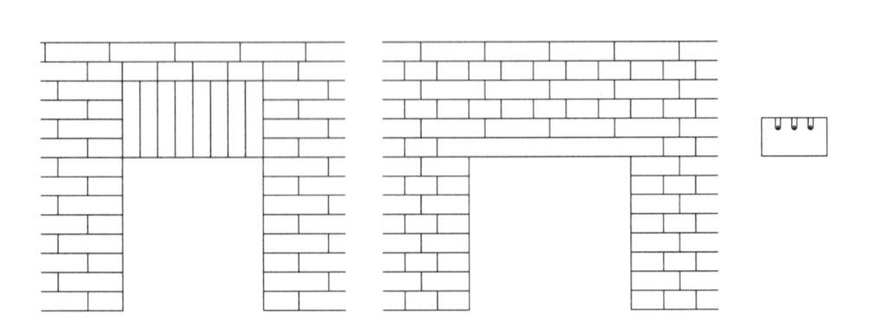

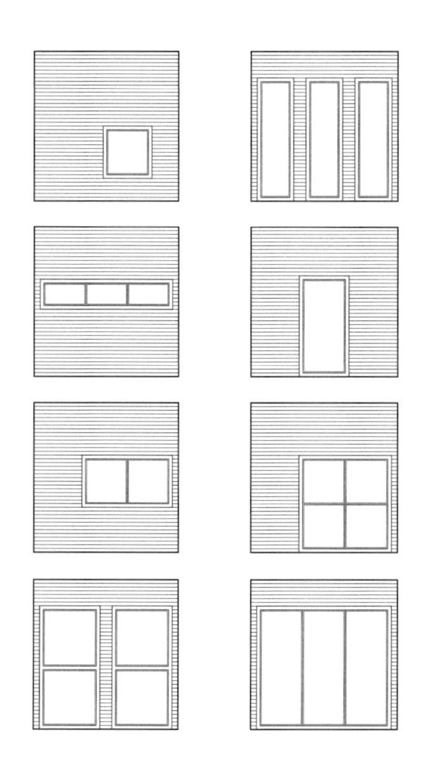

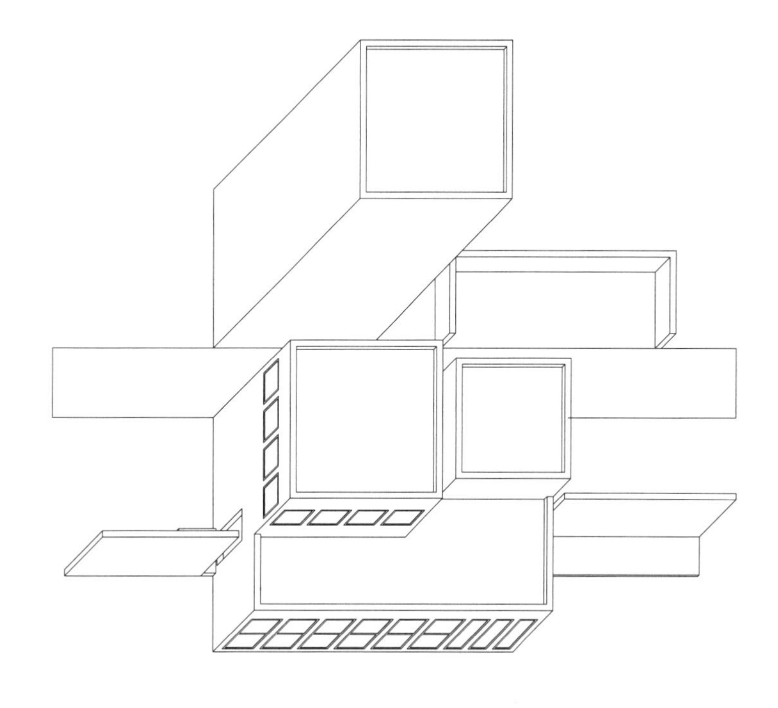

体量被定义为3m边长的立方体。外围护也由模数规范，由此得到了可以在整个画家工作室立面上通用的元素。顶部呈现的立面元素可以如上图那样适用于整个设计。

The volume measures 3m in length. The envelope (the facade) is also modularized. This results in elements that can be used for the overall facade of this sculptor's studio.
The drawings on the top show the facade elements, which are then used for the overall design, as shown above.

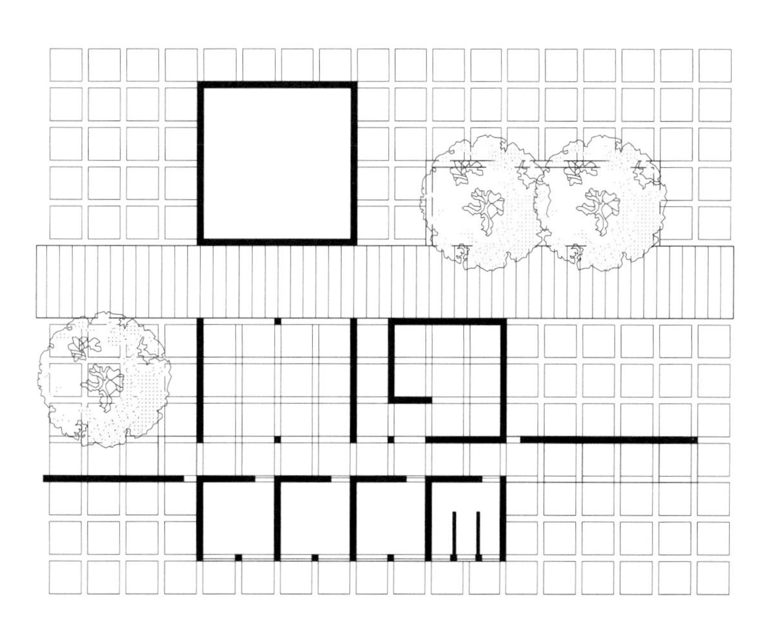

模块二：住宅

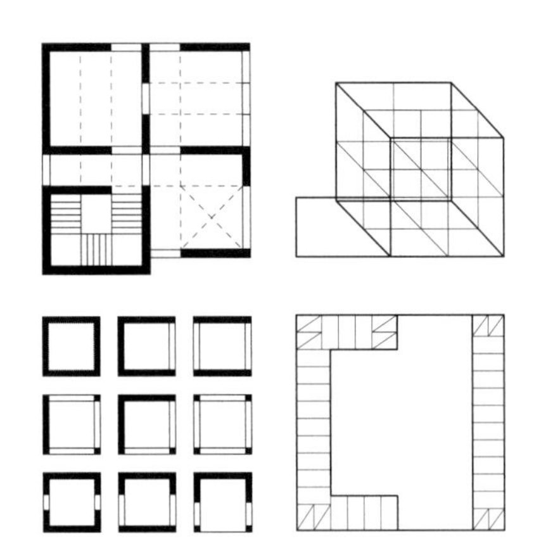

“建成”的环境 | A “constructed” environment

在“结构有序的基本建筑设计”教案中，我们试图让每一步的复杂性“可控”。第一个模块开启了这种编排，现在尝试在第二个模块中继续。模块一是雕塑家的工作室。应当清楚，标题本身并不重要，它只是一个框架；模块二将其延展。在同一场地上为雕塑家营造一个居所。我们以基本建筑元素/单元作为学习工具。模块一通过堆积单元（6个不同大小的立方体）来形成建筑体量，模块二则以由8个相同的基本单元组成的边长6m的立方体为基础进行操作。尽管建筑相当小，但仍不失为一个“教与学的装置”，用以扩展复杂度。

对于模块二，我们重点讨论结构/材料的问题。因为该范畴相当庞杂，必须一开始就加以限制。关键的限定来自材料：使用24cmx24cmx48cm的混凝土块，其尺寸规范了尺度及结构的选择。我们在非洲的项目中[1]使用过这种砌块。它还限定了墙壁和开口的类型，在某种程度上定义了一个复合建筑系统。

正如开头提到的，我们运用“类型”的概念。基本型或基本结构单元来自边长3m的立方体，它类似于混凝土砌块的角色，可以借助它达到各种功能要求。

[1] 1985年前后，克莱默教席受委托为坦桑尼亚的莫罗戈罗市设计建造一个学校（包含教室和宿舍）。团队以方形纸圈模型来规划场地，并在设计中运用模数（轴线距离3m）规范基本构件的尺寸：混凝土砌块尺寸如正文（考虑了抹灰厚度）；墙、柱厚25cm。

Module 2: The House

In our program of "structural basic architectural design", we try to build "controlled complexity" in every step. The first module kick-started this program, which we now try to continue with the second module. Module I dealt with a building—a workshop for a sculptor. It should be clear that the title in itself is not important. It serves only as a "functional framework". In Module 2, we expand the framework. We want to build a habitat for our sculptor on the same site, using a basic architectural element/unit as the learning tool. While we dealt with an architectural volume as a congregation of BAUs (6 cubes of various sizes) in Module 1, Module 2 deals with a cube of 8 original basic units. While the intended object is rather small, this "habitat" of 6mx6mx6m should serve as a teaching-learning device. With it, we want to increase the complexity of our work.

Module 2 shows that we want to deal specifically with structure/material. Since this field is quite complex, we have to introduce limitations even from the beginning. The key limitation for us should be the material; the use of concrete blocks of 24cmx24cmx48cm. Specifically, its dimensions will have to provide scale and structure. The use of these blocks (that we tested in Africa [1]) defines also the type of walls and openings, which to some extent, defines a complex building system.

As mentioned in the beginning, we want to use the concept of "type". The basic type or basic architectural unit for our work comes from a 3mx3mx3m cube. From this, similar to the concrete blocks, we develop our various functional requirements.

[1] Around 1985, Kramel and his team were commissioned the non-profit Morogoro Complex project in Tanzania. The team organized the site with cubic loop models and designed with modulus. The modular was 3 meters from axis to axis, unifying the dimensions of basic components (cement blocks as mentioned, 25cm for wall/column thickness).

1住宅基本平面的生成 2 可能的结构方式 3 空间定义元素——墙体的可能形式

1. Development of the preliminary plan 2. Two possible structural systems 3. The possible construction of walls as space-defining elements

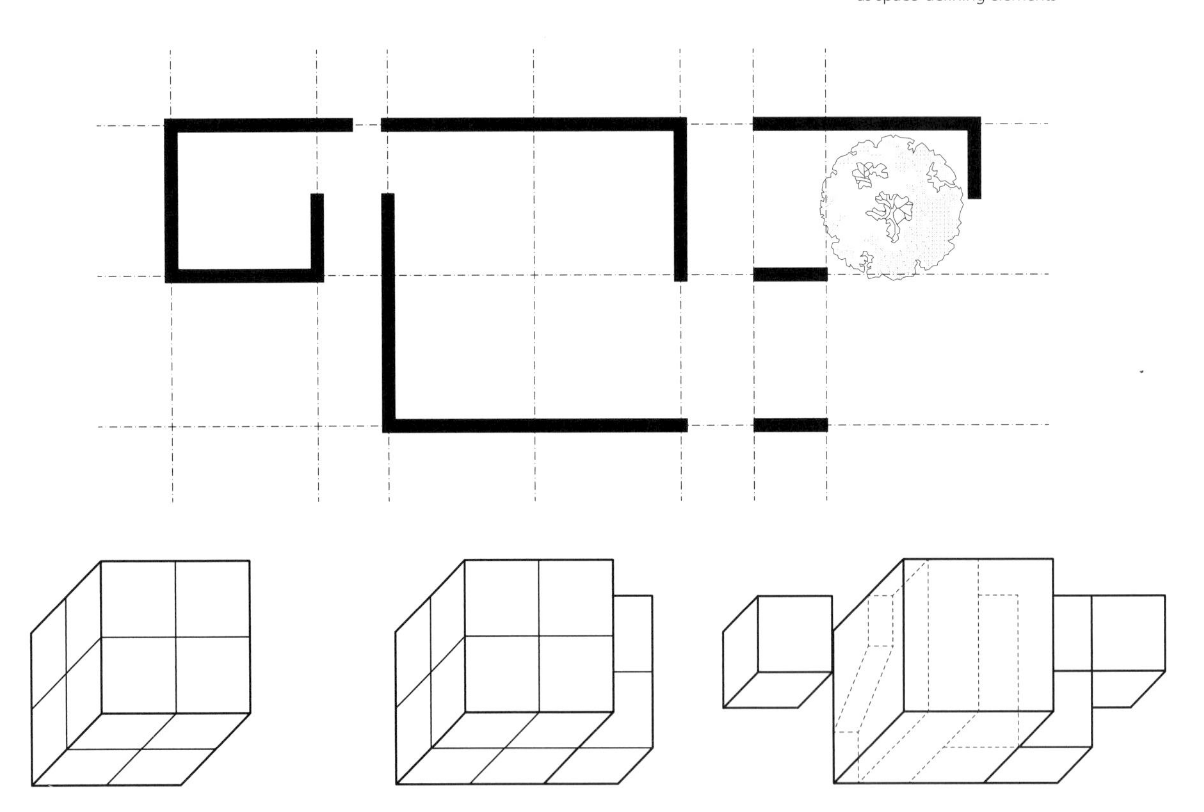

第一步：基本建筑单元的组合

这一步要设计一个供两人使用的住宅。基本建筑单元再一次成为设计序列的起点，形成边长为6m的立方体。它初始为一个几何实体，而后转化为建筑结构，再形成建筑概念。

其后扩展了两个元素——入口及花园。

Step 1: The composition of BAU

The task at hand is to create a "habitat", a house for two persons. The BAU is again the point of departure for a design sequence, giving rise to a 6mx6mx6m unit.

The BAU is first a geometric solid. Then it is transformed into an architectural structure and later, an architectural concept.

This core "habitat" is then extended by two elements—an entrance and a garden.

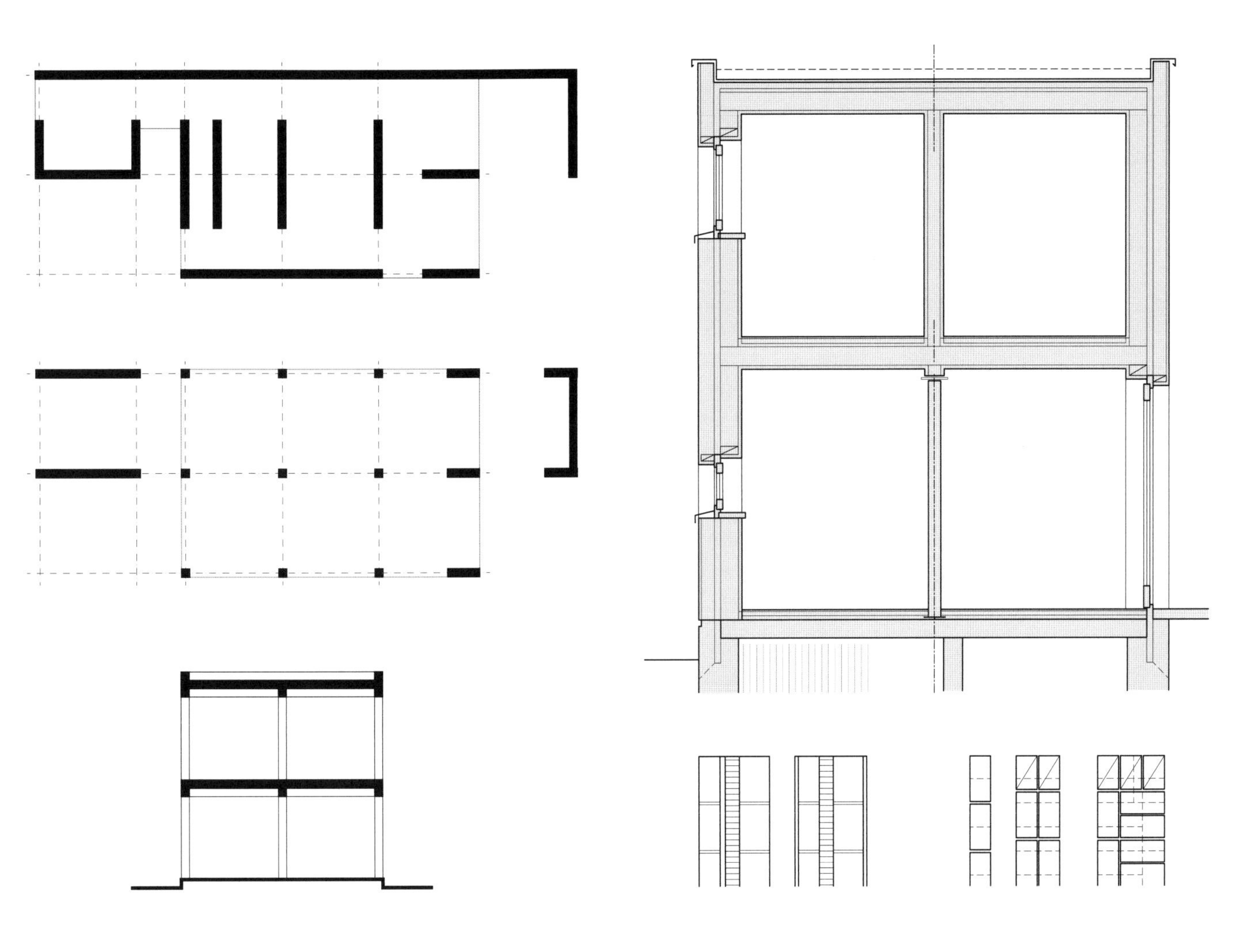

第二步：空间、结构、围护

首先处理结构问题，空间中的重力组织。左图显示了两种结构体系的可能性。

墙的建造——建造意指建筑中的工程部分，但同时与构想相关。

Step 2: Space, Structure and Envelope

First, we deal with the structural problem, the organization of gravity in space. Two possible structural systems are shown on the left.

Construction of walls—Construction refers to the engineering part of a building but relates also to the conceiving.

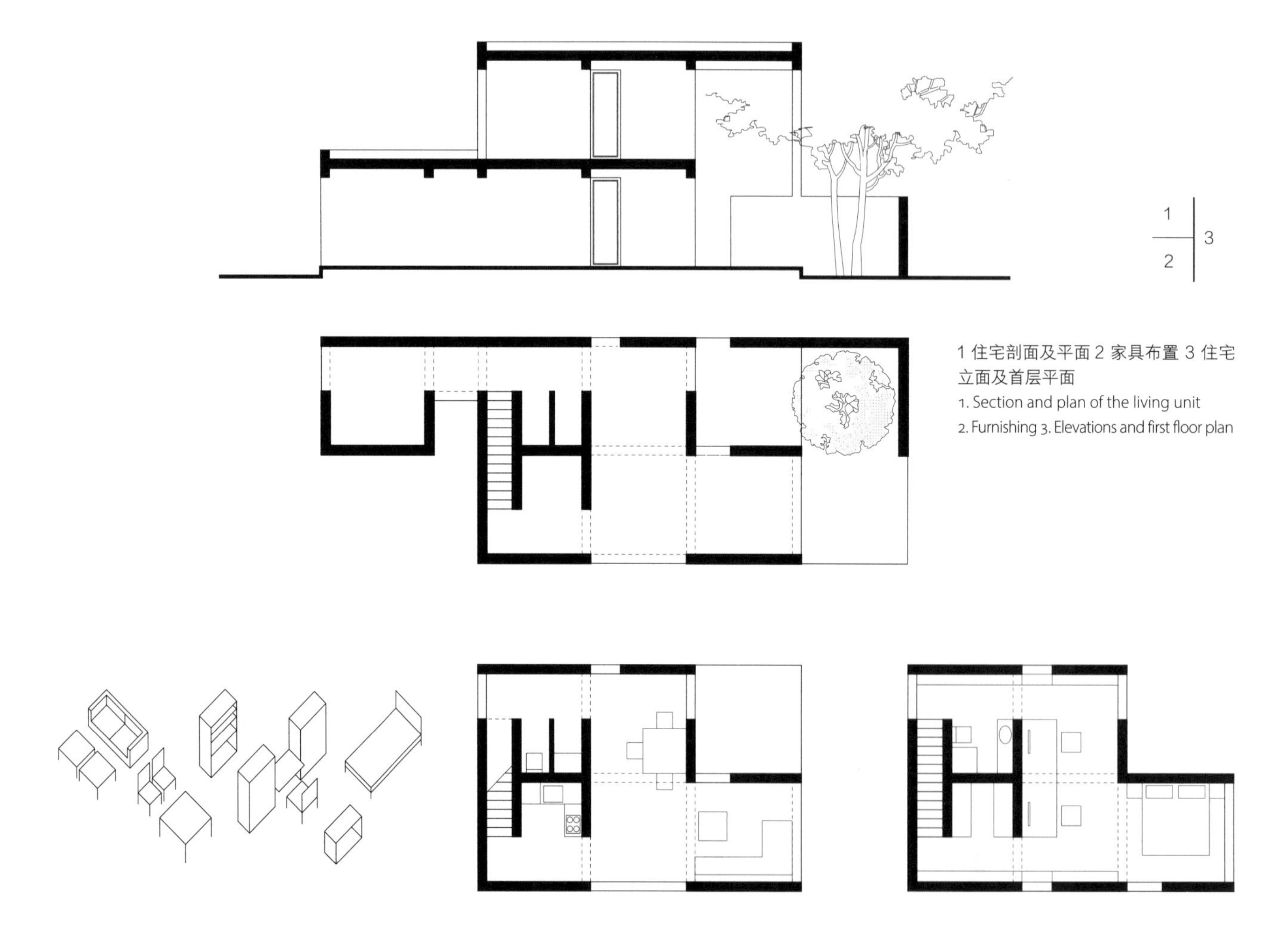

1 住宅剖面及平面 2 家具布置 3 住宅立面及首层平面

1. Section and plan of the living unit
2. Furnishing 3. Elevations and first floor plan

第三步：空间与使用

根据功能组织进一步深化平面。平面组织体现在剖面上，剖面体现垂直的层次及空间域。

平面呈现了9个空间类型中的一些。家居布置与人的生活相关；家具与生活及文化相关。

Step 3: Space and Program

A further development of the floor plan according to functional organization. The planar organization is reflected in the section, and the section reflects the vertical level and spatial zones.

The plan presents some of the 9 space types. Some of the spatial definition prototypes can be recognized here. The furnished plans relate to the life of the person. The furniture relates to living and culture.

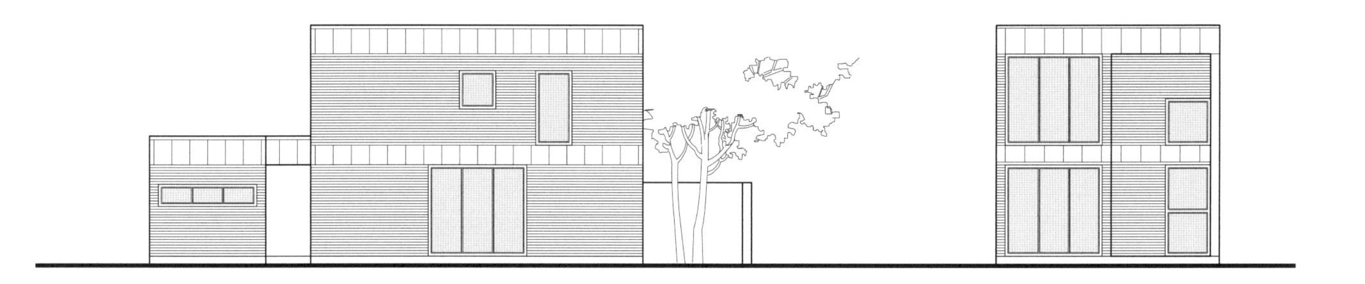

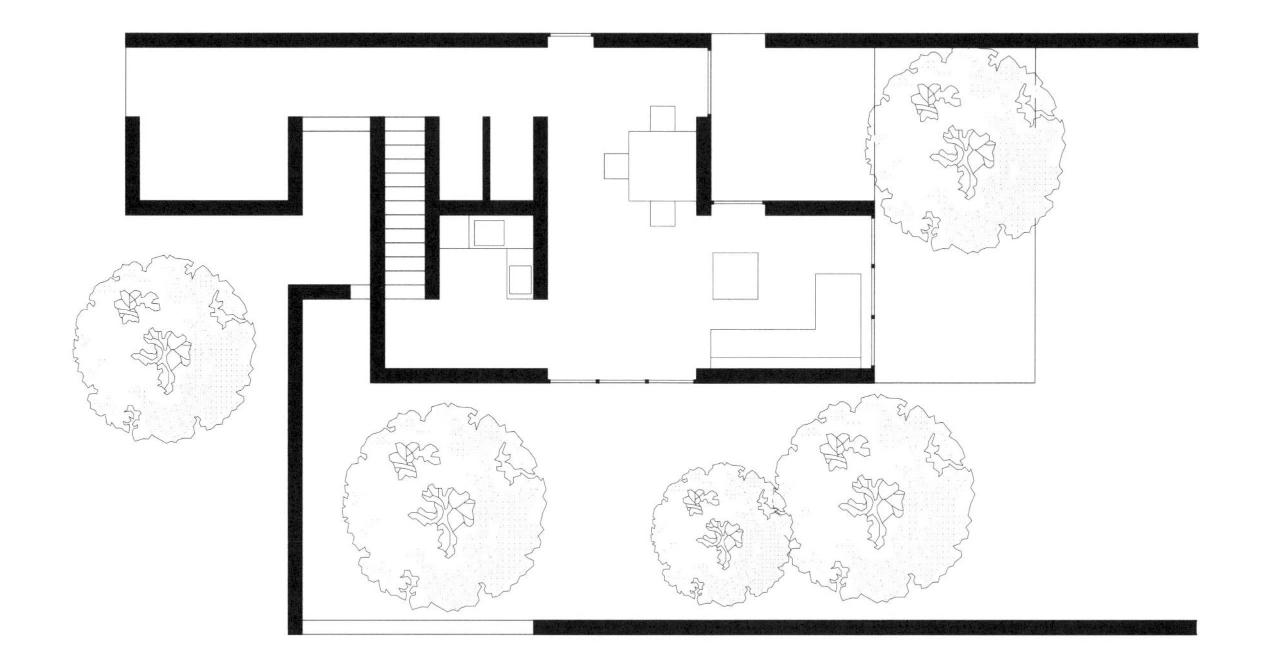

第四步：场地与场所

空间限定不仅包含了内部，还有外部的空间。从6m边长的立方体概念出发，定义了住宅结构，最后将其置于场地之中，并避免产生消极的空间。

通过逐步发展出来的建筑逻辑、材料、尺度及几何形态，进一步产生了建筑的表情。

Step 4: The Site and Place

Space definition covers not only the internal space but also the outside space. We define the core (6mx6mx6m) as our concept. We develop the structure (the construct) of our habitat. We then try to emplace the habitat and avoid any leftover space.

Through introducing the logic of the object, its material, scale, and the geometry, we then further develop the building's expression.

模块三：集群

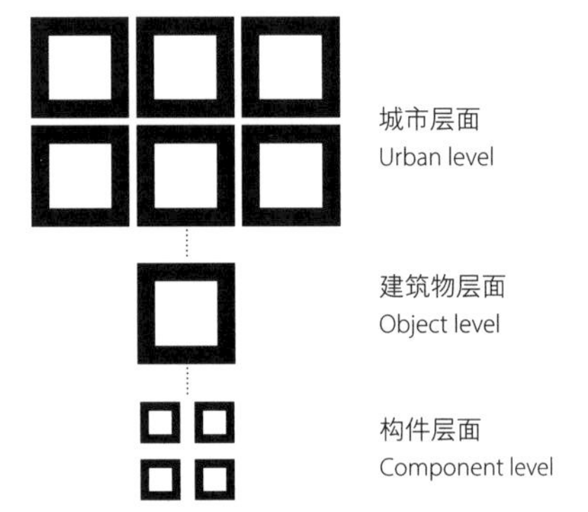

三个操作层面 | Three levels of operation

我们引介基础设计教程的同时，显然也是在讨论建筑设计初步。在模块三，相比前两个模块涉及更大的尺度，更多的建筑数量，并带入城市层面的思考。从“整体”出发，从系统到部分。引入“类型”的概念，尝试处理组织模式。最后一步才是构件，即“细部”的层面。

前两个模块处理的是建筑单体，而模块三关注的是建筑体和功能的组群——强调重复与个体、多个与单一的对比。三个操作层面分别是：

1.城市层面（多个建筑物）；

2.建筑层面（单个建筑物）；

3.构件层面（局部及细部）。

因为这3个层面的相互作用，对“概念”及“建造”的理解，以及对“建筑系统”、整体、子系统的协作理解就很重要。新的复杂性出现了：部分与整体；“常量”与“变量”；合作与竞争；群组与“村庄”；社会结构与混乱；空间类型；私密与公共空间；主空间、剩余空间及过渡。

还有另外一个考虑层面——“类型”。我们知道城市设计中的类型，但我们现在将其理解为传统的一部分。传统代表着对于成功解决之道的集体记忆。密度、私密性、安全、领域感及身份的问题随之进入视野。

Module 3: The Village

In our attempt to introduce a basic design program, it was clear to us that we were dealing with an introduction to architectural design. Compared to the last two, the third module of our documentation introduces a much larger scale and larger number of buildings. Following the introductory lines, we have introduced the urban concern into our work. We have progressed from the "whole", the system, to the parts. We have introduced the notion of the "type" and have tried to deal with patterns of organization. The last step introduces the level of the components, the "details".

While Modular 1 and 2 have dealt with individual objects, Modular 3 is concerned with the grouping of objects and functions. In this phase, the repeated versus the individual, and the many versus one are addressed. Three levels of operation come into play:

1. The urban level (many)
2. The level of the object (one)
3. The level of the component (the part and the detail)

Because of this interplay of three levels of work, our understanding of "conceiving" and "constructing" is important. And again, the understanding of "the architectural system", the whole, and the co-ordination between sub-systems is significant. Also, new levels of complexity are involved: the whole and the part; "constants" and "variables"; cooperation and competition; the group and the "village"; social structures versus chaos; different types of space; the private and the public space; formal spaces and the leftover, and the in-between.

Another aspect of concern involves "type". Historically, we have known about typology in urban design. But we understand "the type" as part of tradition. And tradition in this framework represents "the collective memory of successful solutions". With it, questions of density, privacy, security, territoriality and identity enter the picture.

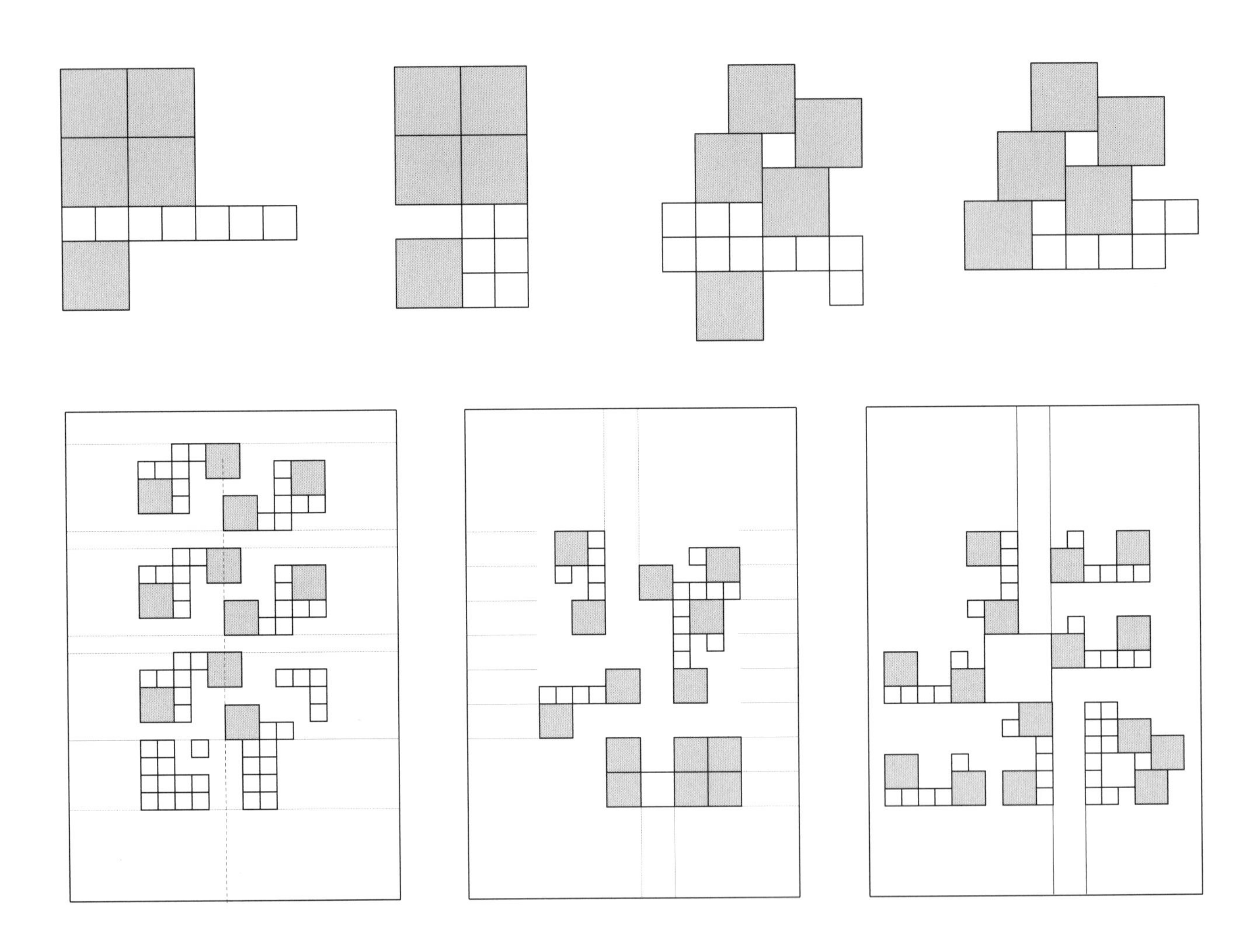

单个与众多，组群与组织，结果应是“井然有序”的。基本建筑单元形成统一整体及多样的局部变化。如何重复一个体量，又不造成单调。要考虑单元本身，但更需注意单元之间的空间。

本页展示了多种场地组织形式或模式，应理解它们的结构及“边缘”，它们用不同的关系组织了若干个居住单元和一个画廊。

The one and the many. The group and the organization. The result should be “well structured”. The BAUs should form a unified whole with diversity in the parts. The question is how to repeat a volume without creating monotony. The units are a problem, but more so is the question of neighbouring and in-between spaces.

Each one of the fields presents a different form of organization or pattern. Their structure and the question of the “edge” have to be understood. They relate to our organization of various living units and a gallery.

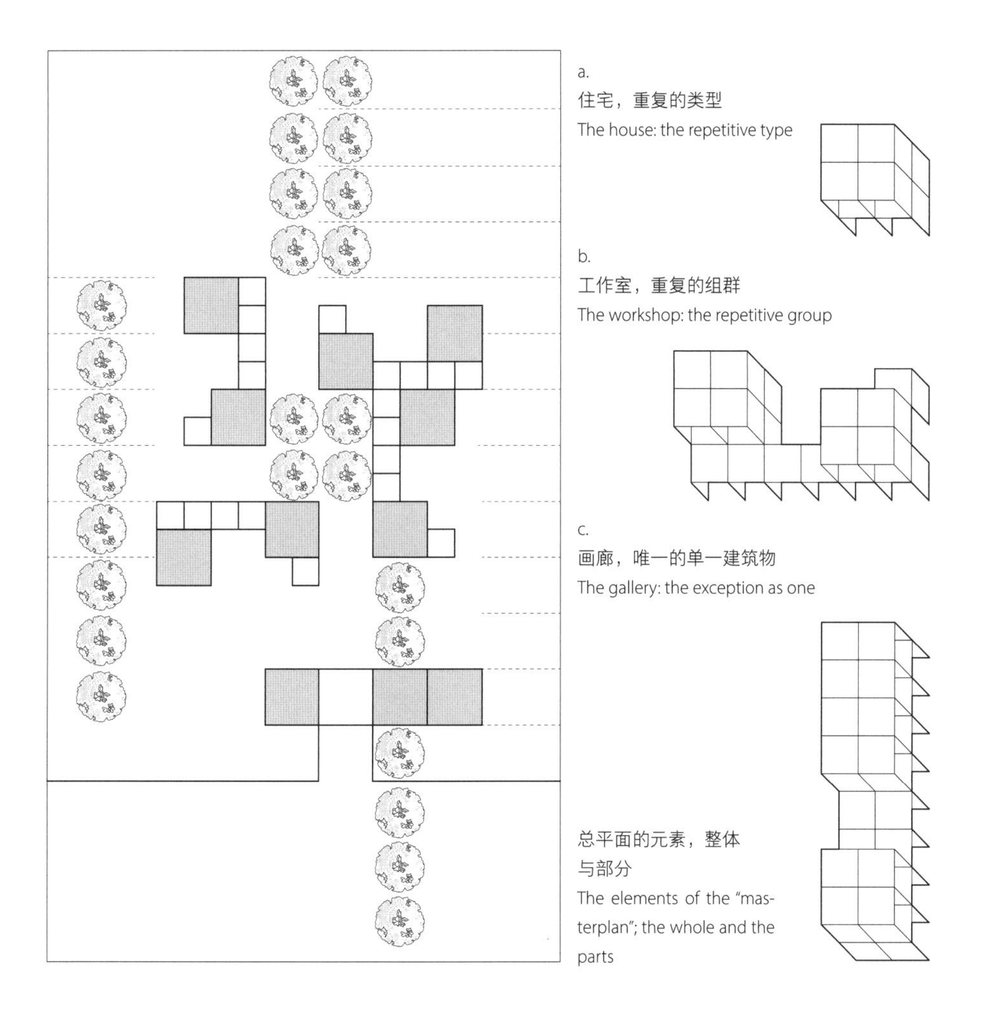

1 几种画廊平面组织 2 对应不同画廊平面的总平面组织 3 一种可能的总平面及与之对应的建筑单体

1. Possible plan organization of the gallery 2. Site planning according to galleries 3. Site plan and individual buildings

1 | 3
2 |

从单一的问题转向多个问题，对类型的研究是共同点。用袖筒模型代表大量的基本建筑单元。它们不但代表了体量，而且代表了空间。相同又不同。类型、类型学，是组织的工具。

基本建筑单元的使用以及“类型”之间的互动。多种功能类型构成一个集合体，一个社区。建筑之间的空间、体量、植物、光影。

We want to move from the problem of one to the problem of many. The study of types serves as the common denominator. We use "sleeve" models to represent a large number of BAUs. We want to present the elements not only as volumes but also as space. All the same but different. The type, the typology, is a tool for organization.

The use of the BAU and the types in interaction. Various functional types render a unity, a community. The in-between space, the volumes, the plants, light and shadow.

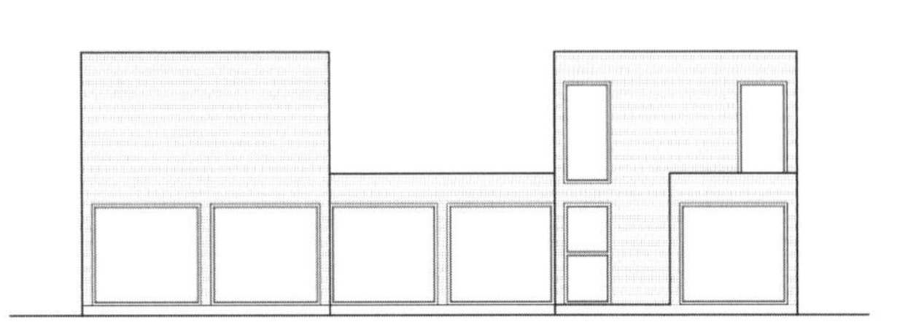

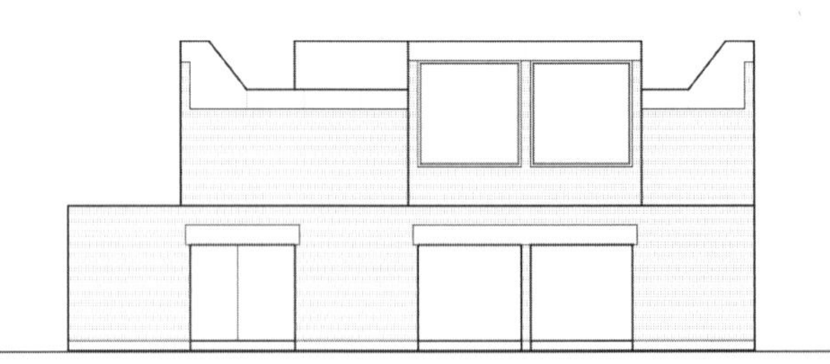

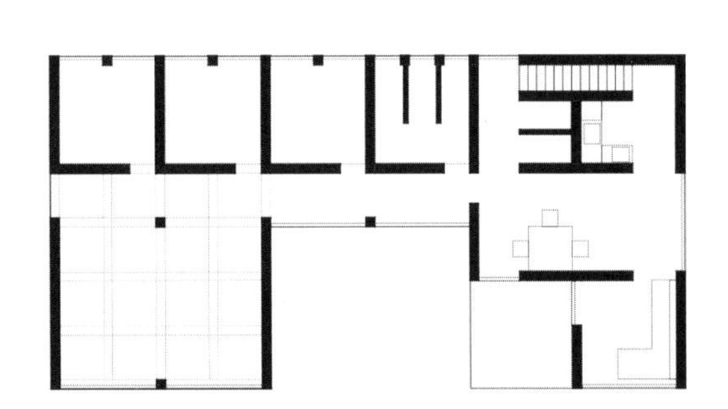

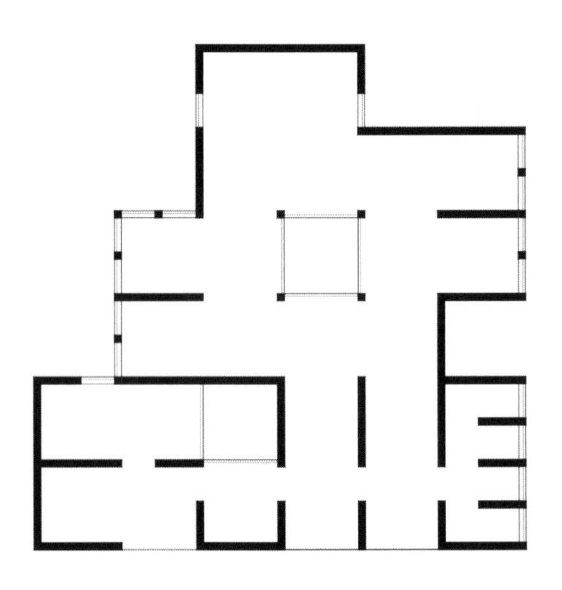

尝试总体组织后，转向局部设计：住宅、工作室，以及作为特殊元素的画廊。统一和变化，基本建筑单元成为多样化的工具。

作为一个特殊的单体，画廊运用与社区共同的建筑语言。体量由光照、空间及环境所决定。平面及基本建筑单元的组织符合该建筑的特殊角色。入口将人们引向中心空间。

After attempting to develop the organization, we moved to the parts: the habitat, the workshop, and as a diversion from the pattern, the "art gallery". Unity and diversity. The BAU becomes a tool for diversification.

The art gallery is a very different object, but it shares the same "language" of the community. The volume of the gallery is defined by light, the space, and the environment. Plan and the organization of the BAU support the specific role of this building. The entrance leads to the central space.

1 住宅与工作室组合体的立面与平面 2 画廊的立面与平面 3 总平面
1. Elevation and plan of the living unit and studio combination
2. Elevation and plan of the gallery 3. Site plan

1 | 2 | 3

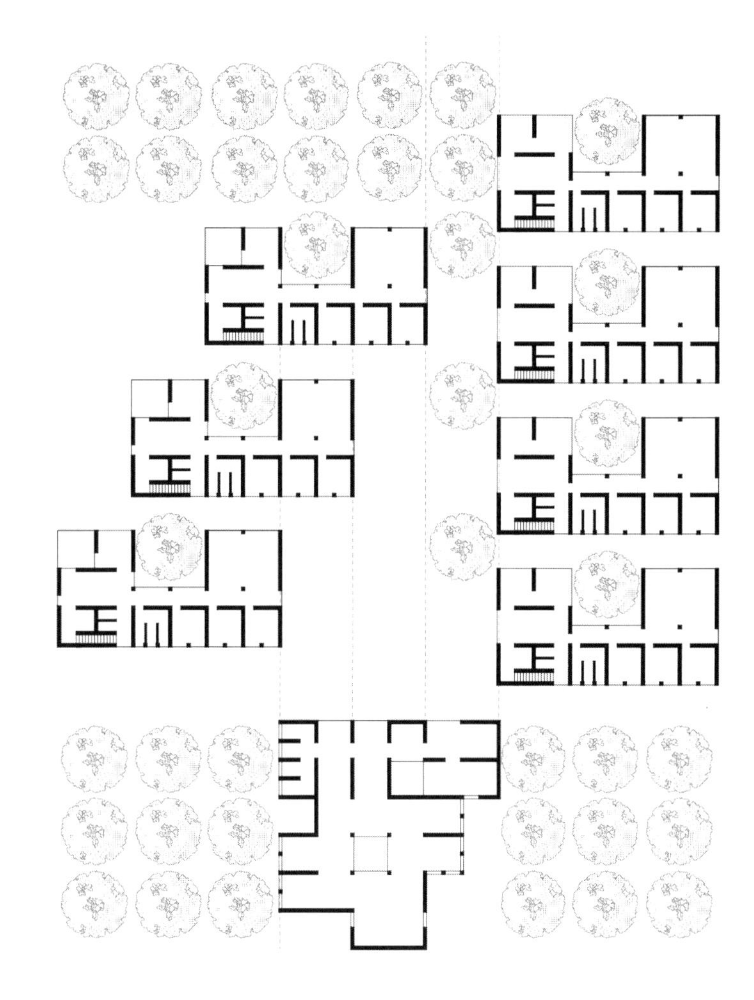

由于在模块一和模块二中我们已经反复地讨论了建筑构件层面的问题，在模块三中我们则不再深入。

社区作为一个整体。不同的群组形成了不同的空间。道路、中心，以及绿植。社区的总平面示意图中，体量被有力地组织起来，绿植与空地遵循着统一的秩序。

Since we have discussed the building component level repeatedly in Module 1 and Module 2, we will not go into it in Module 3.

The community as a whole. Different groups forming different spaces. The roads, the centre, and the plants. In the schematic plan of the resort, the volumes are strongly organized, the plants and free space follow a unified order.

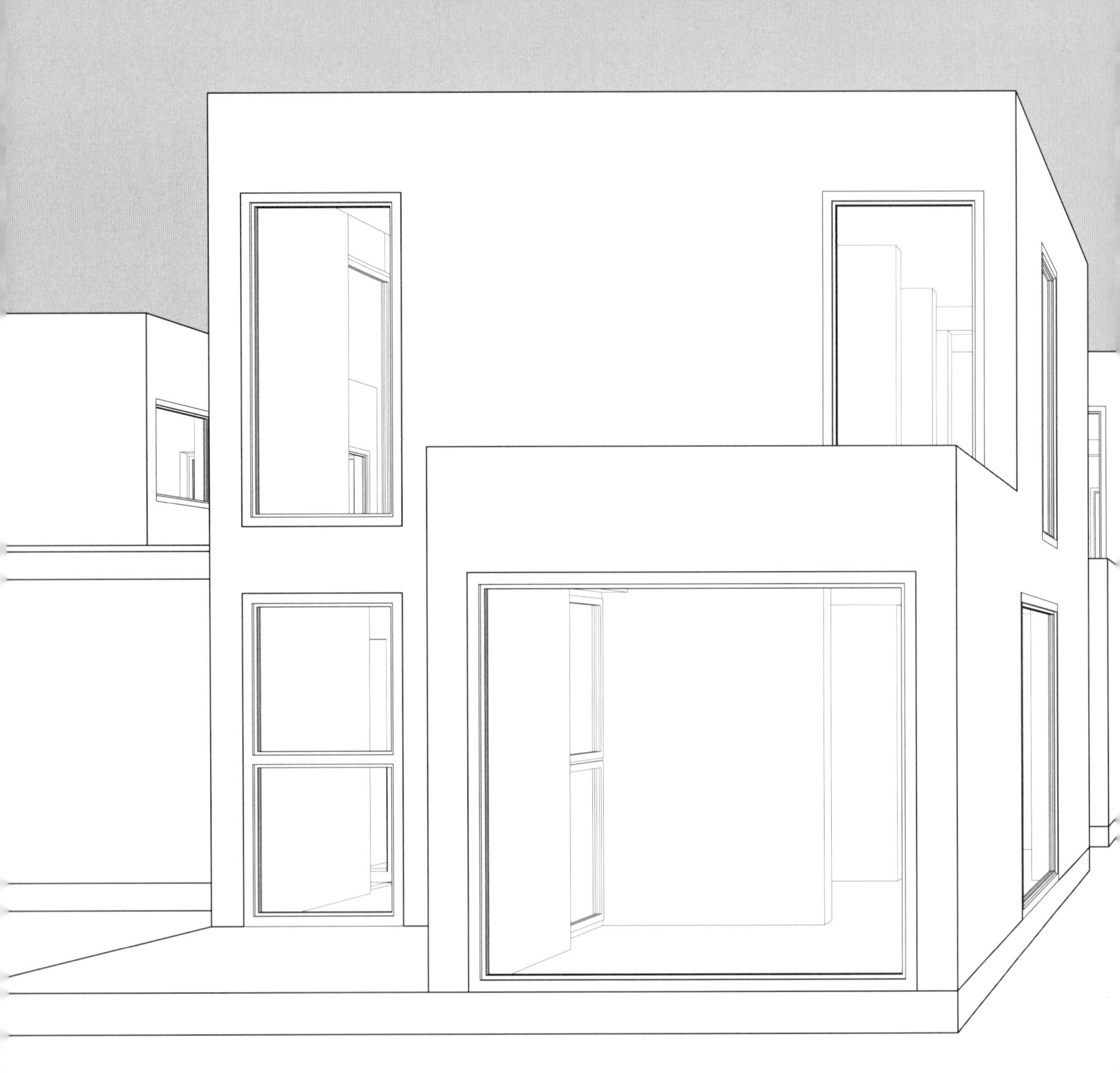

建筑群透视 | Perspective of the complex

3 三要素 | The Three Determinants

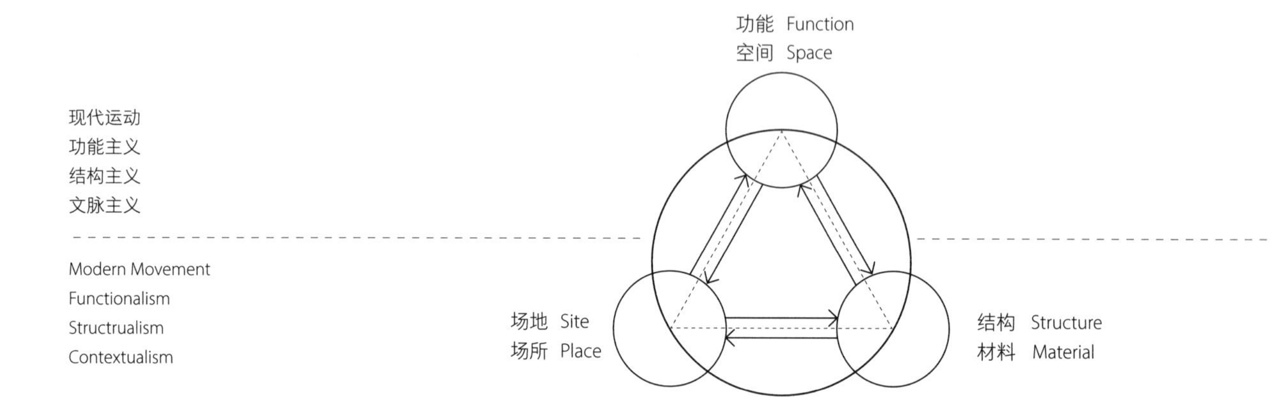

核心观念系统 | The system of vital ideas

“现代建筑运动”常常被视为功能主义，“形式追随功能”，其后出现了新的思潮。功能主义无疑是我们的思维基础，接着是“结构主义”，而“文脉主义”几乎不可避免地紧随其后。作为现代运动的一部分，应当把这些发展阶段理解为将现代建筑推向更高复杂度的进程。

这一思路的内涵是建筑形式首先是各种内力作用的结果，正如我们的建筑系统图解所示。它帮助我们理解建筑形式的决定因素：空间/功能，结构/材料，以及场地/场所。因为每个要素本身已经是复杂的集合体，我们将它们作为一个更大的建筑系统的子集去理解。

同时，“建筑系统”随着时间推移而变化。“核心观念系统”改变了我们对建筑的思考。历史上以风格来谈论建筑，如哥特风格、罗曼风格……如果我们认同整体的“设计决策的筹划”定义了一个系统，我们就不能将其中任何一个阶段或计划称为“风格”，而必须讨论“建筑系统”。

建筑三要素是建筑设计的基础，通过每周的练习来逐一强调各个要素，使学生形成对单个概念及复合情况的理解。这个图解也是学生掌握其他建筑要点的基础，如实体、比例等。

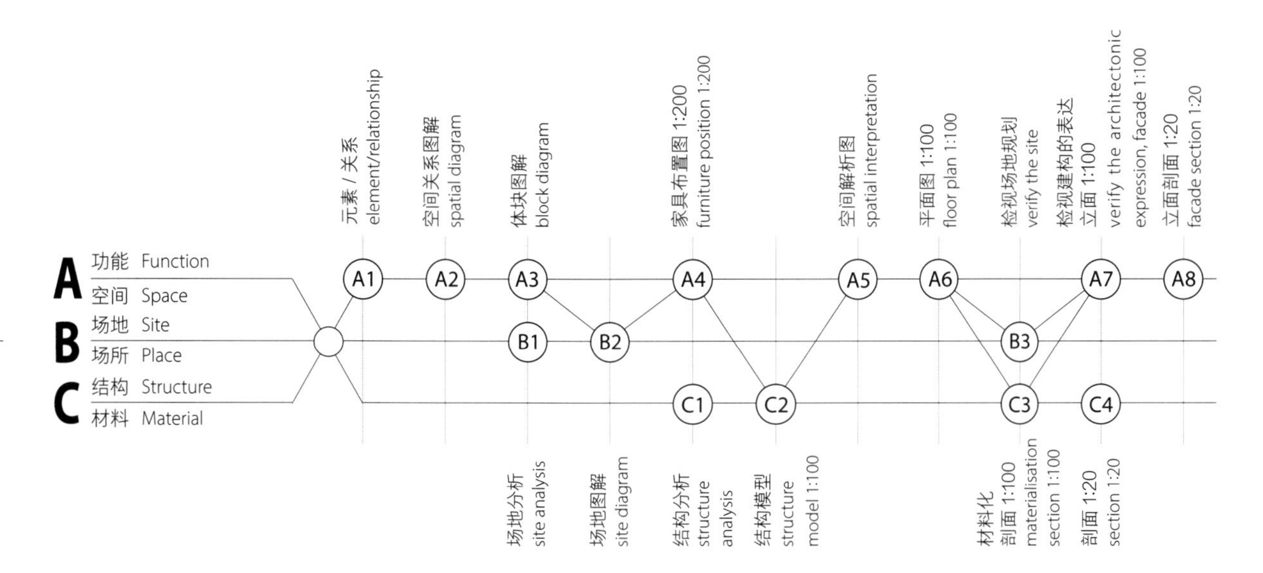

1985学年设计问题四：练习序列与设计方法的内在联系 | The relationship between the exercise sequence and design method, Phase IV, 1985

While the "Modern Movement" had often been identified with functionalism, "form follows function", new forms of "thinking" later emerged. Functionalism is without question the base of our thinking. "Structuralism" was the next step, and was almost inevitably followed by "contextualism". It is important to understand these developmental phases as steps leading the Modern Movement to new levels of complexity.

The insight inherent in this thinking is that architectural form is, first of all, the result of internal forces, which we have addressed in our "system diagram". It helps us understand architectural form and its determining factors, which are space/function, structure/material, and site/place. Since each of these determinants is in itself a rather complex unit, we can understand them as interacting subsystems within a complex larger architectural system.

At the same time, "the architectural system" transforms in time. The "system of vital ideas" has transformed our thinking about architecture. Historically it was possible to talk about styles. The gothic style was followed by the Romanesque style…If we accept that the "programming of design decisions" in its entirety defines a system, we cannot identify any phase or program as "style", instead we must talk about the "architectural system".

This triad of concepts is fundamental to the Basic Design Program. Each concept is emphasized separately in the weekly exercises so that students may build an understanding of the implications of each, alone and together. This triad also is the base from which students learn to explore and gain knowledge about other architectural notions such as mass, scale, etc.

空间/功能　Space / Function

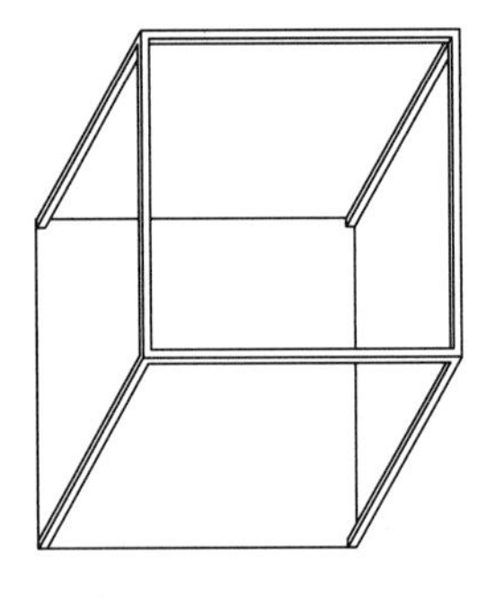

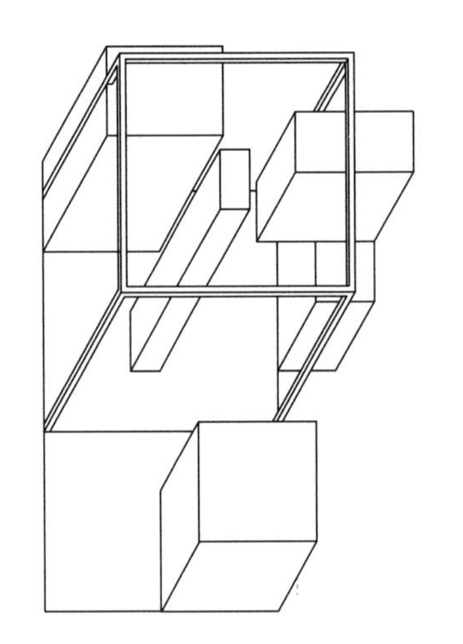

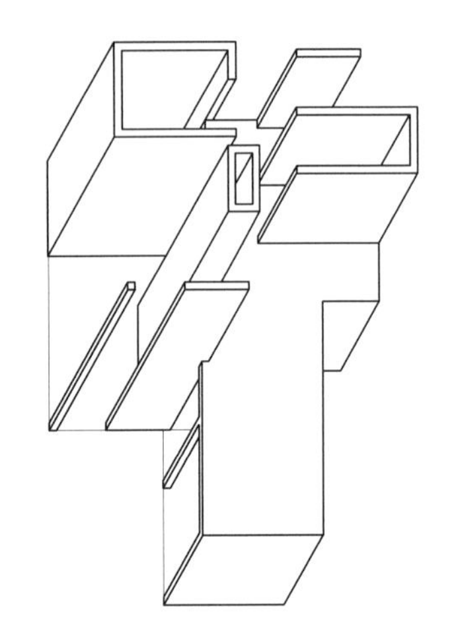

1	2
	3

1 居住单元空间组织的形成 2 居住单元的组合 3 平面及结构的模数关系

1. Spatial development of the living unit 2. Grouping of the living units 3. Modular of the plan and structure

为双人居住单元设置一系列限制条件（元素、结构形式），没有场地限制。首先通过给定元素组织限定元素之间的空间。然后拓展已有的空间组织，加入新的空间定义元素。

原有的空间域扩张了，超出了初始边界（结构框架），从而赋予周边环境一种结构。内部与外部从而联系起来。由此引入了（风格派的）“开放平面”。

A set of limitations (elements, structural forms) have been established for a "living volume" for two persons, and a zero environment (no limitations) has been given. The "givens" have to be organized; space as an "in-between" (between space-generating elements) has to be generated. Then, the originally defined spatial configuration is expanded; additional space-defining elements have been added.

The original spatial zones have to be expanded, by going "beyond" the original limits (the structural frame), thus structuring the surrounding environment. The "inside" and the "outside" become integrated. The "open plan" (De Stijl) is thus introduced.

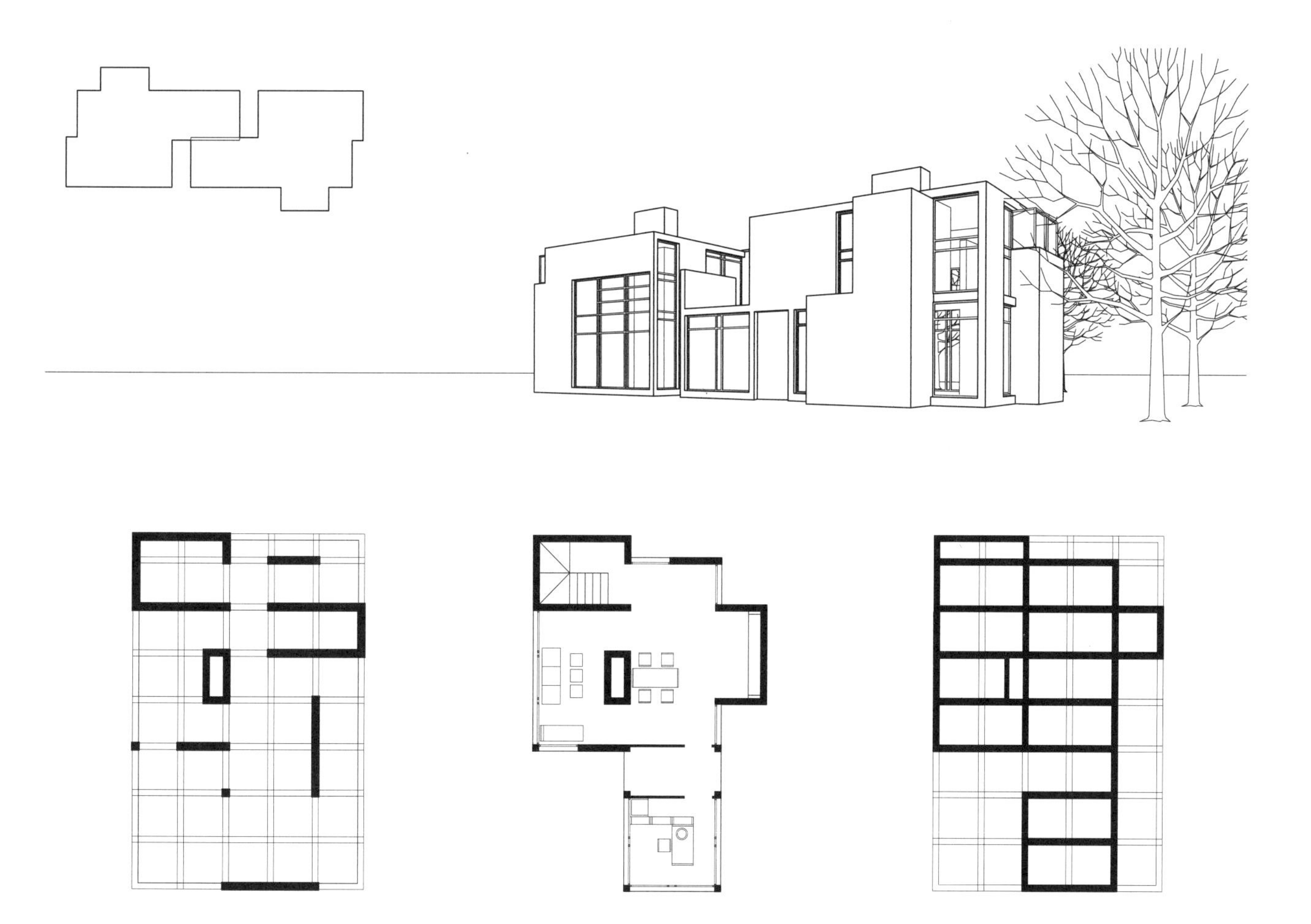

首先构想“结构的空间”，后加入“闭合空间”的概念。这两者形成一个建筑体，再分析其空间域（功能域）及建筑层。结构发展过程：平面；确定空间与功能（家具）；确定结构框架；产生可能的体量；建筑外围护；开敞与封闭；控制开口。住宅基于6mx6mx6m的立方体框架，并以1.5m为模度，因而可以进行体量的叠加、组合，形成一个新的单元。

Defined space was first conceived as “structural space”. The notion of “confined space” was then added. These two formed “an architectural construct”, which was then analyzed in terms of architectural zones (functional zones) and architectural layers. Developing the structure of the object: the plan, space and function (furniture), the structural framework, a possible volume, the envelope, open and closed, and the openings. Our “habitat” is based upon a modular structure of 6mx6mx6m and can be controlled by 1.50m units. The integration of volumes is therefore possible. The fusion of two units into a new unit is also an option.

材料/建造　Material / Construction

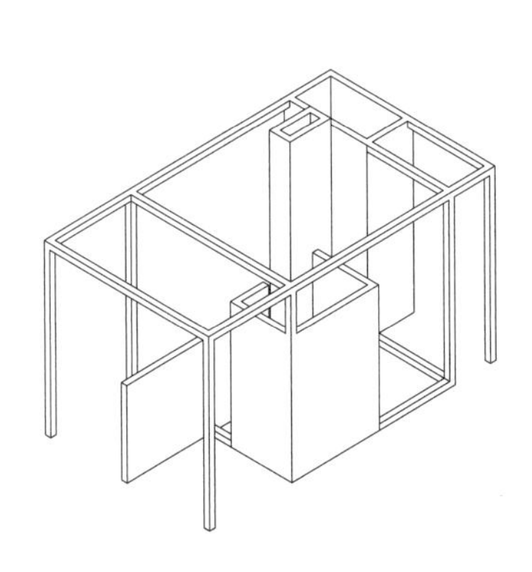

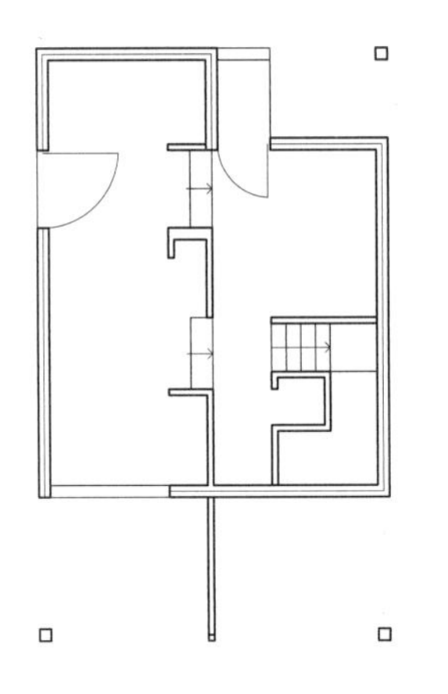

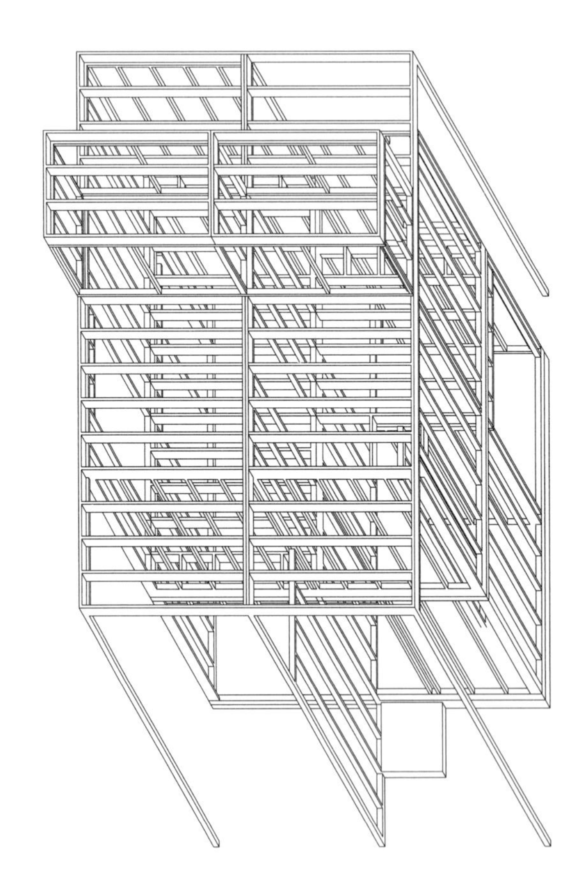

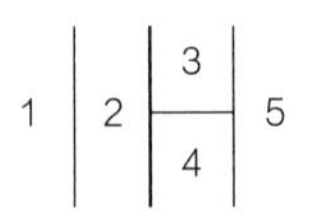

1 居住单元 2 由平台木结构系统建造居住单元 3 墙板的构造层 4 居住单元剖面 5 屋面、楼板及基础细部

1. Living unit 2. Constructing the living unit with timber platform system 3. A wall panel in layers 4. Section of the timber living unit 5. Detials of the roof, floor and base.

按照“三要素图解”的逻辑，下一个问题有关物质化。通过木构带入从抽象结构到建造的转变。一个结构框架，承重系统墙板厚度为12cm，隔热层、内外覆层固定在此核心板上。整体结构是木质的，但外表皮依然可以使用其他材料。随之引入尺度的观念和现实感，如人和家具，使单元更为宜居，同时验证了原有的设计。

The question of materialization is a logical next step. The transformation of an abstract construct took place through the introduction of the wooden technique. The structural skeleton of a load- bearing system consists of a 12-cm-thick layer. The insulation, the inside and outside layers are fixed upon the 12cm core. While the overall construction is made of wood, outside cladding can be of other materials. A sense of scale and reality ensues, allowing the introduction of people and furniture, rendering the volume more habitable. It serves also as a verification of the original design hypothesis.

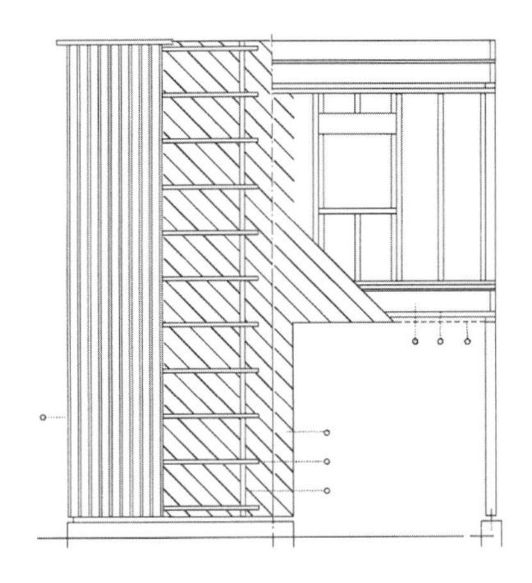

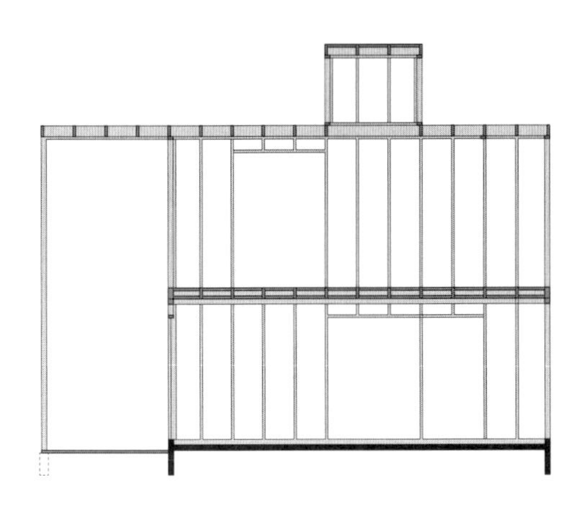

木结构系统本质上是一种叠加式的营建。细部表达了生产过程的逻辑。这一系统可以现场搭建，也可以用吊车现场拼装预制墙板。

内部空间为功能服务，而外表、体量则是一种建构的呈现。（还可见美国2x4模数系统）。

The wooden system is in essence an additive construction. The details illustrate the logic of the fabrication process. The system can be produced "in place" or produced as wall panels and put in place with a crane.

While the interior space served functional purposes, the exterior appearance and the form of the volume were manifested in tectonic terms. (See also: two-by-four system, USA)

场地/场所　Site / Place

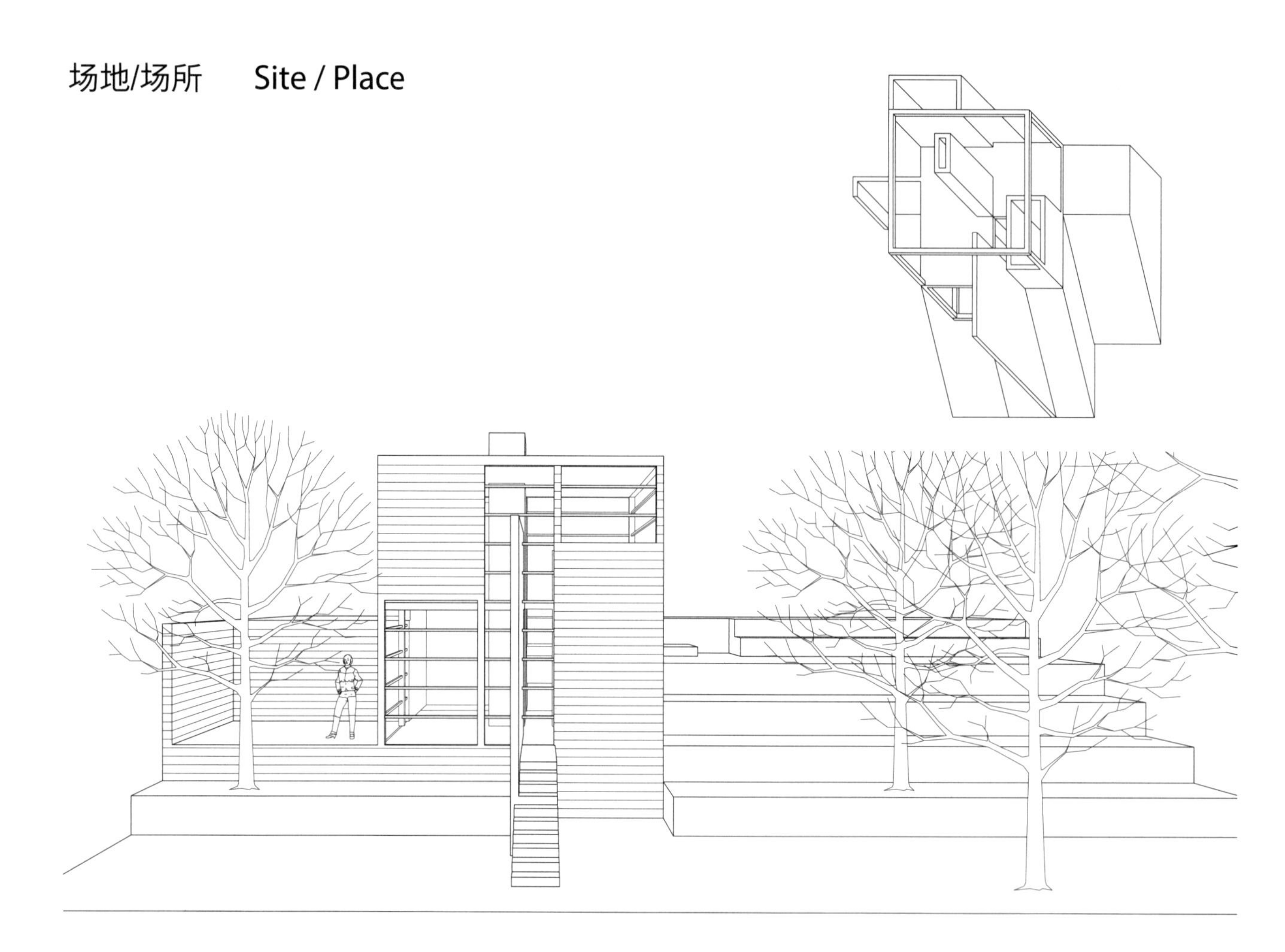

这个试验序列的最后一轮转化由“环境”的引入来激发，意即引入树木、尺度、朝向、色彩和光照作为设计的问题。意图将场地转化为一个场所。阶梯状的地形便于我们“阅读”坡地。

The final transformation of the sequence of experiments was generated through the introduction of "the environment", meaning that trees, scale, orientation, colour and light as design issues were introduced. The intention was to transform the site into a place. The terrace enabled us to understand and to "read" the slope.

1 居住单元与坡地结合 2 透视图 3 居住单元形成组群
4 组群内的外部空间
1. Living unit on a slop 2. Perspective 3. Grouping of the living units 4. Space formed between living units

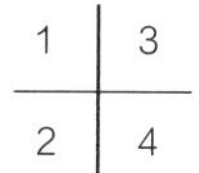

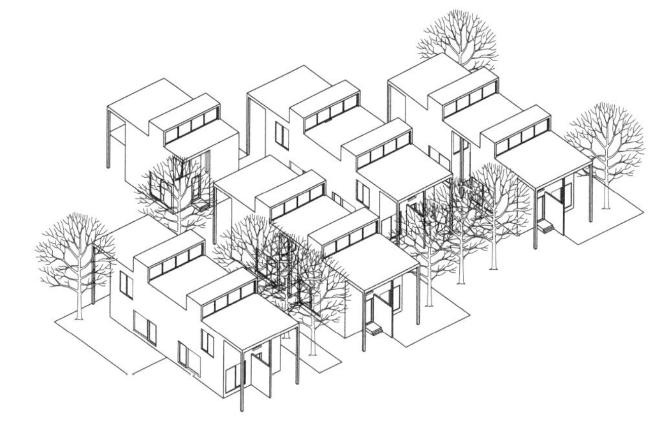

接下来的例子（本页上图和下图）展示了对某个物体或类型形式的重复使用。物体的组合，或者具体的“放置”形成了多样性、区别及总体上的不同结果。下图的例子用相同的基本单元（6mx6mx6m的立方体）形成组团。这里虽然未能显示光线及色彩的差别，也未能看见材质的肌理，但建筑单元的组织足以让我们窥见众多设计的可能性。

The next examples show the repetitive use of one form of object or type. It is thus the combination of objects or the specific “placing” of objects that creates diversity, difference and in general, various results.
The bottom example shows how grouping, or the formation of clusters can be formed using, again, the same object (6mx6mx6m). The documentation does not show the difference in light or colour, we do not see the materials and their surfaces, yet the organization of objects give us an initial idea of the immense range of design options.

赫伯特 · 克莱默给学生评图，1988/89 学年 | Herbert Kramel reviewing students' work, 1988/89

2 教学 | Pedagogy

1 教学模型 | The Educational Model

教学环境 | The Environment
教师，学生 | The Teacher, the Students
基础课程 | The Basic Courses

2 教学过程 | The Teaching Process

要点 | Issues
过程 | Process
方法 | Means
教案设计清单 | Checklist for Program Design

1 教学模型[1] | The Educational Model[1]

教案 | The program

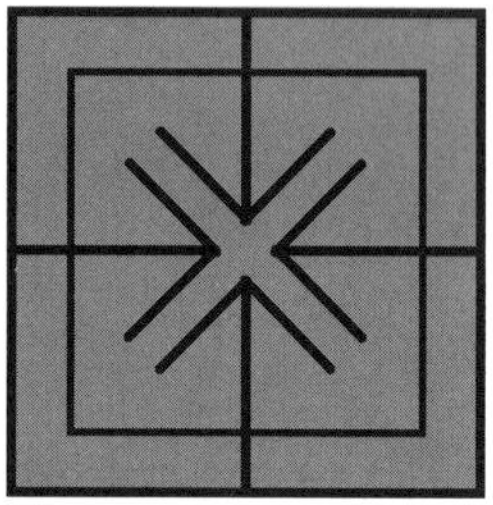

教师 | The teacher

学生 | The student

教学环境 | The environment

教学模型 | The educational model

按内容来理解一个课程或教案似乎顺理成章，却忽略了同等重要的问题，即如何向学生们呈现这些内容，如何用相关资料辅助课程，以及最后如何评价课程或教案的成效。以上这些构成了教师的常规职责。此外，应以更广的视角去审视教与学的过程。教案的内容和结构无疑是最核心的，但教师、学生、教学环境在决定“什么可教”、“如何在教案框架下组织教学”时起到非常重要的作用。

教师，每个教师有其自身的经历，具有一套自己的价值观及自觉的操作模式。工作的职业环境决定了价值与奖励机制。

学生，同样必须考虑学生的经历及现实状况。一名美国大学生的角色、期望及行为模式与一名以色列海法市的学生必定有天壤之别。

环境，首先，学院有自己的定位、礼仪与仪式，但学校的地理位置也很重要。此外，职业环境及建筑业的角色决定了教育的外部气候。

在理解教学模型时将这四点视为相互作用的整体至关重要。

[1] 本章节内容主要来源为 1994 年由克莱默指导、助教卡罗 · 马克欧文完成的设计教学 NDS（Nachdiplom，相当于硕士后文凭）毕业作品《构建一个操作基础》及由后续修读硕士后的学生所不断充实完善的关于克莱默教学的系列册子（均未出版）。

[1] The main sources of this chapter are Carol McEowen's thesis for *Postgraduate Program in Design Education 1993-1994: Building an Operational Base*, instructed by Kramel, and a series of booklets on the same topic that have been enriched by subsequent students (none have ever been published).

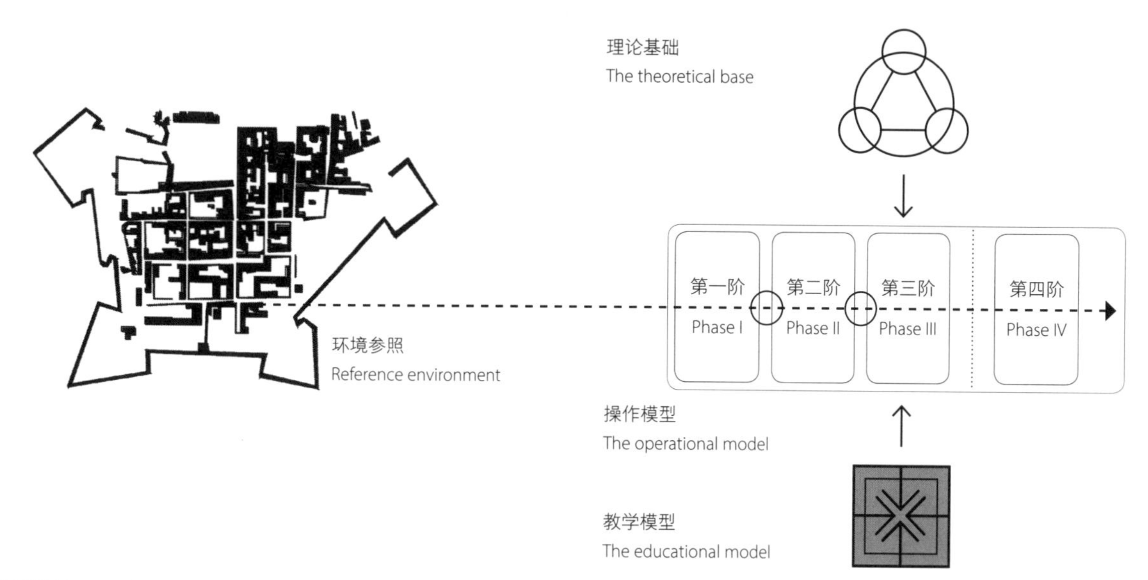

教学结构模型 | Diagram for educational structure

It is normal and logical to try to understand a course or program based on its content. However, this leaves out the equally important questions of how to present it to the student, how to support it with materials relevant to the course and finally, how to evaluate the success of the course or program. All this constitutes the normal work of a "teacher". Yet at the same time, it is necessary to look at the learning/teaching process in broader terms. The content and the structure of the program are no doubt essential, yet the teacher, the student, and the environment play very important roles in determining what can be taught and how the teaching process is organized within the program:

The teacher. He or she has their own history, carrying a set of values and operating more or less consciously within their own models. They operate in a professional environment that determines values and rewards.

The student. Again, the history and the reality of the student have to be considered. The role, expectations and behavioural patterns of an American student are drastically different from an Israeli student in Haifa.

The environment. First, there is the school with its identity, its rites, and rituals, but the place in which the school is located is also essential. In addition, the professional environment and the role of the (building) industry define the climate in which an educational process is taking place.

It is important that an understanding of the educational model considers all four of these subjects as interactive parts of a whole.

教学环境

建筑教育通常被认为是一种职业教育，它的使命由建筑的目的或仅仅由建造业所定义。一方面，建筑教育必须提供适应各种目标的知识、技能与能力的专业训练。另一方面，作为备受争议的课题，建筑研究在教育内部进行，而且研究成果应运用于专业实践，并反过来支持建筑教育。ETH（及其中所有的教学计划）肩负双重职责，首先，作为瑞士法定设立的联邦学院，ETH必须服务于国家需求。它是整个教育体系中的一环。它的使命包含各种高等教育及研究形式。另一方面，为求卓越，ETH必须展现国际水准。

场所的物质结构是另一个考虑层面，还有学校所处的大环境。例如位于贝德福德广场的英国AA建筑学院自然与位于洪格堡的ETH建筑学院不同。前者周遭繁杂的城市环境与ETH的郊野环境传递着不同的信息。

基础设计教程正是学校整体系统中的一个子集。只有当我们对所有其他层面的现状、运作及职能有所知觉，才能专注于基本教程。

［专业体系］

建筑学专业系统包括教育机构、建筑行业和社会环境。在瑞士，学生修读建筑学专业之前并未经过筛选考试，生源素质的控制在他们进入高中之前就已完成。也没有执照考试。瑞士建筑行业或许可以说是未被规范的。行业的素质很大程度上与竞赛制度相关。建筑竞赛由私人或公共实体组织。这一机制为年轻建筑师提供了迈入行业的契机，同时维持了较高的专业水平。专业合作在确立建筑业运作的基本框架中起了关键作用。建筑业的结构基于各个组成部分间的紧密协作。建筑师、工程师及施工方起核心作用，而社会力量也参与其中。

瑞士的社会、政治、文化和经济环境已经发展为紧密联系、相互依存的结构体。它的基本单位是社区。在社区层面通过复杂的相互关系、功能、成员的角色及职责将社会环境结构化。在国家层面，政治体系赋予社区诸多决定权从而支持了“社区原则”的实行。这一权利赋予所有社区成员通过直接或间接参与来定义环境的可能性。

瑞士的建筑教育并没有与某个社区关联。然而，建筑师在受教育后回归了社区体系。建筑师首先通过受教育初步参与了国家层面的行业环境。行业环境再被定义为与建成环境相关的建筑师、学者、技术人员和设计师等多个团体。因为与公众利益紧密相关，建造业直接受到社区系统的影响。瑞士建筑师清楚社区环境的优势在于它提供了国家层面影响建筑思想发展的框架。

The Environment

Architectural education, in many cases, is identified as a professional education, and its aim is defined by architectural purposes, or narrowly, by the building industry. On the one hand, architectural education must provide impart the knowledge, skills and abilities for different professional purposes. On the other hand, architectural research as a contentious subject has to take place in education, and the results of the research should be used in professional practice and as support to the education. ETH (and any program within it) has a dual responsibility. First, the ETH as a Swiss federal institute is established by law to serve the needs of the country. It is thus an element in the overall education system. Its mission encompasses all forms of higher education and research. On the other hand, for the pursuit of excellence, it has to perform on an international level. Yet another concern is the physical structure of the place. One should look carefully at the physical setting, its quality, and its spirit, as well as the location of the school in its larger environment. The AA at Bedford Square is by nature different from the ETH on the Hönggerberg. Comparing the intrusive urban environment of the former with the rural conditions faced by ETH, the message conveyed by each of them are very different.

The basic design program is a subset within the overall system of the school. Only when we have created awareness of all the other levels of existence, operation and responsibility, may we focus on the Basic Program.

Professional System

The professional system of architecture includes educational institutions, the building industry, and the social environment. In Switzerland, there are no examinations for the selection of students; the process of quality control is implemented before students enter high schools. There are also no license examinations. It might be said that the architectural profession in Switzerland is not regulated. The professional quality is very much related to the competition system. Architectural competitions are organized by private or public entities. This system provides an access to the building industry for new architects and the maintenance of a high professional level. Professional cooperation plays a vital role in defining the framework of architecture as an activity. The structure of the building industry is based on a strong interaction and interrelation of different components. Architects, engineers, and builders take up the central positions but society participates directly as well.

The social, political, cultural and economic environment in Switzerland has developed into a strongly interrelated and interdependent structure. The primary units of this structure are the communities. Social environments at the communal level is structured by a complex system of interrelations, functions, roles and responsibilities of the members. At the country level, this "community principle" is supported by a political system that gives communities the right to make decisions in many aspects. This right provides communities the possibility to define an environment through the participation, direct or not, of all members of the community.

Architectural education in Switzerland is not related to a certain community. However, architects who influence the Swiss environment return to the communal structure after their education. The architects first receive an education to join primarily the national professional environment. The professional environment is then defined as the group of architects, academicians, technicians and designers who are related to the built environment. As an activity strongly related to public interests, the building industry is directly

EUROPE 欧洲
SWITZERLAND 瑞士
ZURICH 苏黎世
ETH 联邦理工学院

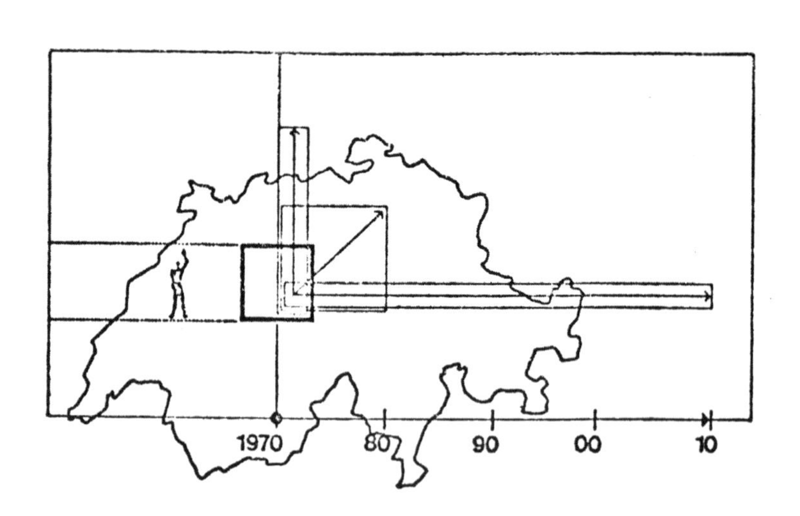

时间与空间中的ETH建筑教育。下个跨页中的图解描述了多个观察及思考瑞士教育过程的方面及层级。教学计划实际上处于大至苏黎世、瑞士乃至整个欧洲，小至学院本身的多个经济、社会及政治框架之中，决定了自身的潜力、局限及特点。| The architectural education of ETH in space and time. The diagrams delineate various layers or levels on which the process of education can be observed and considered. The fact that a program exists in the economic, social as well as political frameworks of first, Europe, then Switzerland, Zurich, and finally, the ETH defines its potentiality, its limitations and characteristics.

瑞士建筑教育与建筑行业。建筑学的教学与实践不可能脱离于社会、政治、经济及文化环境的强烈影响。| The architectural education and profession in Switzerland. The teaching and practice of architecture can never be fully detached from the strong influence of the social environment, as well as political, economic and cultural forces.

[1] 瑞士大学学制原先使用学时计法，并且没有本科与硕士阶段的划分。1999 年瑞士与欧洲其他 28 国签订《博洛尼亚宣言》后，开始致力于建立以两阶段模式为基础的高等教育体系及学分体系。2004 年开始，ETH 的学制及学时由此经历了一段转变时期。从 2007 学年至今，学制改为 3 个学年的本科课程及 1.5 个学年（3 个学期）的硕士课程。学生要获得硕士学位，必须在此基础上完成毕业设计以及 12 个月的实习（其中有 6 个月需在修读硕士课程之前完成）。

[1] The Swiss university system originally used the class hour system, and did not have the undergraduate-master stage division. Following the signing of the Bologna Declaration with 28 other European countries in 1999, Switzerland began to work on the establishment of a system of higher education and credit system based on the two-phase model. Since 2004, ETH have undergone a period of transformation. From the academic year 2007 to date, the system has been converted to comprise 3-year undergraduate courses and 1.5-year (3-semester) master's courses. To obtain a master's degree, students must complete their graduation design and a 12-month internship (6 months accomplished before the commencement of the master's program).

［建筑学校］

建筑学校是组织在一定结构和计划中的实施教育的工具。瑞士联邦理工学院和州立大学的建筑院校代表了本国教育的最高水平。他们的目标是培养不仅具有建筑设计能力，而且具有专业理论背景的建筑师。大学教育的国际背景提供了对建筑灵活多元的理解。而建筑技术学校则以为地区建筑业提供专业人才为宗旨。对于这些地区，快速实用为导向的技巧是解决短期问题所必备的。

理解院校结构是接近社会及国家导向模式的途径。联邦理工学院可以视为国家及科学界宏观背景中的一个缩影。这缩影由学部或学院组成，建筑学院就是其中一员。建筑系是苏黎世联邦理工学院最早设立的学部之一，并遵循工科大学的模式，而不是艺术院校的传统。

20世纪90年代，建筑学院的教学计划分为3个阶段：基础课程，核心课程（包括实训）以及文凭（或硕士）课程[1]。第一年是基础课程，旨在建立对设计和建造的基本认识，然后进行一次预文凭考试。接下来的两年是核心课程，补充和加强建筑形式发展的知识。毕业需要额外一年的建筑实习，让学生在教育框架内体验职业实践。最后，学生需要通过一系列考试并在10周时间内提交毕业设计以获得文凭。随着文凭的授予，学生便具有了成为专业建筑师的执业凭据。

influenced by the communal system. Swiss architects understand that the communal environment offers the advantages of a framework that influences the development of architectural thinking at the national level.

The Institution of the School

The Institution is an operational instrument for education, which has been organized within a structure and program. The departments of architecture at ETH and the State Universities represent the highest educational levels in Switzerland. They are meant to cultivate architects who have not only professional abilities but also a theoretical background for architectural design work. The education at the university imparts a flexible and manifold understanding of architecture oriented towards and embedded in an international context. Whereas the technical school of architecture is formed to train professionals who will be part of the regional building industry where fast and practically oriented craftsmanship is necessary to solve problems on a short-term basis.

Understanding the institutional structure is a way to approach models of social and national orientation. ETH can be defined as a microcosm inserted into the larger context of the nation and the scientific world. This microcosm is organized in departments or schools, one of them being the school of architecture. It is an original department at the university and follows the tradition of architecture in technical and polytechnic universities rather than art academies.

In the 1990s, the curriculum of the school can be divided into three phases: the Basic Courses, the Core Courses (including practical training), and the Diploma or Master's Courses [1].

The Basic Courses take place during the first year of studies. It is oriented to build up a fundamental understanding of design and construction, and is followed by a pre-diploma examination. The Core Courses take up the following two years. It complements and reinforces the knowledge about the development of architectural form. An additional year of studies required for the diploma provides the practical training that allows students to experience the profession within the framework of education. The last year is the Diploma Courses, in which the students are required to take exams and submit a design project within ten weeks. With the awarding of the diploma, students are given the credentials to practice as professional architects.

教师，学生

ETH的教学延续了教席制的传统，由一至两名教授与麾下若干助教组成一个教学团队。对于建筑设计课，教授负责制定教学目标及计划，主持讲座；课程设计分组进行，每组由一至两名助教辅导——讲解任务要求、演示绘图技法、与学生讨论设计等。助教通常是毕业于本校的青年建筑师，通过申请获得校内的兼职职位。这种等级化系统合理地配置了人力资源，高效地保证了面对大量学生时的教学质量。

制定教程的教师（教授）必须厘清将要执行的课程内容——模式及目标。这在一个学生数目庞大的课程中尤为重要——如克莱默教授的“基础设计”每年面向超过300名学生，有20至25名助教参与。助教来自瑞士的不同地区或国际交流，因此一开始不仅学生是“新手”，许多教师也都是“新人”。所以明晰可行的教学计划是整个教学团队合作的基础。

重要的不只是那些建筑问题，除了建筑教学所特有的问题外，还有一般教学的固有问题，如沟通、督导和评价方法。由于“基础设计”以简化组件的形式呈现建筑概念，即便新手教师也能够利用各个阶段构建交流的方式。

正如学生奠定建筑学习的基础，教师亦在奠定教学的基础。通常，学生被称为教育的受体，但事实并非如此。教育过程中，学生可以处在主动的位置，因而学生的行动，作为个体或学习团队中的一员，都可以成为教育过程的主体。

学生有着不同的兴趣和经历。通常他们对建筑教育为何物是茫然的。在瑞士，中学阶段成绩优异的学生享有选取大学专业的高度自由，就读建筑学并无附加条件。尽管如此，作为课程目标，我们依然假设每个学生都是以进入建筑行业为最终目标的。关键是呈现核心概念，因为学生对这些概念的驾驭构成了其建筑教育和职业的基础。他们应该通过“基础设计”的初步经验建立技能、感知力、操作和评判及工作习惯。

The Teacher, the Students

The organization of teaching at ETH is based on Professorship, or Chair (Lehrstuhl). One or two full professors appoint a group of teaching assistants to form a basic operation unit inside the department. In the case of design courses, a professor's role is equivalent to a course coordinator who is responsible for setting up the goal and the program for the course and giving lectures, while teaching assistants are responsible for the implementation of the program in his/her studio group, such as distribution of assignments, discussion with students on the design, demonstration of drawing and modelling skills, etc. Normally, part-time assistants are young alumni who work mainly as architects. Such a hierarchical system is an efficient composition of teaching personnel that guarantees the quality of pedagogy when a great amount of students are to be taught.

It must be assumed that a teacher understands clearly the content - the model and its goals - of the program he or she is presenting. This is especially important in a program as large as the Basic Design Program taught by Prof. Kramel at ETH with over 300 students and 20 to 25 assistant teachers. The group of assistants constantly changed. The group members came from different parts of Switzerland or via international exchanges. Therefore, at the beginning of every round of the Basic Design program, not only were the students novices but also many of the teachers. A clearly defined and workable plan for the program was necessary as a practical solution to the cooperative mode of the group.

Important issues are not just architectural ones; there are issues inherent in the teaching profession in general - such as communication, monitoring and evaluation methods - in addition to issues specific to the task of teaching architecture. Because the Basic Design Program presents architectural ideas in a simplified component form, a novice teacher can use each phase as an opportunity to structure an approach for communicating ideas.

As the students are laying the foundations for their education in architecture, the teacher is laying the foundations for a way of teaching. Normally, students are considered the object of the education, but this is not true. In the educational process, the student can be in a positive position, and the action of the student, as an individual or a member of a learning team, can become a major part of the educational process.

Students enter the Basic Design Program with diverse interests and experiences. In general, their ideas of what an architecture education is about are vague; students who demonstrate a specified high level of achievement in secondary education have the highest latitude to enter the university program without additional requirements specific to architecture. Nevertheless, for the purposes of the program, it must be assumed that each student has the goal of entering the architectural profession.

The presentation of the core concepts is a key issue because students' mastery of these concepts forms the necessary foundation of their architectural education and career. Technical skills, perceptual abilities, operational and critical, and work habits are also established during their initial experiences in Basic Design.

基础课程

建筑学系的基础课程最早由霍斯利设立于1961年，包括为一年级学生设计的三门课——视觉设计，建筑设计和构造课。一开始，三门课分别由艾斯、霍斯利及罗纳执教。克莱默从1971年开始负责构造教席，从1985年到1996年，将建筑设计与构造课合并为“基础设计”。它着重于一种建筑及结构的设计方法。因评分需要，建筑设计及构造课会分别打分，但设计练习是合并在一起的。

继艾斯之后，视觉设计课由彼得·耶尼为学生提供了更为感官和概念性的设计入门，并不一定涉及建筑形式。虽然这两种设计类型之间的联系是内在的，但两个课程并未整合；两者的关联留待学生们自行解读。

“基础设计”的目的之一是为学生在往后的学习及未来作为专业人士将发展的多种兴趣方向做准备。他们将理解建筑形式、现代建筑运动及其对当代建筑理论的重要性；还必须熟练掌握操作和评判方法以及相关技能。因此教案的关键在于保证这种学习的发生。教案旨在整体、跨学科、可自我调节的学习过程。该教案的编排适应了新生的多样性，以直观方式呈现建筑概念——在开始的几周几乎是过于简化的——但在这一简单之上不断强调建筑设计、构造、理论、历史及敬业精神等概念之间的联系，从而慢慢建立起复杂性。

The Basic Courses

Bernhard Hoesli first established the Basic Courses at D-Arch in 1961, including visual design, architectural design, and construction for the first-year students. At the beginning, the three main courses were separately chaired by Hans Ess, Bernhard Hoesli, and Heinz Ronner. Kramel had chaired the construction course since 1971. Then from 1985 to 1996, Kramel combined architectural design and construction to form the Basic Design course. It concentrated specifically on an architectural and structural approach to design. For evaluation purposes, the students were graded in design and construction respectively, but the studio exercises were merged to formulate the grades.

After Ess, the Visual Design course taught by Peter Jenny provided students with a more sensual and conceptual introduction to design that was not necessarily related to built form. While a connection between both types of design is intrinsic, the two programs are not actually integrated; the strength of the connection is left to each student's own interpretation.

The Basic Design Program is intended to prepare students for the different interests they will pursue in the following years within the department and afterwards as professionals. It is expected that students will have knowledge of architectural form and knowledge of the Modern Movement and its importance to contemporary architectural theory. They must also have a command of operational and critical methodologies and of technical skills. It is, therefore, essential that the program ensure that this learning is taking place. As a program, it aspires to a learning process that is holistic, interdisciplinary and self-regulating. The Program is structured to accommodate the diversity of the entering students by presenting notions of architecture in a straightforward manner - almost simplistic in the first few weeks - but with an increasing complexity and a continual stress on the interrelation of the notions of architecture design, construction, theory, history and professionalism.

2 教学过程 | The Teaching Process

操作模式是时间维度上的课程结构。它用以组织课程内容，还有相关的辅助活动及资源。理想情况下，模式的结构反映了理论基础的逻辑，并顾及教育模式中的固有资源。参考环境的选取也是教育结构关系的辅助因素。

课程基于四个相对独立又相互关联的学习阶段，时间分段对应于一个学年内的假期[1]。每个后续阶段都建立在上一个阶段的知识基础之上，意在加深学生的理解。

从首个练习，亦即首个设计问题开始，学生在积累建筑知识的同时，便也不断提高应对问题的能力及对设计过程的理解。不同于二年级教学以开放的设计问题作为核心，一年级的教学主体是被严格定义的练习。练习的难度逐步增加，并具有明确的教学目的。例如，分析作业加深了学生的理解，另一个练习则训练草图和绘图能力。还有一些练习训练学生的组织能力，帮助他们发现设计中的问题。

教程不是为了给学生提供某种诀窍，而是形成一种基本的方法和工作模式。如此，所有学生在第一年建立起的共同的知识背景及经历，成为后续学习的基础。

[1] 当时学年的上学期由十月份开始，至次年二月份结束，中间为圣诞假期；下学期由 4 月开始至 7 月份。另，每个学期的第六周为研讨考察周。

[1] At that time, the first semester of the academic year began in October and ended in February, with Christmas holidays in between; the following semester began in April and ended in July. The sixth week of each semester was the seminar week.

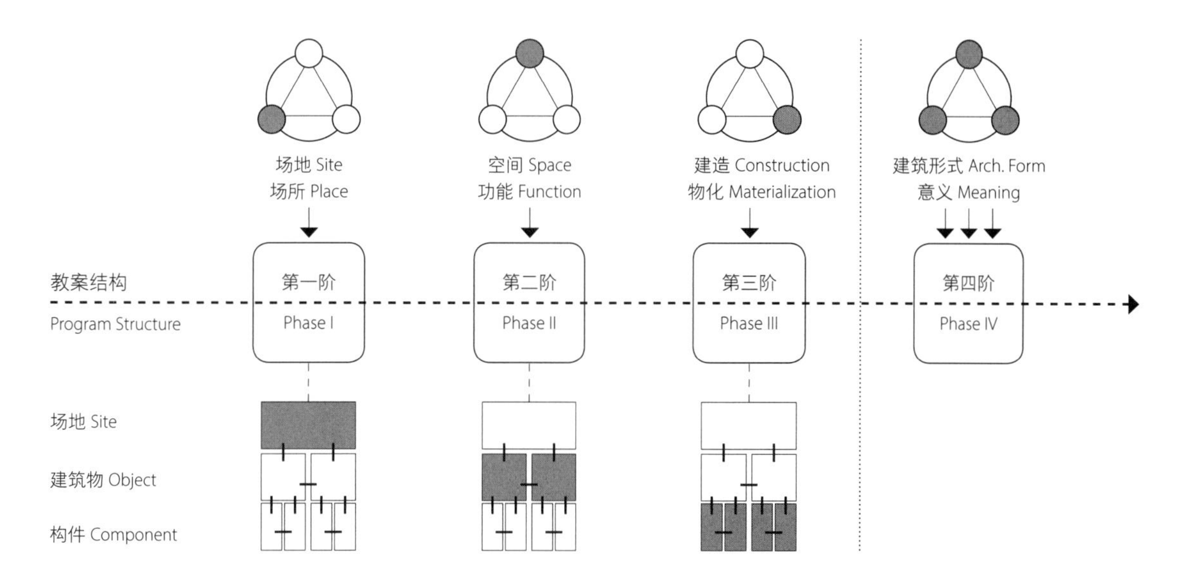

教案结构与设计要素、操作层面的关系 | The relationship between program structure, design elements and operation levels

The operational model is the course structure in time. It is used to organize not only course materials but also relevant support events and resources. Ideally, the structure of this model reflects the logic of the theoretical base and considers the resources inherent in the educational model. The choice of reference environment is also a factor that supports the interrelation within the educational structure.

The Basic Design Program is based upon four relatively autonomous yet interrelated phases that correspond to breaks in the school year [1]. Each successive phase builds on knowledge gained in the preceding phases, intentionally increasing students' depth of understanding.

Through the exercises - that is, from the first design problems - the students not only accumulate concrete facts but also develop their competence and improve their insight into the design process. In contrast to the second year in which an open design problem forms the core of the educational program, the first-year exercises are strictly defined and constitute an important part of the overall teaching program. The exercises themselves are constructed in such a way that the students follow a sequential process in their work that increases in complexity and difficulty. Each step also addresses specific teaching goals. For example, an Analysis serves as a means of improving the student's insight. Another step is intended to develop the student's drafting and graphic abilities. Other exercises develop organizational abilities and aid the students in their efforts to locate problems within their work.

The goal is not to give recipes to the students but rather to develop a methodological basis and system of work. In this way, all students have a common background and experience that forms the foundation on which their education in later years can be built upon.

要点

［知识的问题］

“基础设计”旨在创造令学生能够批判性地应对某个具体“建筑情境”的知识基础。该教案的基本前提是将建筑视为建造之艺。知识基础通过设计、构造和建筑课程的讲座来建立，帮助学生掌握建筑和设计的概念及建造技术。

［能力/技能的问题］

建筑上的能力和技能就是以专业方法解决建筑问题的能力。在“基础设计”中，这些技能在课程设计中传授。教学引入工作方式，并鼓励学生多做尝试，从而获得技能。学生通过关联的练习尝试一系列设计及表现的方法，初步掌握并有意识地运用建筑模型的二维、三维表达方式。关键是学生学会将设计决策与设计方法联系起来。

［方法/媒介的问题］

马素·麦克卢汉提出：“媒介即讯息[1]”，在此处则是“媒介即方法”。建筑师最重要的媒介是绘图。如今自然包括了手工和电脑制图。定义“绘图”这一术语，可以说是“对计划中客体的视觉表达”，还可加上“对建成环境的可视化”。这种可视化已有多种形式，皆可纳入绘图的范畴。绘图方式随着建筑理论基础的演变而变化；反之，绘图方式又是建筑演化所借助的思维模型。

尺度的概念伴随着多种形式的绘图及模型——它本身就可在理论上视为一种建模的技法。

设计课引入了一系列建筑视觉交流的媒介与方法。方法和媒体体现了各个练习的特点。学生从中理解到建筑的几大交流媒介：图解、草图、制图和模型。

图解是某个想法、过程或一系列数据的抽象视觉表达。例如泡泡图、平面构想图、分析图、图－底关系图、流线图等。教学中同样利用图解来组织设计过程及建筑理论。

类似于图解，草图是研究及厘清初始设计概念的工具，但能够进一步将图解转化为有形的建筑空间和实体。设计过程中使用草图来思考光线、形式、空间、实体细节、色彩等。草图直观地记录了学生整个设计发展的过程。

学生必须熟练掌握建筑制图规范——平面、立面、剖面和透视图，但重要的不仅是画得正确，而是如何呈现设计概念。成功的建筑制图应该具有草图或者分析图一样的表现力。

计算机辅助设计作为具有潜力的新技术，“基础设计”为之开设了一个平行教

[1] 马歇尔·麦克卢汉，《理解媒介：论人的延伸》（纽约，1965）。

[1] McLuhan, Marshall. *Understanding Media: The Extensions of Man.* New York: McGraw-Hill, 1965.

Issues

The Question of Knowledge

The Basic Design Program is oriented towards the creation of a knowledge base that enables students to react critically in an "architectural situation". The premise of the program is that architecture is the art of building. The knowledge base is established through lecture courses in design, construction, and architecture. These courses contribute to students' grasp of design and architectural concepts as well as building techniques.

The Question of Abilities/Skills

In architecture, abilities and skills are related to the capacity to deal with an architectural problem in professional terms. In the Program, these are taught in the design studio. Students experiment with a series of design and presentation methods through interrelated exercises. The goal of the Program is to present methods, to encourage students to use and experiment with them in order to hone their skills. Thus, students acquire the preliminary ability to consciously create 2- and 3-dimensional representations of an architectural model. It is important that students learn to relate design decisions to a design methodology.

The Question of Methods/Media

It was Marshall McLuhan who coined the phrase "the medium is the message [1]" and here the media is the method. The most significant medium for the architect is drawing. Today, of course, drawings might just as often be drawn by computers as by hand. To define the term "drawing", one could say that it is the visual representation of a planned object, or the visualization of the built environment. This visualization has assumed many forms, all of which can be subsumed under the idea of the drawing. The drawing changes as the theoretical base of architecture changes; it is the thinking model upon which each step in the transformation of architecture is made. Along with the various forms of drawing and the existing notion of the model is the notion of scale, which can be seen theoretically as a modelling technique.

A series of media and methods for the visual communication of an architectural problem are introduced. Methods and media are used to characterize exercises of the program. Students achieve an understanding of significant media of communication in architecture: the diagram, the sketch, the drawing and the model.

The diagram can be seen as an abstract visual representation of an idea, a process, or a set of statistics, e.g., the bubble diagram, the parti diagram, the analytic drawing, the figure/ground drawing or the circulation diagram. Diagrams are also used as teaching tools to organize information presented in the design process and architectural theory.

Like diagrams, the sketch is a tool for studying and clarifying initial concepts of a design, but it can also further the design process by transforming the former into tangible architectural space and mass. The sketch is useful throughout design process for investigating notions of light, form, space, mass detail, colour, etc. It is the visual evidence of the thought process that leads to a student's final result.

It is important for a student to master the conventions of architectural drawing—plan, section, elevation, perspectives, not purely as ends in themselves, for the way one draws is just as important as what one draws. Successful drawings can be just as provoking as a sketch and just as structured as a diagram.

Harnessing a technology with new potential, the Program uses computers to run a parallel CAD program. Students acquire knowledge by using computers in the modelling, the manipulation, the

场地组织，1985/86学年 | Site organization, 1985/86

单元与地形，1987/88学年 | Units in different situation, 1987/88

程。学生通过运用计算机进行建模、操作，可视化及专业演示特定的建筑条件。

模型或许首先被视为缩小版的建筑物——设计思维结果的三维表达。然而作为教学手段，它具有不同的角色。它是在落实到图纸之前，便于设计者进行三维操作和思考的具有灵活性的草图。模型的空间表达优势正是“基础设计”将它作为第一步操作媒介的原因。它迫使学生运用立体思维模式，重视平面和剖面之间的紧密关系。

[经验/领悟的问题]

“基础设计”可被视为一块试验田。更具体地说，设计工作室如同实验室，学生通过学习过程获得经验。它的目的是为学生尝试不同的空间、方法、媒介和表现技巧的可能性。这一过程基于边做边学的基本思想。除此之外，应考虑其他活动所能带来的经验和领悟。例如郊游、田野调查、考察，尤其是研讨周——面向特定的主题。

“基础设计”将研讨周统筹考虑进设计与构造课程的大纲中。鼓励一年级学生在研讨周选修钢结构、砖混结构或木结构建造课程。这些课程与瑞士的建造业紧密相连，并获得了ETH工程学系的帮助。课程希望学生获得实际搭建的经验，了解砖石结构、钢结构和木结构建造技术的最新发展。

最后必须重申，设计课上产生的经验仍是学习过程中最关键的反馈。心到、眼到、手到成就了设计和设计者。

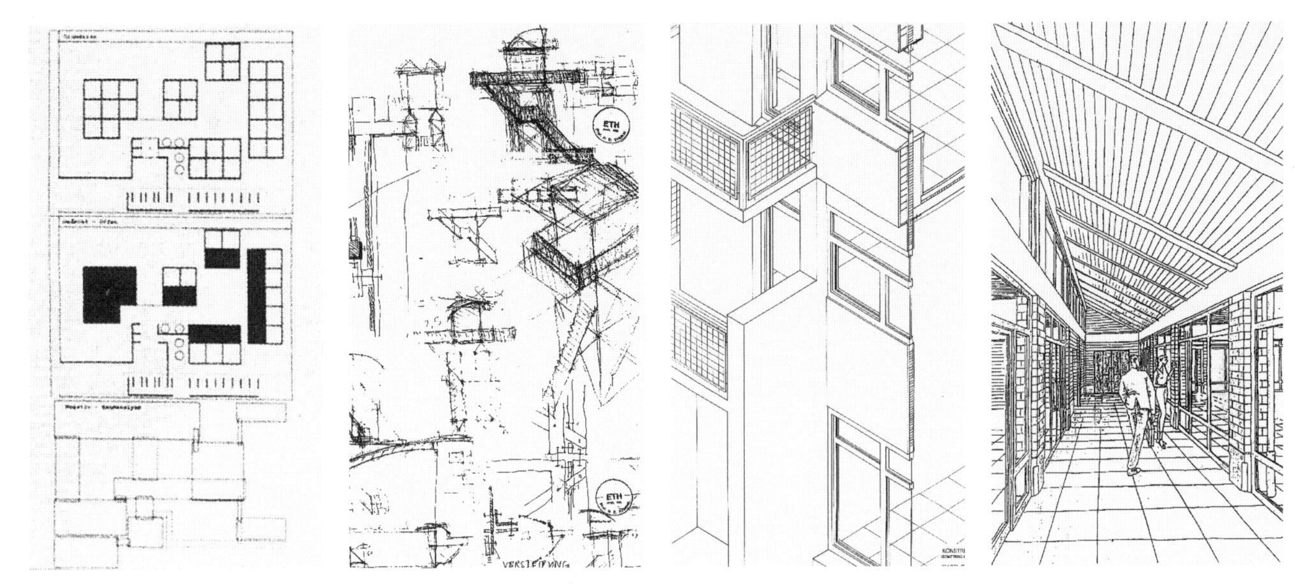

不同的表达媒介，左起：1985/86，1985/86，1991/92，1985/86学年 | Different medias, from left: academic year 1985/86, 1985/86, 1991/92, 1985/86

visualization and professional presentation of specific architectural conditions.

The model is perhaps primarily understood as the scaled end product of an architectural object—a three-dimensional representation of a final idea. But as an educational tool, the model takes on a very different role—that of the pliable sketch, a diagram that enables the designer to study the three-dimensional consequences of ideas in progress before transferring them to the drawing. This advantage in spatial representation is why the model is used in the Basic Design Program as the first step in a new investigation. It forces students to solve problems in three dimensions and to look at the complex integration between plan and section.

The Questions of Experience/ Insight

The Basic Design Program can be seen as a field of experimentation. More specifically, the design studio functions as a laboratory where students gain experience through the learning process. The objective is to offer students different possibilities to experiment with space, methods, media and visualization techniques. The process is based on the fundamental idea of learning by doing. It is also important to consider the experience and insights that additional activities can bring. Excursions, fieldwork, study trips and especially the seminar week are oriented towards specific topics.

Seminar week is integrated into the goals of the design and construction course. Students are encouraged to take one of three building courses in steel, masonry or wood. These seminars are run integrally with the related construction industries in Switzerland and with help from the Engineering Department at ETH. The intention of each building course is to give students both hands-on experience with the materials and knowledge of recent technological developments in masonry, steel and wood construction.

Finally, it has to be mentioned that experiences generated in the studio provide an essential feedback mechanism in the learning process. It is important that the mind, the eye and the hand bring forth the design and the designer.

过程

计划

在可供选用的设计教学方法中，“小练习”的模式似乎是最具成效的。学生在独立的步骤中制作模型，探索、试验、发现并逐步形成一套包含了知识、技能、经验及态度，以及方法论的认知体系。相应的，教师筹备练习。这一过程可称为应用设计研究；教师基于研究构建学习过程，间接地决定了学习内容及目的，而将途径留待学生去探究——促发主动与创新。

设计的设计，即学习过程的设计（一种“元设计”形式）是对教师的真正挑战之一。因此，试做是筹备教案的重要部分。通过实际操作获得对本质的理解，对练习目的及关键问题的反思。每年教案都会根据环境和行业的变化，以及学生特点的不同而调整。重要的是测试这些调整的人对练习的目的有所意识，同时又不会对练习的预期结果过于熟悉。如此才能纠正含混的指示，发掘练习的深层潜力。

试做过程还提供了切实的信息，例如制作模型所需材料量，完成练习所需时间等。此外，试做的结果常常成为告知学生作业要求及标准的范例。

实施

教案的实施是教师工作的关键。教案中信息传递的有效性如同信息本身一样重要。教案的实施直接影响学习过程，以及学生的表现。

不论是按照现成教案教学，还是研发新的教案，一名新任教师应明白教案本身不是结果，而是一个框架、教学过程的指引。教案指出学生应掌握的技能，并将方法与信息组合起来展示给学生，便于他们增长知识、掌握技能。但决定学生的学习体验及表现的是实际在教室中进行的讲座、讨论、设计课及辅助内容（设计实验室）。

实施建筑设计教案的重点有二：其一，学生不仅要学习知识，还要掌握应对专业实践的技能；其二，学生要发展出对建筑形式的判断能力。因此，教师不但要注意学生在具体设计中的表现，还要留意他们对建筑学的广义理解。同理，教师应该全面地呈现主题——把学习和设计的过程锚固于专业目标。

Process

Planning

Within the scope of methods available for the teaching of design, the "exercise" mode seems to be one of the most productive. The student makes models in discrete steps - exploring, testing, discovering and, thus, building a "repertoire" that combines knowledge, skills, experience and attitudes, as well as methodology. In turn, the teacher prepares the exercises. One might call this process applied design research; based on research, the teacher structures the learning process, defining the "what" and "why" by indirect means, leaving the "how" to the students—encouraging their initiative and inventiveness.

The design of the design, or the design of the learning process (a form of meta-design) poses one of the real challenges to the teacher. Thus, the test run is a vital part of the planning of a program. Through hands-on experience, an intrinsic understanding is gained, and reflection can be made on the key issues and purposes of the exercises. Each year, modifications are made to the program, taking into consideration environmental and professional changes as well as the character of the students. It is important that these modifications are tested by someone generally aware of the nature of the exercise but not too familiar with the expected outcomes. Unclear instructions can be refined or modified. Insights into further potential of the exercise can be discovered.

The test-runs also provide practical information such as how much material is required and how much time will be needed for students to complete the exercises. Results of the test-runs are usually used as visual examples for students to better understand what is required of them.

Implementing

The implementation of a program is a crucial task for a teacher. The effective delivery of information embedded in a specific program is as important as the information itself. The implementation directly affects the learning process and students' performance within the program.

A new teacher following an established program, or developing and implementing a new program, needs to understand the program plan not as an end in itself but as a framework—a guide for the teaching/ learning process. The plan identifies the skills and abilities that students should acquire and presents to students a combination of methods and information so that they can achieve these abilities and increase their knowledge. It is, however, the actual events in the classroom—in the lecture, the seminar, the studio and related support material (the laboratory)—which determine the students' experience and facilitate their performance within the program.

Two important concerns when implementing a program in the discipline of architecture are, first, that students should be learning not just facts but knowledge and skills that will enable them to practice professionally, and second, that students are developing an ability to make meaningful judgments about architectural form. Because of this, it is important that a teacher remain constantly aware not just of the performance of students as evidenced by their works within the program, but also of how their performance reflects their larger comprehension of the discipline. Along the same vein, it is necessary for a teacher to present the subject matter in a holistic way—continually anchoring the learning and design processes to the goals of the profession.

督导

督导是实施教案不可或缺的一环。教师可以借此发现学生是否掌握了教学计划中设定的内容。以天为单位，确保学生的理解已经达标，可以向更高层次推进，或者某些地方需要再次强调或以其他形式呈现。因此，教师有必要保持教案的灵活性以便适时加入评图，或为学有余力的学生加码。

对学生理解程度及表现优劣的细致监察有利于教师在整个过程中对教学的有效掌控。这有利于新任教师提高和学生的沟通及互动能力，以及学习如何感知特定学生或群体的需求。

有效的监控可以通过结构化的考试、论文或评图来实现，但也可以通过在个人评图或小组讨论时仔细的观察来达到。

从更广的角度来看，督导的重要性不仅在于确保学生对教学计划的掌握度，同时使教师可以将教案放在更大的教育、建筑教育背景及不断变化的专业要求中进行比较。

评价

结束“基础设计”教程学习后，学生应该对建筑形式、建造及设计过程有基本的理解，以便顺利地进入高阶学习，并在日后成为称职的建筑师。因此，在一年中，尤其是“基础设计”完结之时对学生进行评估至关重要。

评估的手段与督导类似，但督导是在教学期间持续进行的，而评估则是对获得信息价值的整体衡量——不管是对一名学生，班级整体，课程教师，抑或是整个教案。

要对学生作出评估，必须首先在教案中明确指出学生应掌握的技能、能力及知识，然后根据平均理解水平来衡量学生个体的表现。评估方式要与被评价的技能和能力有逻辑关系。考试及论文仅能确定学生在建筑学习过程中的部分能力，因此要确定评价视觉技能、感知及判断能力、知识及经验的方法——通常借助于设计展示、评图及学生作品集。

Monitoring

Monitoring is an invaluable part of implementing a program of education. Through monitoring, a teacher can discover whether or not students are mastering the material identified in the program plan. On a daily basis, monitoring provides the means to ascertain that students have reached a level of understanding necessary for moving on to a more advanced level, or that specific points need to be reemphasized or presented in a different way. For this reason, it is important that a teacher maintains some flexibility within the program framework in order to accommodate needed reviews or challenge students who are able to achieve above average.

Careful monitoring of the student's understanding and performance enables a teacher to be aware of the effectiveness of his or her teaching throughout the program. This is an important point for new teachers expanding their communication and interaction skills or learning to anticipate the needs of a particular student or group.

Effective monitoring can be achieved through structured exams, papers or reviews, but monitoring can also be accomplished by careful observations during individual reviews or group discussion sessions.

From a wider perspective, the monitoring of a program is important not just to assure that students are mastering the material as stipulated in the program plan but also to enable a teacher to compare a program to the larger contexts of education, architecture education and the ever-changing needs of the profession.

Evaluating

Students completing the Basic Design Program must have a fundamental understanding of architectural form, construction, and the design process in order to successfully continue in the architecture program at ETH and proceed to practice as professional architects. The assessment and evaluation of students throughout the year and, most importantly, at the end of the program are essential.

The means used for evaluation are similar to those used for monitoring, however, monitoring is an on-going process during the program, whereas evaluation is the placement of value on the information obtained—be it about a student, the class as a whole, the teacher of the program or the entire program.

In order to evaluate students within a program, it is essential to identify the skills, abilities and knowledge a student should obtain through the program and rate each individual student's performance against an average level of understanding. The assessment needs to be logical in relation to skills and abilities being evaluated. Exams and papers are able to ascertain only part of the capabilities a student must develop in a program of architecture; means of assessing and evaluating visual skills, perceptual and judgmental abilities and the use of knowledge and experiences also need to be determined–normally through formal presentations, reviews and final portfolios of students' works.

方法

讲座、研讨会及课程设计可视为学习过程中传授知识及技能的三种途径。任课教师必须熟知它们各自的优、劣势，以便在教学中有效运用。一个教案如果能够将三者结合起来，就能够扬长避短，发挥最高效用。“基础设计”的课程模式是以早上的构造、建筑及设计讲座来引入并激发下午的课程设计。

［讲座——一般知识］

讲座是一种有组织的展示形式，讲者一般在约45分钟的时间内将经过结构组织的信息呈现给听众。课程讲座传达了一般性的知识以及某一专题的背景信息，并为学生建立知识框架。讲座支撑了进阶学习递增的复杂度。讲座的时间有限，相关信息必须尽量简明扼要。视觉辅助，如照片、图形图像、图表和图解等非常有助于说明重点。

从本质上讲，讲座主要是一种被动的学习模式。可是，它促成并鼓励了进一步的独立研究，并激发了相关的讨论和课程设计中的主动学习。因此在“基础设计”中，设计课讲座对练习的讨论不是将其视为单个的孤立问题，而是将其置入更大的系统中。构造课讲座以建构的方式呈现砖石结构和木结构的基本知识。诸如基础、墙身、屋面、洞口等元素被视为整体的一部分，是承载着技术与建筑意义的构件。设计课讲座则为现代建筑的发展提供了历史文脉并为建筑形式提供了当代理解。这些讲座的重点更多地放在了从现代建筑的经验中提取操作程式，而非历史陈述。思考具有现代特征的空间概念，尤其是流动空间的概念。由此，现代建筑的思维发展被纳入教学。这些关于建筑如何被构想出来的例子启发了学生思考生成建筑的可能方式。可以由每组的助教为期中测验、命题小论文及期末考试进行评分。

［研讨会——重点探究］

相比讲座，研讨会能为讨论及深入研究提供更多时间。基本的主题、参考资料及研讨会框架是由组织者或小组提供的，但是，学生应能够积极地参与到讨论中，并主导讨论的重点及最终成果——论文及演示。互动促使学生快速且批判性的思考，并意识到不同的观点。教师在研讨中的角色是引导讨论，确保理解的深入与结论的形成。

［课程设计——综合实践］

课程设计教学是设计类专业独特的学习形式。设计练习直接或间接地与学生从讲座及研讨课中获得的知识相关，但需要学生更高一级的智力参与，即判断并转译练习给出的信息，从而解决特定的问题。学生通过设计练习习得传达信息的约定俗成方法及媒介。

Means

Lecture, studio, and seminar can be seen as three different means of conveying and acquiring knowledge and skills within the learning process. A program that utilizes all three in coordination can take advantage of the strengths of each method and reinforce information acquired in each, thus minimizing disadvantages. The Basic Design program at ETH uses morning lectures in construction, architecture and design to introduce and provoke a reflective thought process required in the afternoon studio course.

Lecture -general knowledge

A lecture is an organized form of presentation, in which the "teacher" presents information in a structured way to an audience usually via a 45-minutes time frame. In a course situation, a lecture is used to convey the general knowledge and background information about a topic, and to give structure to the knowledge acquired by the student. Lectures complement the rising complexity of ideas throughout the phases.

The lecture is limited in duration, thus pertinent information needs to be given as succinctly as possible. Visual aids such as photographs, graphic images, charts and diagrams are invaluable for clarifying important points.

By nature, the lecture is predominantly a passive learning situation. However, it enables and encourages further independent research and provides inspiration for active learning in related seminar and studio courses. Consequently, in the Basic Design Program, the lecture course in design discusses ways of seeing the exercise not as the study of an isolated problem, but as a part of a larger system. The lectures on construction present the fundamentals of masonry and wood construction in a tectonic format. Elements such as foundation, wall, roof, opening, etc. are seen as part of a whole, components that carry technical and architectural significance. The lectures on architecture provide a historical context by tracing the development of Modern Architecture and offer a contemporary understanding of the notions of architectural form. They place less emphasis on the history of Modern Architecture but rather seek to draw operative programs from the experiences of the Modern Movement. Significant concepts of space—and, in particular, the concept of continuous space—are considered. In this way, the conceptual developments of Modern Architecture are included into the teaching program. Given these examples of ways in which architecture "is conceived," students are provoked to consider ways in which architecture "can be made." A mid-term exam, a limited topic paper, and a final exam can be given with a teaching assistant grading each section.

Seminar - focused inquiry

The seminar allows for more discussions and more in-depth research than is possible within the lecture. The basic theme, reference material, and seminar structure are provided by the person or group organizing the seminar, however, it is assumed that students will participate actively in discussions and direct the emphasis and the end product—through papers and presentations—of the seminar. Interaction challenges students to think quickly and critically and to be aware of differing opinions. A teacher's role in a seminar situation is to guide discussions and assure students' progress towards understanding and conclusions.

Studio - synthetic practice

Studio teaching is a form of learning unique to design professions. Studio exercises can be developed to be directly or indirectly related to the knowledge students acquire through lecture and seminar

教师有多种增进学生理解的方法——灵活性正是课程设计的优点。讲评、评图及指导可以面向整个小组、学生组合或个人。师生、学生间的互动是课程的重要一面。

［参考读物］

参考读物可以作为讲座的辅助手段。资料取材于讲座的参考书目。为使每个听众都能翻阅这些参考书，可以放置在图书馆的“保留”书架上，限制每次借阅的时间。这类阅读巩固了讲座内容，增进学生理解。同时，参考书目所包含的知识系统将来可供学生索引，读物将激发那些有求知欲的学生进一步阅读和研究。

保存在计算机数据库中的材料可以让更多的学生同时阅读，这弥补了“保留”读物的不足。

研讨会中学生的角色更为主动，因此需要范围更广的文章、书籍、电影及其他音像媒介。读物——摘选了相关文本及图像的资料集，可以提供研讨会参与者共同的讨论及演示的基础。但这些材料的有效性可能会受到研讨会组织者筛选信息的主观因素的限制。

［旅行］

旅行弥补了阅读的不足，使学生能亲自观察及发现建筑形式、结构及场所文脉。旅行使学习过程丰满起来，为课程设计专门制定的考察活动可以增进知识与观察力。

具有很强目的性的短期调研或许是讲座最好的搭配。适时选定的适宜区位、场地、建筑或建筑群可以一针见血地指明问题并且比讲座图片更为有效地传达信息。学生通过实地调研把被动的学习转化为亲身的视觉经验。

研讨周是对某一个问题的集中研习。许多教授将其视为整体教学的一部分，利用这段时间进行建筑旅行。成功的研讨周旅行需要细致的筹划：主题的选取、辅助材料的选编、旅行及导览规划。整个研讨周可以高度有序地安排日程，考察并记录多个地点；同样也可以由学生自己独立设定研究的地点和主题。

跟研讨会一样，旅行一周结束时的成果很重要：参与者记录、演示、讨论并得出结论。目标是形成有深度的观察。记录形式可以是文本或图像，记录的方法和内容尤为重要。调研旅行的另一个优点是，在集体完成知识积累之外，每个人都会获得自己对建筑的独特认识。

courses, but the exercises require an additional level of participation from students in that a student must use his or her judgment to interpret information to "solve" a specific "problem". Through studio exercises, students learn the conventions and media used for conveying information.

A teacher has a variety of methods available to further students' understanding; it is this flexibility constitutes an advantage in the studio. Criticisms, reviews, and instruction can be delivered either to the entire group, to small groups or to individuals. Interactions between students and teacher and among students account for an important aspect of a studio course.

Readings

In support of a lecture, reading assignments can be given. Logically, these will be from books and articles that the lecturer has used as sources for the lecture. Books can be put "on reserve" in the library so that they are available for everyone who attends the lecture. These assignments reinforce and clarify the content of the lecture, thereby improving students' understanding. At the same time, source books represent a specific bibliography which students can draw from in the future. Thus, readings can also provide an inquiring student with stimuli for further reading and research.

Reading materials put on a computer database would allow access to more students at one time, thus eliminating one of the major problems with reserved materials.

The seminar allows for a more interactive role for students. Because of this, a broad-based selection of articles, books, films and other audio/visual media including computer data can be incorporated into the structure of the seminar for research and reference purposes. A "reader"—excerpts from relevant texts and collected visual material—provides seminar participants with a common source of materials from which to draw inspiration for presentations and discussions. The effectiveness of these materials can be limited by the editorial decisions of the seminar organizers.

Excursions

To complement the support that reading can bring to a program, active support is found in travel—a visit to a location where each student can experience and discover through observing architectural form, structure and context. Travel, in general, can be a source of enrichment for a student and an important part of the learning process. But travel especially tailored to the goals of a program can be a direct way to increase a student's knowledge and insight.

An excursion—a short trip for a specific purpose—is perhaps best paired with lectures. A well-timed excursion to a well-chosen location, site, building or buildings can underscore specific points and illuminate knowledge being presented in lectures more efficiently and is more memorable than photographs and images, thus increasing the value a student can gain from lectures. An excursion gives students the opportunity to directly apply passive learning to first-hand visual experience.

Seminar Week is a week of intensive study of a single topic. Many professors use it as an opportunity for travel as part of an overall education process. A successful seminar week travel program, like a regular seminar, requires extensive planning: choice of theme, compilation of support materials, and travel and touring arrangements. The week can be highly structured with a detailed itinerary to visit and document a number of sites. However, the week can also be dependent on students' initiative for research and discovery within the given location and theme.

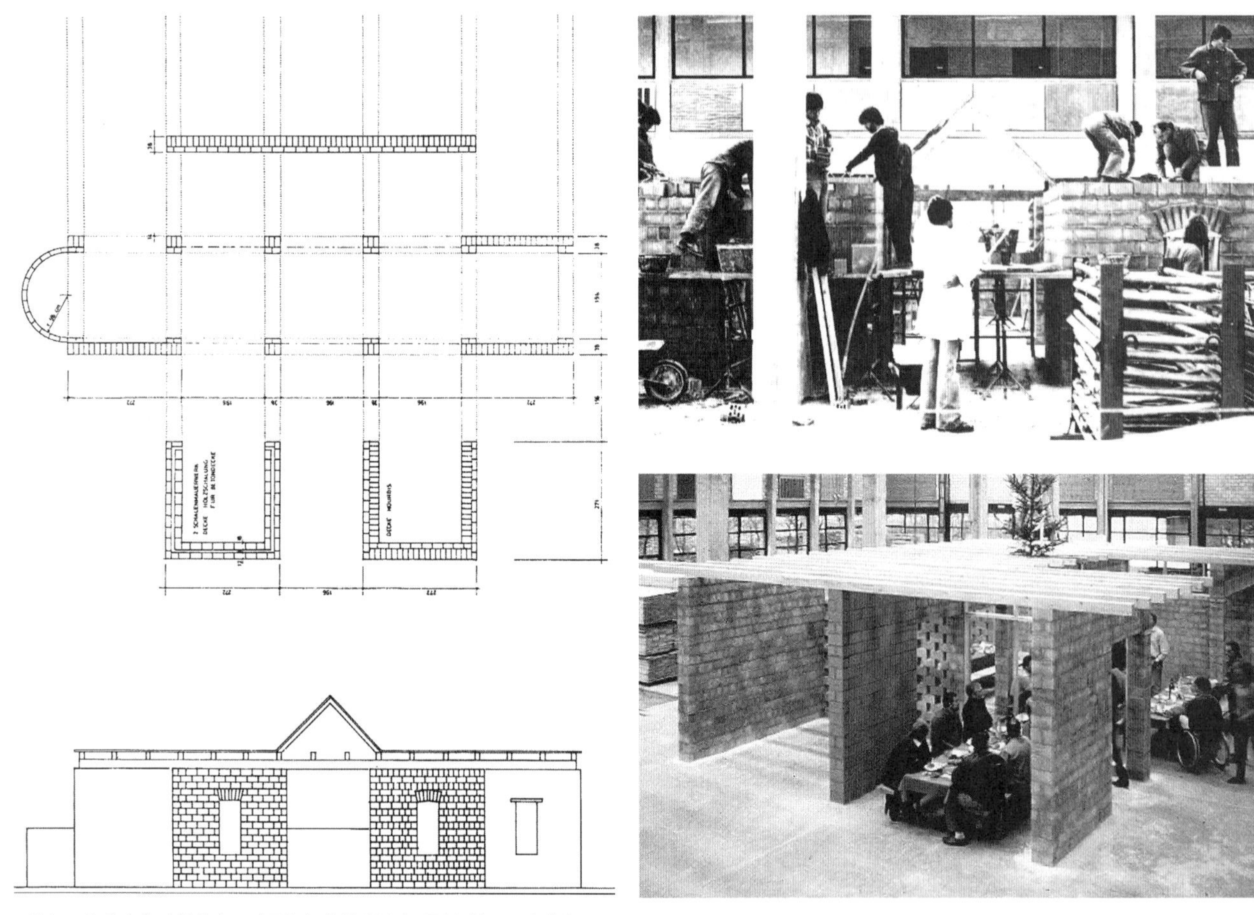

20世纪80年代克莱默教席在研讨周组织的搭建活动：图纸，施工及庆典 | Building activities organized by Kramel's Chair in the 1980s: drawings, construction, and celebration

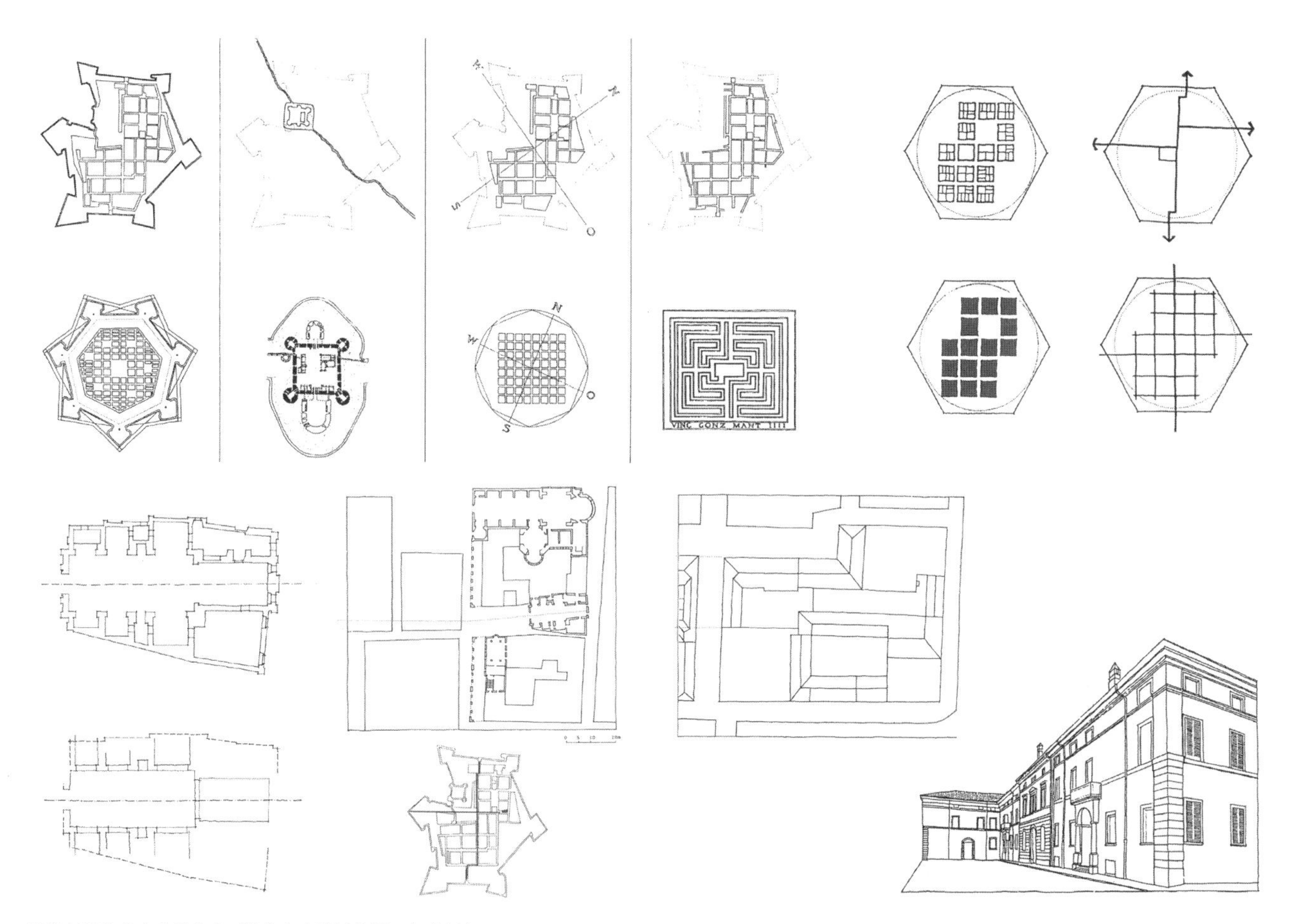

20世纪90年代克莱默教席对撒比奥内塔城的调研与分析 | Survey and analysis of Sabbioneta by Kramel's Chair in the 1990s

As in a seminar, an important aspect of the week of travel is the end product; documentation, presentation, discussion and conclusions about what have been learned by the participants. The documentation of in-depth observations is the goal. Documentation includes written as well as visual records, and knowledge of materials and methods for making such records are especially important. An advantage of the study trip is that in addition to the "group knowledge" represented in the documentation, each individual participating in the trip gains valuable personal insights and understanding about notions of architecture.

教案设计清单

在教育过程中应强调教学法，但更为实际的技术性问题却常常被低估或忽视。有经验的建筑或设计教师有一套备课的基本程序。

以下给出一套按步骤总结的教案设计要点。这套程序积累了多年的经验，同时面向新的诠释、调整和定义，可以被视为构建课程或教程的纲要。

1.定义教程 教程的主题和目标分别是什么？	· 知识　· 技能　· 体验　· 态度　· 过程性知识
2.策划教案 现有的限制 边界条件	· 环境　· 学生　· 教师　· 教案　· 时间结构　· 课程结构　· 配套活动 · 辅助材料　· 理论基础　· 参考人物　· 参考研究
3.实施教案	· 过程模式（评图，课程设计）　· 媒介/方法　· 学习过程　· 学生/教师互动
4.过程督导	· 预期的结果　· 反馈分析　· 实际结果　· 灵活性
5.评估教案	· 表达与展示　· 评分　· 期末评图　· 实际情况　· 价值尺度
6.应用教学成果	· 存档　· 展示　· 总结　· 反馈回路　· 洞察/结论　· 出版/论文

Checklist for Program Design

While in an educational process, the role of pedagogy has to be emphasized, the more pragmatic issues of technicalities are often underestimated or neglected. There exists an underlying set of procedures that every experienced teacher of design or architecture follows in his or her preparation of a course or a program.

A step-by-step procedure as a check-list for teachers to follow when designing a program is made. This procedure reflects years of experience yet still remains open for interpretation, adjustment and definition. It can be seen by those who use it as an outline for structuring a course.

1. Definition of the program What are the theme and the goal?	· Knowledge · Know-how · Skills · Experience · Attitudes
2. Planning the program Existing limitations Boundary conditions	· Environment · Students · Teachers · Program · Time structure · Course structure · Supporting events · Supporting materials · Reference persons · Reference research · Theoretical base
3. Implementing the program	· Process mode (reviews, studio) · Media/methods · Learning process · Student/teacher interaction
4. Monitoring the program	· Expected results · Feedback analysis · Actual results · Flexibility
5. Evaluating the program	· Presentations · Critics · Value scales · The real picture · Grading
6. Utilising the program	· Documentation · Presentation · Debriefing · Feed-back loops · Insights/conclusions · Publications/papers

第一阶评图展板局部，1988/89 学年 | Students' pin-ups for Phase I review, 1988/89

3 教案 | Programs

1 教案结构 | Program Structure

第一阶 | Phase I
第二阶 | Phase II
第三阶 | Phase III
第四阶 | Phase IV

2 演化 | Evolution

历年教案汇总 | Summary of Teaching Plans
教案选例 | Examples

3 实例 | Cases

1987/88 划艇俱乐部 | Rowing Club
1991/92 游客中心 | Tourist Centre
1998/99 旅馆 | Hotel Complex

1 教案结构 | Program Structure

本页及后两页：1992/93学年“基础设计”发给学生的课程安排及部分任务书 | This and the next spread: the handouts distributed to students in Basic Design program, 1992/93

教案结构是向学生呈现信息的方式。它是学生赖以增长知识、能力及技能的框架；是教师借以考察、评价学生进步的参照。它决定了设计课练习、讲座的节奏（以天为单位）。虽然“基础设计”的基本结构大体不变，课程的安排却会为确保学生适时地掌握方法、能力及技能不断调整。

四个进阶中的每一段均研习建筑集群中的一个建筑组群。随着学生对建筑形式理解的不断加深，每个组群都比前一个更复杂，更具挑战。换言之，工作方法是延续的，而工作内容递进变化。另一个值得说明的是，前一阶的成果作为下一阶的输入条件鼓励了经验的传递。我们坚持这一做法的原因是学生们常出现的“从头再来”的学习习惯——每当他们完成一个设计任务，就会在心理上把它封存起来。因此必须创造出一种有利于他们“取回”经验的过渡情景。另一方面，这些他们所产生的，熟悉的场景避免学生在每一阶都需要不断应对全新的设计条件。

每一阶末尾的小组及个人评图是对学生作业的正式评估。同时要求学生将每周的练习成果整理成作品集，以便教师与学生翻阅设计过程的完整记录。

应该明确的是课程设计的目的并不在于完成一个设计项目，而是一个框架，将一系列有针对性的练习有机地联系起来。因此，我们关注的不是设计的最终结果，而是设计过程；关注的不是学生给出的答案，而是在此过程中他们面对的问题。练习中将应用如下概念：

· 练习相对独立并且步骤明晰；

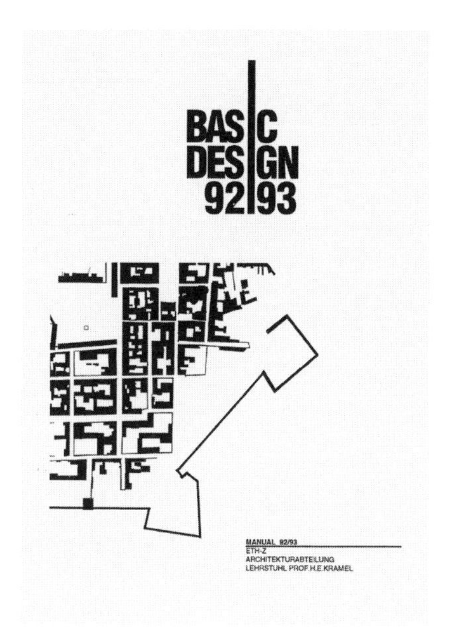

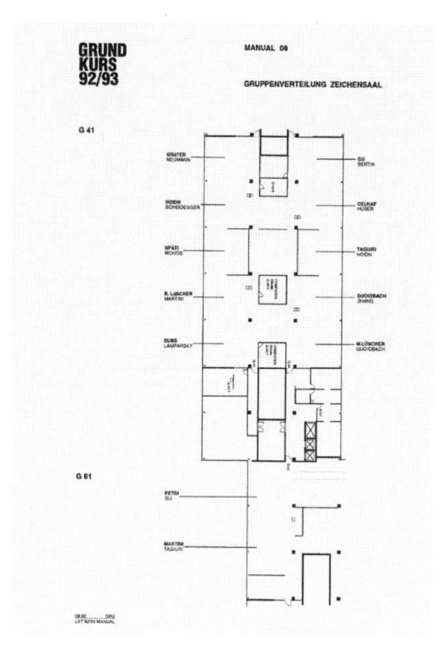

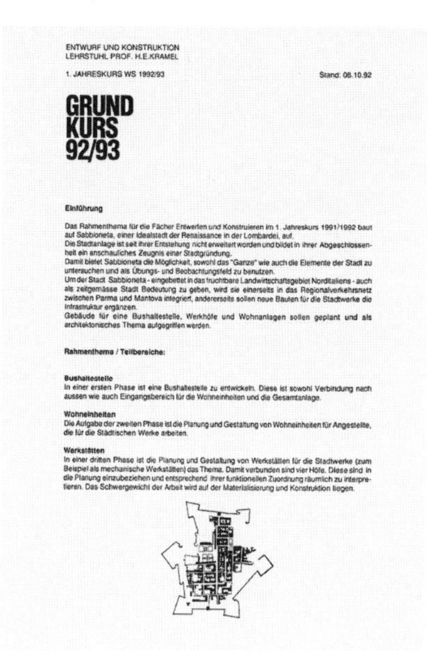
ENTWURF UND KONSTRUKTION
LEHRSTUHL PROF. H.E.KRAMEL

1. JAHRESKURS WS 1992/93 — Stand: 08.10.92

GRUND
KURS
92/93

Einführung

Das Rahmenthema für die Fächer Entwerfen und Konstruieren im 1. Jahreskurs 1991/1992 baut auf Sabbioneta, einer Idealstadt der Renaissance in der Lombardei, auf.
Die Stadtanlage ist seit ihrer Entstehung nicht erweitert worden und bildet in ihrer Abgeschlossenheit ein anschauliches Zeugnis einer Stadtgründung.
Damit bietet Sabbioneta die Möglichkeit, sowohl das "Ganze" wie auch die Elemente der Stadt zu untersuchen und als Übungs- und Beobachtungsfeld zu benutzen.
Um der Stadt Sabbioneta - eingebettet in das fruchtbare Landwirtschaftsgebiet Norditaliens - auch als zeitgemässe Stadt Bedeutung zu geben, wird sie einerseits in das Regionalverkehrsnetz zwischen Parma und Mantova integriert, andererseits sollen neue Bauten für die Stadtwerke die Infrastruktur ergänzen.
Gebäude für eine Bushaltestelle, Werkhöfe und Wohnanlagen sollen geplant und als architektonisches Thema aufgegriffen werden.

Rahmenthema / Teilbereiche:

Bushaltestelle
In einer ersten Phase ist eine Bushaltestelle zu entwickeln. Diese ist sowohl Verbindung nach aussen wie auch Eingangsbereich für die Wohneinheiten und die Gesamtanlage.

Wohneinheiten
Die Aufgabe der zweiten Phase ist die Planung und Gestaltung von Wohneinheiten für Angestellte, die für die Städtischen Werke arbeiten.

Werkstätten
In einer dritten Phase ist die Planung und Gestaltung von Werkstätten für die Stadtwerke (zum Beispiel als mechanische Werkstätten) das Thema. Damit verbunden sind vier Höfe. Diese sind in die Planung einzubeziehen und entsprechend ihrer funktionellen Zuordnung räumlich zu interpretieren. Das Schwergewicht der Arbeit wird auf der Materialisierung und Konstruktion liegen.

The program structure is the means of presenting information to the students. It is the structure within which students expand their knowledge, their abilities and their skills. It is the reference with which the teacher monitors and assesses students' progress. It determines the course of action in the design studio exercises (a daily rhythm) and the lectures on architectural design and technology that support these exercises. Although the fundamental structure for the Basic Design Program remains more or less the same, the arrangement of the course is constantly changing within the framework.

In each of the four phases, one building group of a larger complex is studied. The brief for each group increases in complexity in each successive phase as the students develop a more sophisticated understanding of architectural form. In other words, the "how" remains constant while the "what" is changing, making the learning process gradual but continuous. Another point is worth mentioning. Given our method of teaching, the output of one phase becomes part of the input of the following one. With this, a transfer of experience and knowledge is encouraged. The reason for our insistence on continuity is the common phenomenon of students trying to get things "over with"—they learn something, create a mental compartment for it and store it away. Therefore, transfer situations have to be created to help students "retrieve" their own experience. Besides, notwithstanding a new problem condition in each phase, students get to design for a familiar context that they understand since they were themselves responsible for its formation.

Studio teachers make formal evaluations of students' work at the end of each phase through group and individual reviews. Students are also required to collect their weekly exercises in a portfolio. These portfolios have the advantage of giving both teacher and student an overview of a student's work in a readily accessible format.

It is important to understand that the intention behind all this is not to produce a design project. Rather, it is intended as a framework that ties together a number of exercises that each deals with problems of specific and limited scope. Consequently, it is not the result—the final product—that is of interest but the process of learning. We are

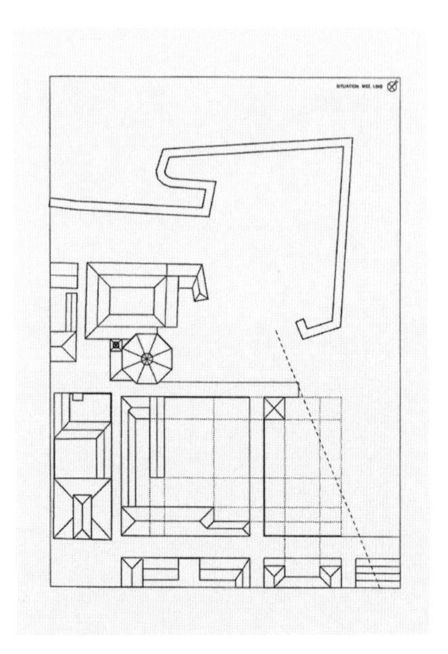

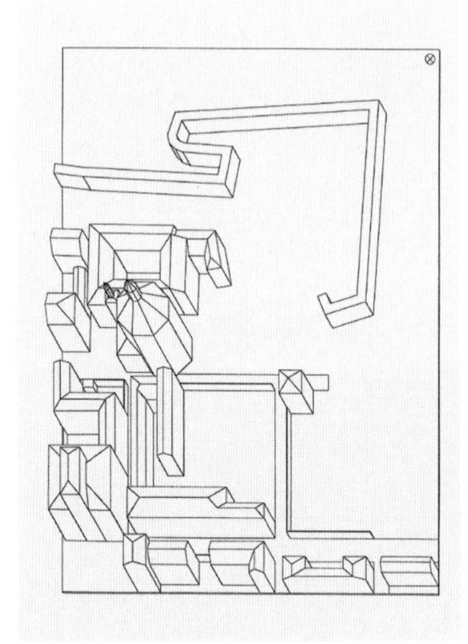

ENTWURF UND KONSTRUKTION
LEHRSTUHL PROF. H.E.KRAMEL

1. JAHRESKURS WS 1992/93 Stand: 03.11.92

**GRUND
KURS
92/93**

Übungsplan

PHASE I	Dienstag:		Mittwoch:	
Arbeitsschritt 1.1/ 1.2 27.10./ 28.10. 92 (Woche 44)	Situationsmodell Ausgangslage	1:200	Schlaufenmodell Gesamtsituation	1:200
Arbeitsschritt 1.3/ 1.4 03.11./04.11.92 (Woche 45)	Situationsplan Zeichnung	1:500	Schlaufenmodell Überarbeitung Situation	1:200
Arbeitsschritt 1.5/ 1.6 10.11./11.11.92 (Woche 46)	Volumenmodell Bushaltestelle	1:100	Funktionsdiagramm Bushaltestelle Volumen	1:100
Arbeitsschritt 1.7/ 1.8 17.11./18.11.92 (Woche 47)	Strukturplan Bushaltestelle	1:100	Strukturmodell Bushaltestelle	1:100
Arbeitsschritt 1.9/ 1.10 24.11./25.11.92 (Woche 48)	Schnittzeichnung Bushaltestelle	1:20	Schnitt / Ansicht Bushaltestelle	1:20

SEMINARWOCHE

Arbeitsschritt 1.11/ 1.12 08.12./09.12.92 (Woche 50)	Grundriss Raumzelle mit Möblierung	1:50	Gesamtsituation Überarbeitung
Arbeitsschritt 1.13/ 1.14 15.12./16.12.92 (Woche 51)	Grundriss Raumzelle Überarbeitung	1:50	Portfolio Schlussabgabe Phase I

· 工作方式由实体模型开始过渡到更为抽象的表现形式（如平面图、剖面图）；
· 有意识地运用相应的表现方式和媒介是设计学习过程的一部分；
· 小组工作（讨论及互动）是学习过程的基础（观察、参与、对比、评论）；
· 通过练习之间的相互联系增加理解的复杂性；
· 设计过程运用预设的基本建筑模块，便于在模数化的基地上进行快速操作；
· 空间、结构、场地的概念是建筑形式的决定因素。

上述方法将直接影响学生的工作成果。虽然对多样性有所期待，但每个设计练习都严格规定了需要解决的问题以及完成的时限。因为，只有让每个一年级学生都从类似的问题出发，才能在整个学院达成共识。

因此，四个相连的进阶各自有严格界定的教学目标。设计教育的基本信条之一是：“形式只能在限制中演化。”这套限制条件一路指引学生完成“基础设计”，为的是令学生明白设计一直是，也将会是一个发展的过程。事实多次证明了设计与建筑技术的结合教学是可行的，而且可以适应学生人数众多的情况。此外还证明了在设计和学习过程中，在不牺牲品质的前提下处理较高复杂程度的问题是可能的。

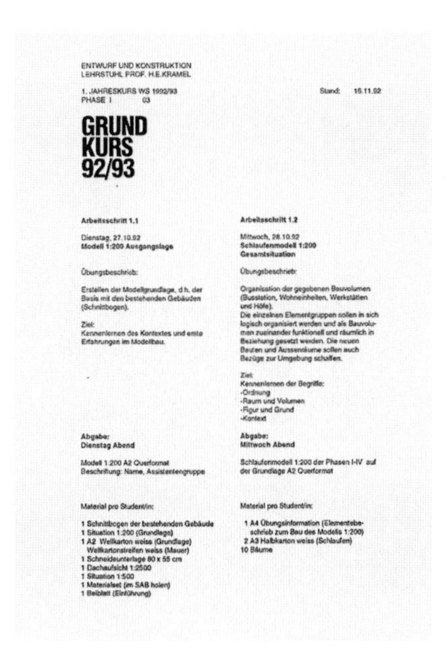

ENTWURF UND KONSTRUKTION
LEHRSTUHL PROF. H.E.KRAMEL

1. JAHRESKURS WS 1992/93 Stand: 16.11.92
PHASE I 03

GRUND
KURS
92/93

Arbeitsschritt 1.1

Dienstag, 27.10.92
Modell 1:200 Ausgangslage

Übungsbeschrieb:

Erstellen der Modellgrundlage, d.h. der Basis mit den bestehenden Gebäuden (Schnittbogen).

Ziel:
Kennenlernen des Kontextes und erste Erfahrungen im Modellbau.

Abgabe:
Dienstag Abend

Modell 1:200 A2 Querformat
Beschriftung: Name, Assistentengruppe

Material pro Student/in:

1 Schnittbogen der bestehenden Gebäude
1 Situation 1:200 (Grundlage)
1 A2 Wellkarton weiss (Grundlage)
Wellkartonstreifen weiss (Mauer)
1 Schneideunterlage 80 x 55 cm
1 Dachaufsicht 1:2500
1 Situation 1:500
1 Materialset (im SAB holen)
1 Beiblatt (Einführung)

Arbeitsschritt 1.2

Mittwoch, 28.10.92
Schlaufenmodell 1:200
Gesamtsituation

Übungsbeschrieb:

Organisation der gegebenen Bauvolumen (Busstation, Wohneinheiten, Werkstätten und Höfe).
Die einzelnen Elementgruppen sollen in sich logisch organisiert werden und als Bauvolumen zueinander funktionell und räumlich in Beziehung gesetzt werden. Die neuen Bauten und Aussenräume sollen auch Bezüge zur Umgebung schaffen.

Ziel:
Kennenlernen der Begriffe:
-Ordnung
-Raum und Volumen
-Figur und Grund
-Kontext

Abgabe:
Mittwoch Abend

Schlaufenmodell 1:200 der Phasen I-IV auf der Grundlage A2 Querformat

Material pro Student/in:

1 A4 Übungsinformation (Elementebeschrieb zum Bau des Modells 1:200)
2 A3 Halbkarton weiss (Schlaufen)
10 Bäume

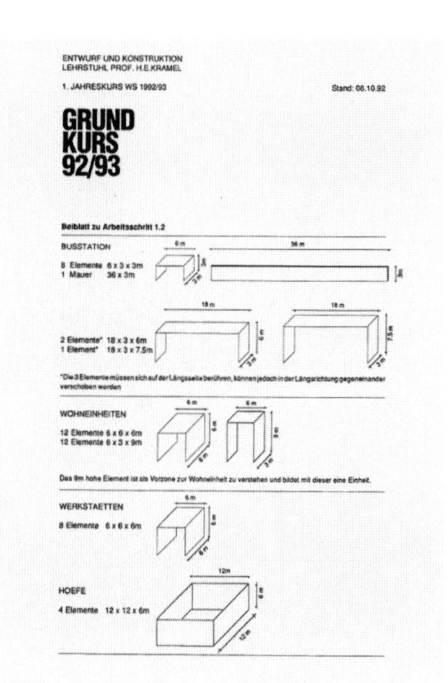

ENTWURF UND KONSTRUKTION
LEHRSTUHL PROF. H.E.KRAMEL

1. JAHRESKURS WS 1992/93 Stand: 08.10.92

GRUND
KURS
92/93

Beiblatt zu Arbeitsschritt 1.2

BUSSTATION

8 Elemente 6 x 3 x 3m
1 Mauer 36 x 3m

2 Elemente* 18 x 3 x 6m
1 Element* 18 x 3 x 7.5m

*Die 3 Elemente müssen sich auf der Längsseite berühren, können jedoch in der Längsrichtung gegeneinander verschoben werden

WOHNEINHEITEN

12 Elemente 6 x 6 x 6m
12 Elemente 6 x 3 x 9m

Das 9m hohe Element ist als Vorzone zur Wohneinheit zu verstehen und bildet mit dieser eine Einheit.

WERKSTAETTEN

8 Elemente 6 x 6 x 6m

HOEFE

4 Elemente 12 x 12 x 6m

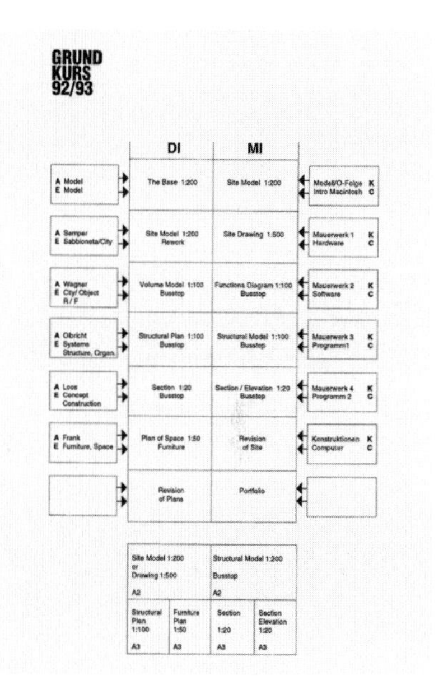

GRUND
KURS
92/93

not concerned with answers that the student develops but rather with the questions that he or she is compelled to consider.

— The exercises move through discrete and clearly defined stages.

— The working method moves from physical modeling to more abstract forms of representation.

— The conscious utilization of specific representational methods and media is part of learning the design process.

— The work-in-a-team situation (discussion and interaction) is fundamental to the learning process (observe, participate, compare, critique).

— The interrelation of the exercises builds a complex understanding.

— The design process utilizes "pre-established" basic architectural units (BAU), which allow the quick manipulation of information on a modular base.

— Concepts of space, structure and site are used as the determinants of architectural form.

This approach directly influences students' work. Although a great diversity is desired, the design exercises are strictly limited in the problems they seek to address and in the amount of time on which the student is allowed to spend. It is believed that only if each student in the first year deals with the same problems, a common denominator for the school as a whole can be established.

Consequently, the didactic goals for each successive phase are rigorously defined. One of the basic assumptions in design education is that "form evolves only through limitations". It is thus a set of limitations that guides the student of architecture through the four phases of our Basic Design Program. This approach seeks to demonstrate to the student that design has been and still is a process of development. In a number of ways, it has been possible to prove that an integration of design and technology is possible and can be executed even with a very large number of students. In addition, it has become clear that a high degree of complexity could be handled without loss of quality in the design and learning process.

第一阶

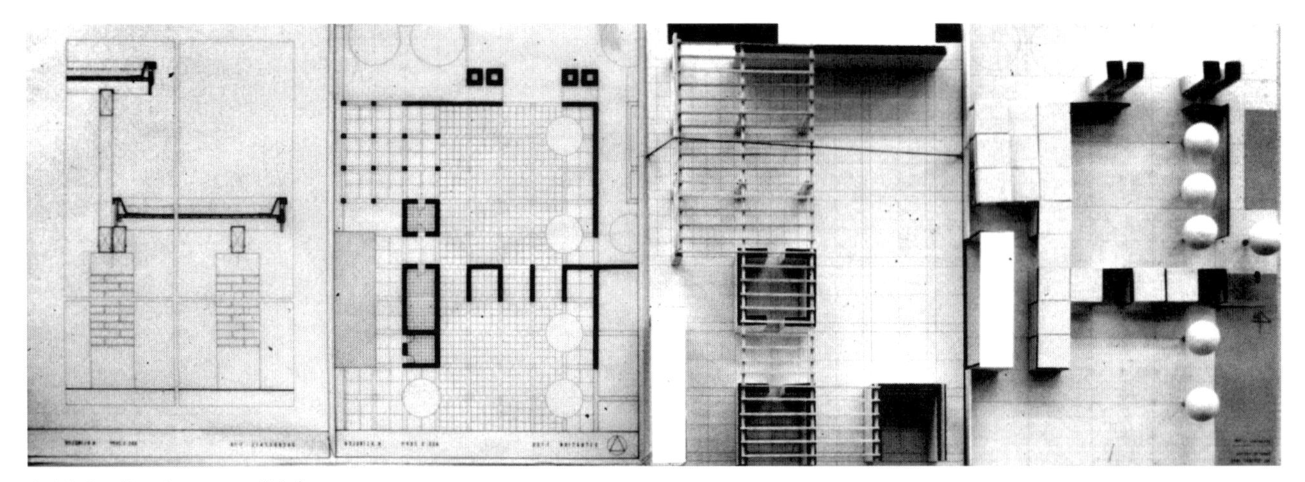

学生作业，第一阶，1986/87学年 | Students' work, Phase I, 1986/87

第一阶引入作为建筑现象的空间，同时强调设计与建造方法的关系。

构思教程的时候，应区分职业建筑师设计的过程和学习设计的过程。学生是设计新手，尽管如此，仍要求他们一开始即用给定的模块组织总平面，也正因为他们专业能力的不足，这个初始的尝试为的是让他们更好地理解问题，而不是真正解决问题。

从场地模型开始，以屋顶节点模型结束。学生由此经历各种尺度的设计问题，即便他们还未能完全驾驭。每周引入新的设计表达方法，使学生逐步掌握模型制作、绘图及表达的技法。学生必须明白尺度的重要性。绘图的比例决定了它所要解决的问题的深度。

第一阶对于新任教师来说，同样是一种对新概念及陌生思维的浸润过程。新任教师虽然对建筑概念并不陌生，却很可能并不熟悉设计教育的概念及教师应有的角色。教师寻找传递知识的途径，而更重要的是，帮助学生形成自己的知识体系。要求新任教师以日记或周记的方式将教学思维过程正规化。学生的进度各不相同，练习固有的相关性和进阶性使教师能够适时引导个别学生或小组回顾相关概念。这种形式架构给予教师试验各种教学技巧的契机。因材施教，某些讨论和解说适合在整组范围内进行，但个别或小组讨论往往更能激发对话，促进交流。

Phase I

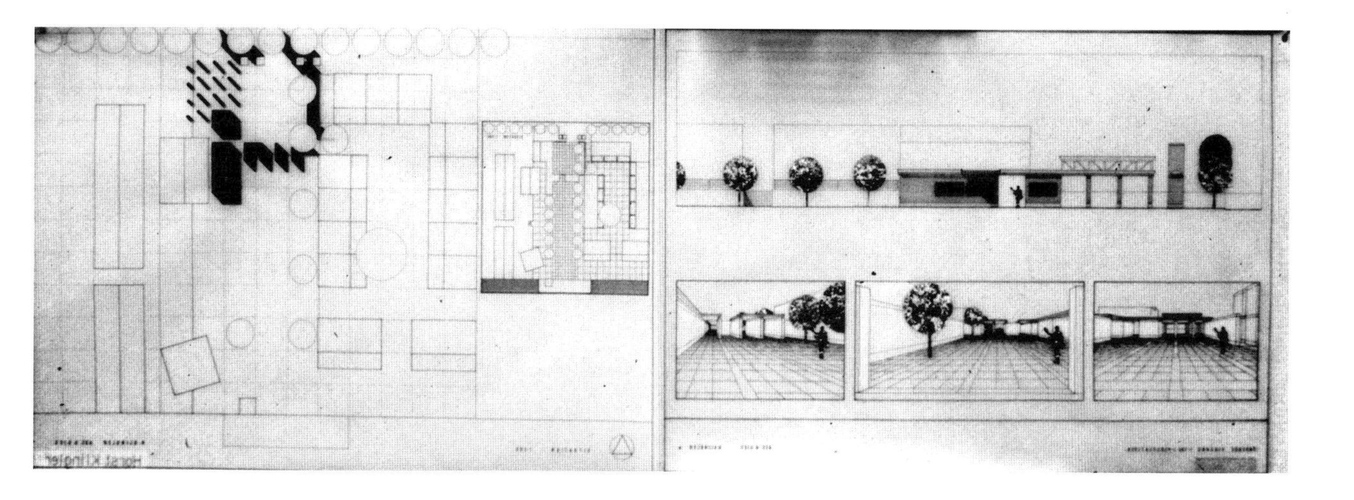

The first phase serves as an introduction to space as an architectural phenomenon. Furthermore, the relationship between design and construction methods was emphasized.

In the planning of the teaching program, the difference between the logic of the professional design process and that of the process of learning became important. Unlike professionals, students approach design problems as beginners. Nevertheless, they are expected to design the whole site based on the given components. Because the student lacks developed practical or conceptual skills, the end product of this exercise constitutes not a solution but rather the launch pad for a better understanding of the problem.

Phase I starts with a site model and ends with a roof detail. Although the student works through the full range of design problems, from large-scale planning to details, we do not expect the student to be thoroughly competent. Each week, new technical skills are also required for presentation, progressively building students' knowledge of the architectural conventions for model making, drawing and presentation. It is important that the students understand the significance of scale. The scale of a drawing defines the issues one can deal with and therefore determines the types of questions that must be addressed in that drawing.

Phase I can also be seen as an immersion for the novice teachers. The teacher is not unfamiliar with the notions of architecture but rather with those of design education and the role of the teacher within the program. The teacher is discovering ways to convey knowledge and, most importantly, ways to help students build their knowledge. A journal kept by the new teacher in preparation for and evaluation of weekly events in the studio forces the teacher to formalize their thoughts. Each student learns at a different pace. The interrelated and progressive nature of the exercises gives a teacher the opportunity to review concepts when necessary for individual students or a group of students. This format allows a new teacher the opportunity to experiment with studio techniques. Some discussions and explanations are best done with the entire group together while discussions with individual students or small groups can stimulate conversation and increase communication between teachers and students.

第一阶 1991/92 学年

目标	引入建筑形式：空间、文脉、体量、系统。	进一步理解概念；绘图与模型技能作为表达概念的途径。	用设定的空间元素生成空间；理解较大系统与较小系统之间的协同关系。	厘清设计想法，测试空间组织；用二维绘图来描述三维空间。
周次	1	2	3	4
讲座	建筑设计 1.1 构造 1.1	建筑设计 1.2 构造 1.2	建筑设计 1.3 构造 1.3	建筑设计 1.4 构造 1.4
课程设计	练习1 场地 场地模型 1:200 场地现状	练习3 场地 总平面图 1:500 4个子系统，包含不同类型的元素	练习5 市场 总平面图 1:500 在总平面中修改集市	练习7 市场 结构平面 1:200 平面及剖面 1:200
	练习2 场地 纸圈模型 1:200 总体规划	练习4 集市 纸圈模型 1:100	练习6 市场 功能图解 1:200 体量草图 1:200	练习8 场地 总平面 1:500 体量、铺地、绿化、整体调整
过程	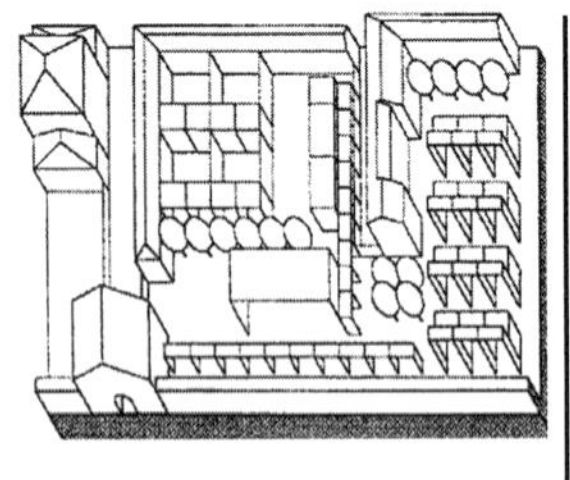	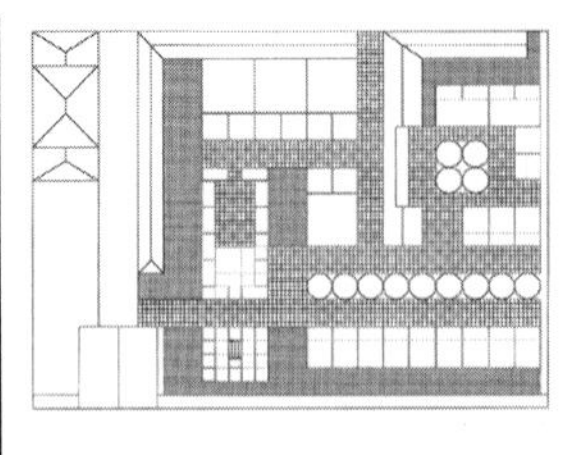 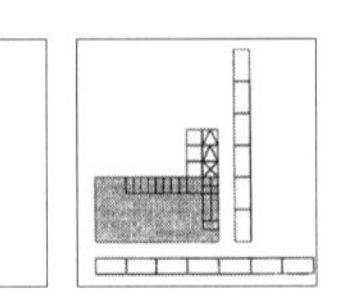	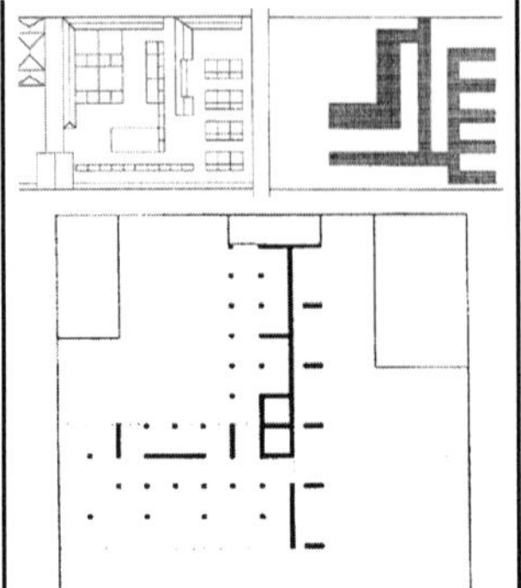	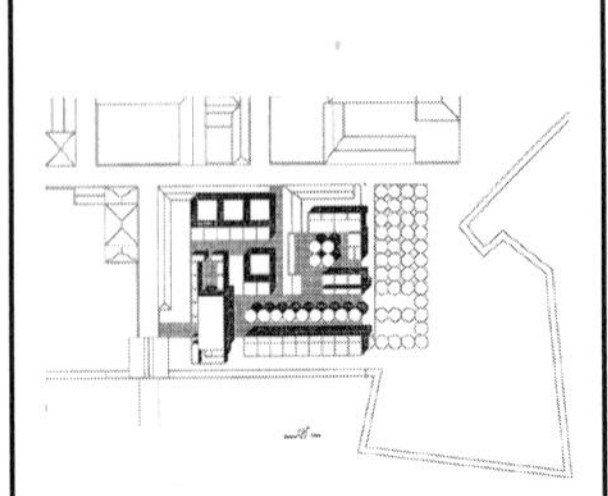

周次	5	6	7	8
目标	理解从想法转化为建造及材料的涵义。	现场体验从概念到建造的建筑过程。	综合前面几周所得完善想法。	为学生总结设计过程，并将学生作品和全班的平均水平相比较。
讲座	建筑设计 1.5 构造 1.5		建筑设计 1.6 构造 1.6	
课程设计	练习9 市场 体量模型 1:100 练习10 市场 用网格纸绘制剖面 1:33	研讨周 建造课：钢/木/砖石结构	练习11 场地 总平面 1:100 表现体量、绿化、铺地及阴影 练习12 市场 首层平面，立面图 1:50 家具布置；重绘结构平面	练习13 市场 发展1.10中的构造剖面 立面图 1:33 第一阶评图
过程	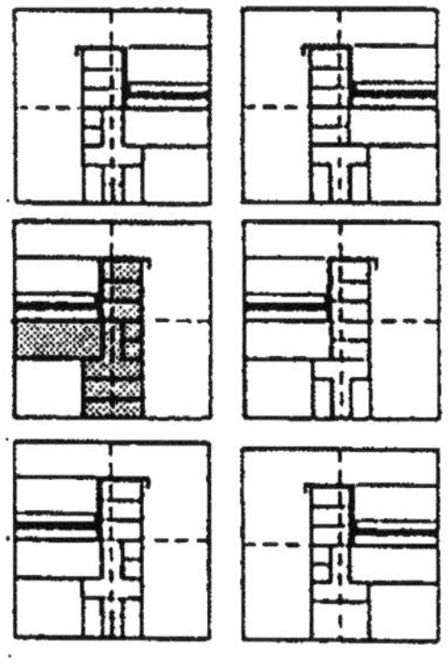		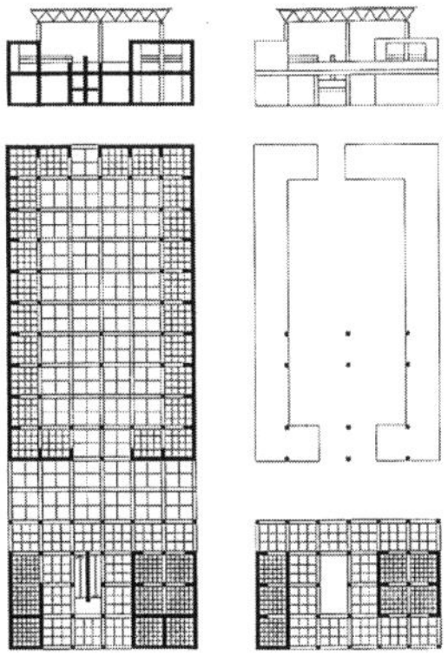	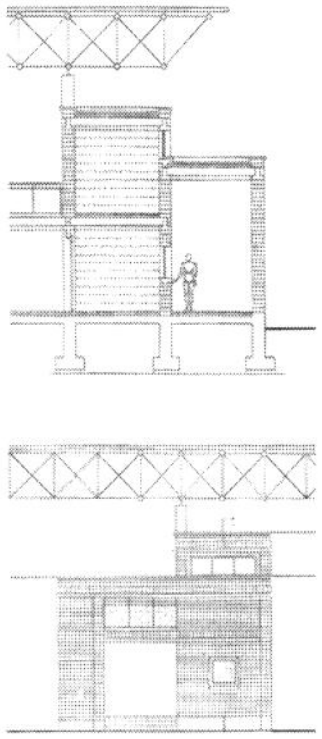

Phase I 1991/92

Objectives	Introduction to architectural form: space, context, mass, systems.	Reworking to clarify understanding of concepts, drawing and model building skills as a way to illustrate ideas.	To use established spatial elements to create space; to understand the interactive relationship between larger and smaller systems.	Clarification of ideas and testing of spatial organization; use of 2D drawings to describe 3D space.
Week	1	2	3	4
Lectures	Architecture Design 1.1 Construction 1.1	Architecture Design 1.2 Construction 1.2	Architecture Design 1.3 Construction 1.3	Architecture Design 1.4 Construction 1.4
Studio	Ex.1 The site Site model 1:200 initial situation	Ex.3 The site Drawing 1:500 4 subsystem	Ex.5 Market hall Drawing 1:500 To modify market hall in overall situation (Rev. 1)	Ex.7 Market hall Structrual plan 1:200 Plan and section
	Ex.2 The site Loop model 1:200 Overall situation	Ex.4 Market hall Loop model 1:100	Ex.6 Market hall Function diagram 1:200 Volumn sketch 1:200	Ex.8 The site Drawing 1:500 volumn, flooring, trees overall situation (Rev.2)
Process	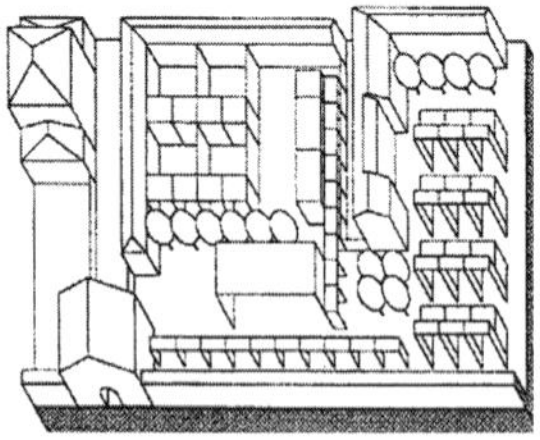	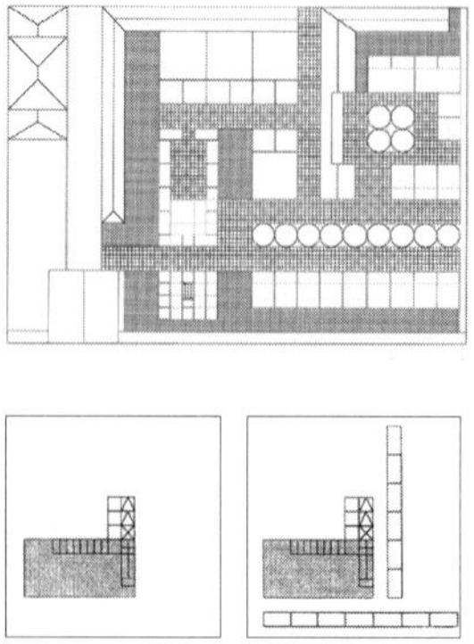	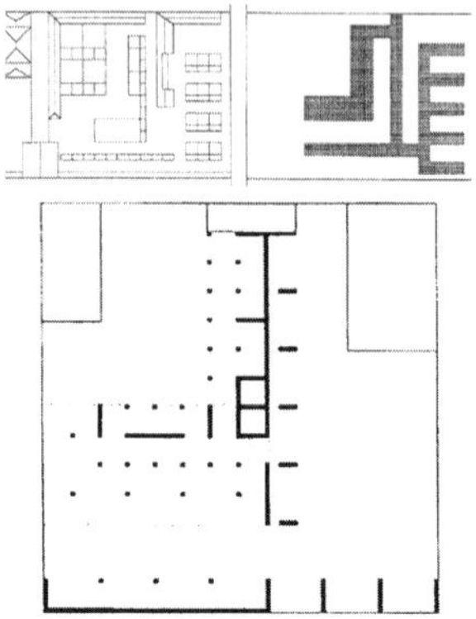	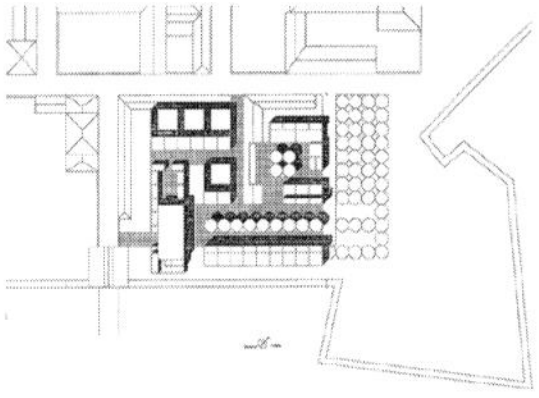

	5	6	7	8
Objectives	Understand the implications of converting ideas into the reality of construction and materials.	Hands-on experience in the architecture process from conception to construction.	Refine ideas -synthesizing understanding gained in the previous weeks into a reworked solution.	For students to hear a summary of their progress and their current position in course work in comparison to other students.
Week	5	6	7	8
Lectures	Architecture Design 1.5 Construction 1.5		Architecture Design 1.6 Construction 1.6	
Studio	Ex.9 Market hall Volumn model 1:100	Seminarweek (construction course: steel / wood /masonry construction)	Ex.11 The site Site plan 1:100 volumn, trees, flooring, shadow	Ex.13 Market hall Develop the section in Step 1.10 elevation approx. 1:33
Studio	Ex.10 Market hall Scheme section with grid paper, approx. 1:33		Ex.12 Market hall 1st floor plan, elevation 1:50 with furniture; rework the structural plan	Review of Phase I
Process	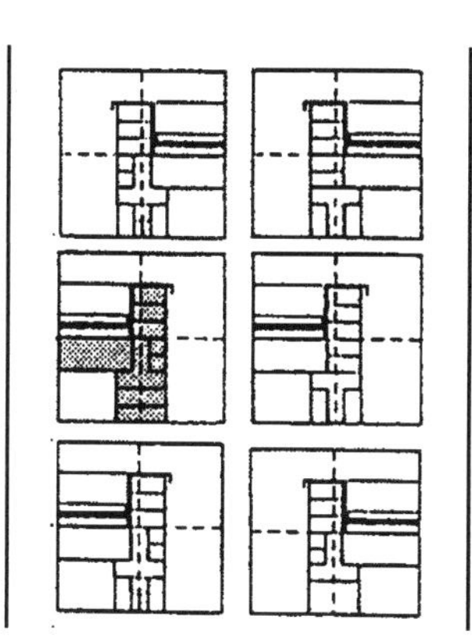		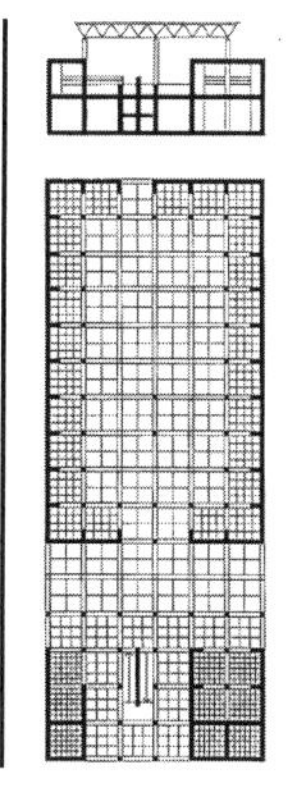 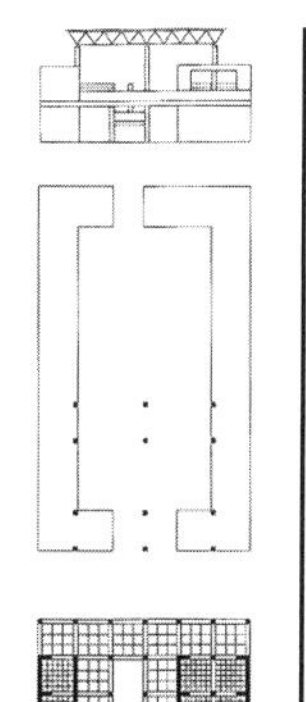	

第二阶

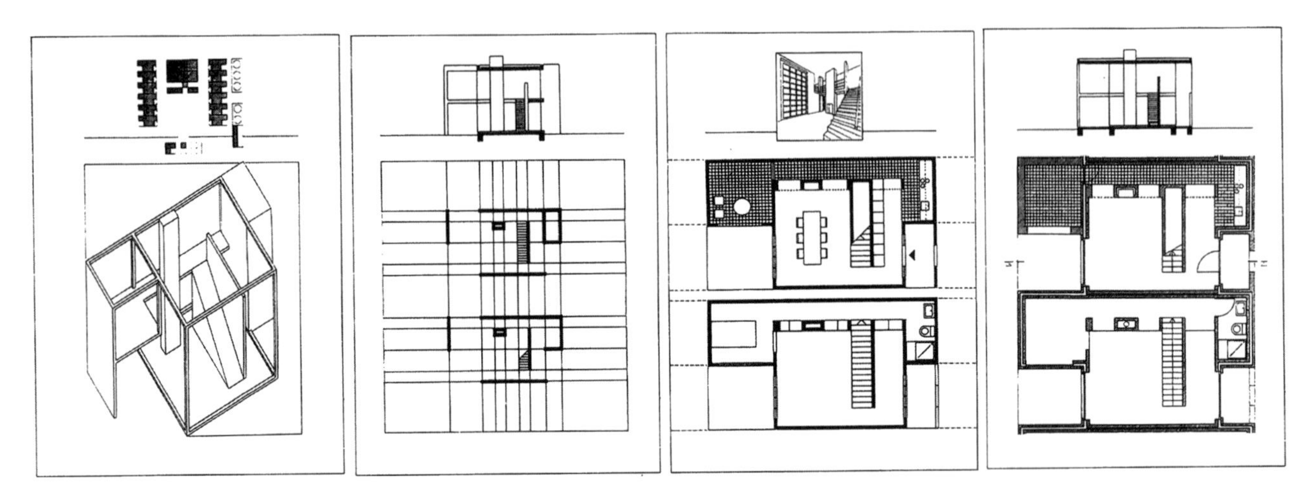

学生作业，第二阶，1986/87学年 | Students' work, Phase II, 1986/87

学生在第一阶中组织的总平面成为第二阶的初始条件。第一阶的设计成果是第二阶的参照。学生将在场地中设计居住单元，扩展对建筑空间的认识。使用需求更为复杂，需要操作的元素增多，作为两层的住宅构造难度也提升了。承重材料在砖和木材以外增加了钢。构造层面涉及热散失、热工性能及防潮问题。

第二阶整体强调的是功能组织。简单的住宅单体有利于多方位的设计考量，例如家具的摆放及使用需求发生变化时的影响。与第一阶一样，练习覆盖了不同比例尺度上的设计问题。

在第一阶的浸润之后，第二阶是技能与能力、所学知识与方法的内化过程。陌生的概念逐步转化为个人的思考工具。学生们已认识到决定建筑形式的基本要素，并熟悉二维及三维的建筑表现常用方法。他们做决定的能力提高了，积累经验及提炼知识的能力也随之增长。一系列更高难度的强调“空间/功能”“建造/物化”“文脉/场所”三要素的循环练习促进深化理解。

到了第二阶，教师已经适应了每周的教学节奏，课程设计的氛围，以及学生能力的个体差异，可以侧重于沟通技巧和感知力的完善。教师能够更敏感地觉察到学生的需要及理解程度的差异，更有效地使用课堂时间。

Phase II

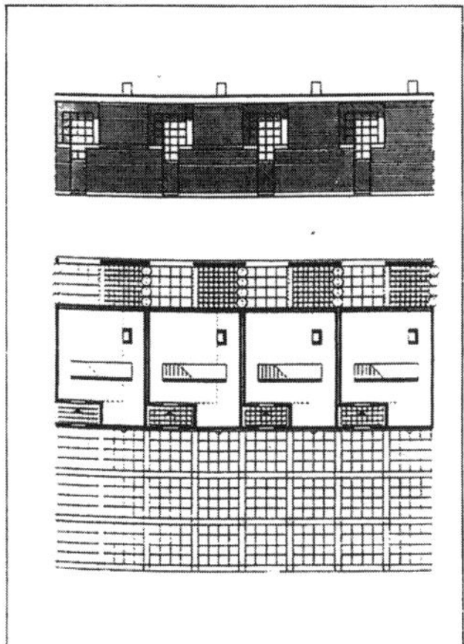

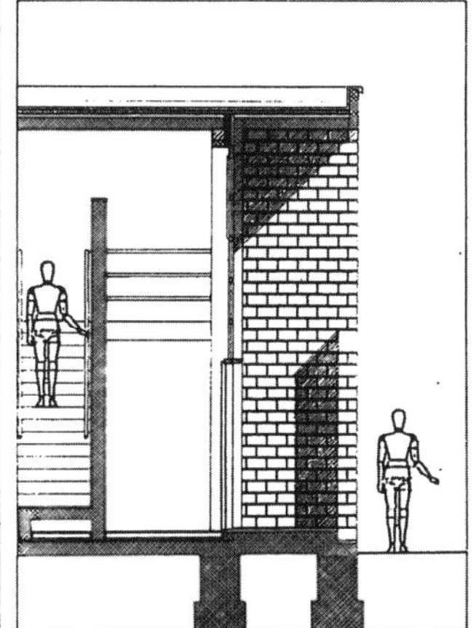

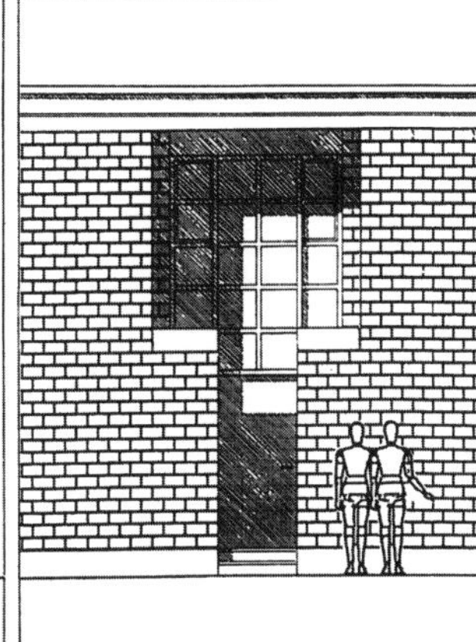

The overall site established by the students in Phase I became the input in Phase II and was further developed. The design that had been developed in some detail in Phase I became a point of reference in the second phase. The students had to develop the design of a housing unit within the context of their site plan. This meant that they had to expand the notion of architectural space, that a more complex program had to be organized and that the understanding of form had to be further developed. On the level of technology, the introduction of two stories and the problem of thermal insulation raised the level of complexity. As an additional material, steel was added to the already existing options of wood and brick. Given this, it became possible to introduce questions of heat loss, thermal performance, and insulation against humidity.

Throughout Phase II, specific emphasis was placed upon the question of functional organization. The simple two-story housing unit allowed the investigation of multiple considerations such as the role of furniture and the impact of changes on conditions of use. As in Phase I, the exercise was executed on various levels of scale, forcing the students to address a range of problems.

After the immersion of Phase I, Phase II can be seen as the phase in which assimilation of the skills and abilities, the knowledge gained and methodologies learned, occurs. Students are aware of the basic notions that determine the architectural form, and they are familiar with many of the conventions used for 2- and 3-D architectural representations. For them, recurrent exercises, which emphasize the triad of space/function, construction/materialization and context/place—at a more demanding and complex level—offer the opportunity to check their understanding and to deepen their investigations.

By Phase II, the teacher is acquainted with the pace of the week and with the various levels of abilities, skills and knowledge the students possess. Attention can be given to the refinement of communication skills and perceptual abilities.

第二阶 1991/92学年

	1	2	3	4
目标	用模块定义抽象空间，阐明空间概念；两人为组——团队合作的重要性。	抽象概念转化为具体功能；空间序列与公共/私密空间的区分；立面的组织。	结构与空间。	结构与使用需求对立面的影响，立面自身的形式逻辑。
周次	1	2	3	4
讲座	建筑设计 2.1 构造 2.1	建筑设计 2.2 构造 2.2	建筑设计 2.3 构造 2.3	建筑设计 2.4 构造 2.4
课程设计	练习1 居住单元 空间组织模型 1:50 体量的形成	练习 3 居住单元 平面兼家具布置 1:50 空间结构平面修正与分析 1:100	练习 5 居住单元 结构模型 1:50 单元组合	练习 7 居住单元 空心墙构造剖面 1:50 立面构成 1:100
课程设计	练习 2 居住单元 空间组织平面、剖面 1:100 轴测图1:100	练习 4 居住单元 立面模型 1:20	练习 6 居住单元 一层、二层平面与剖面（表现空心墙体）1:50	练习8 居住单元 正/背立面 1:50 实墙体与空心墙体的分化
过程	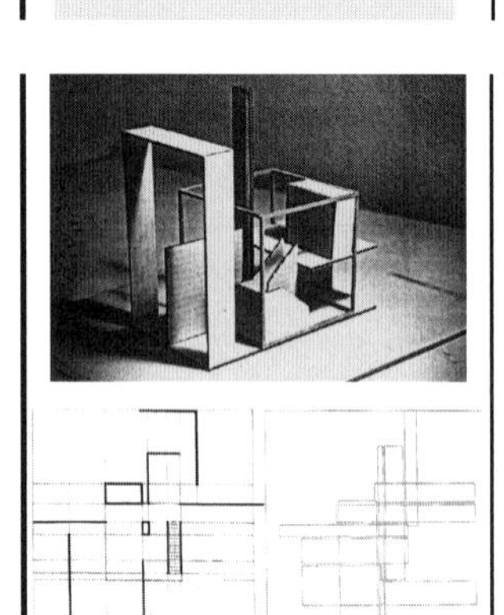	 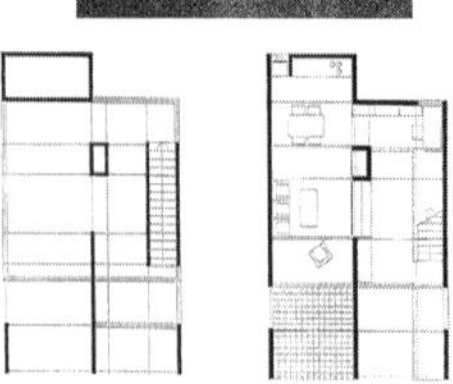	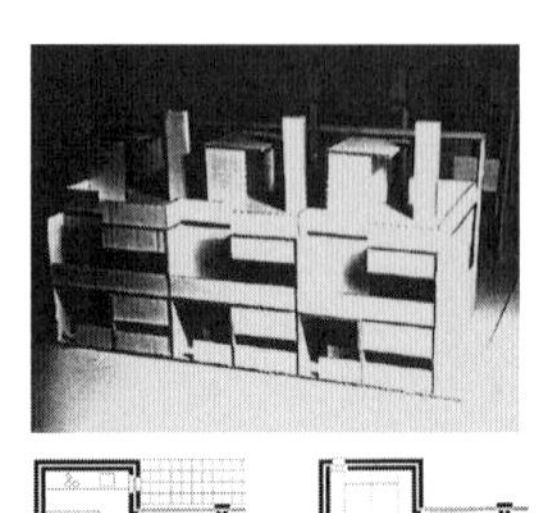 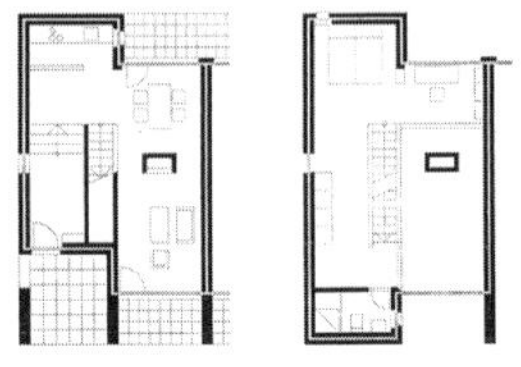	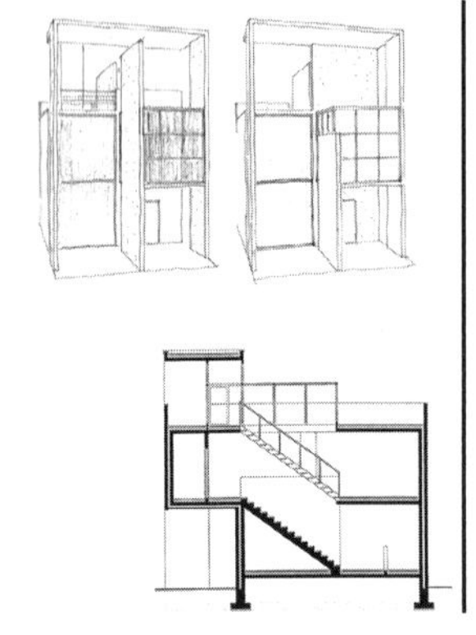

周次	5	6	7	8
目标	深化立面表达。	通过构造细部深化建筑形式，以大比例绘图研究结构与材料；总平面空间关系。	透视及设计表现。	对学生个人及小组成果的总结与评价。
讲座	建筑设计 2.5 构造 2.5	建筑设计 2.6 构造 2.6	建筑设计 2.7 构造 2.7	
课程设计	练习9 居住单元 平面、剖面 1:50 练习10 居住单元 立面 1:50	练习11 居住单元 构造节点轴测图 1:20 练习12 总平面 住宅组群与市集的空间组织结构 1:200	练习13 居住单元 场地中的单元组合透视图/单元组合轴测图/提交正图 练习14 居住单元 场地中的单元组合透视图/单元组合轴测图/提交正图	第二阶评图
过程	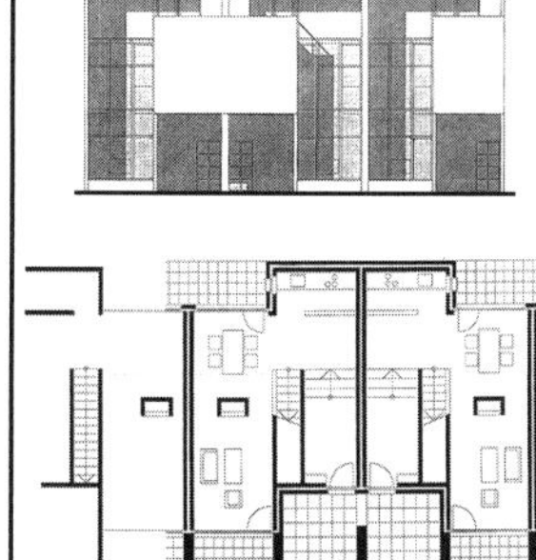	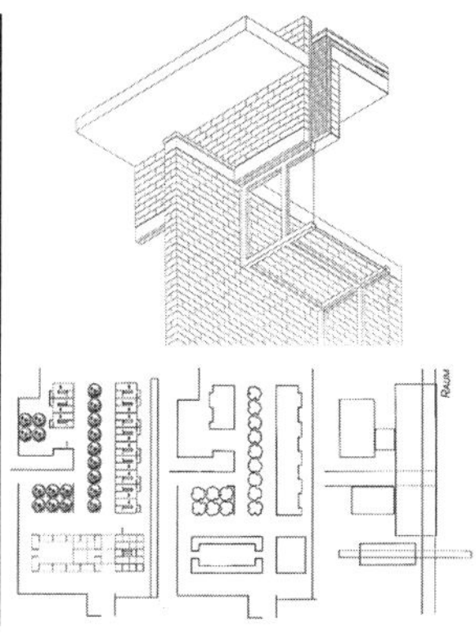	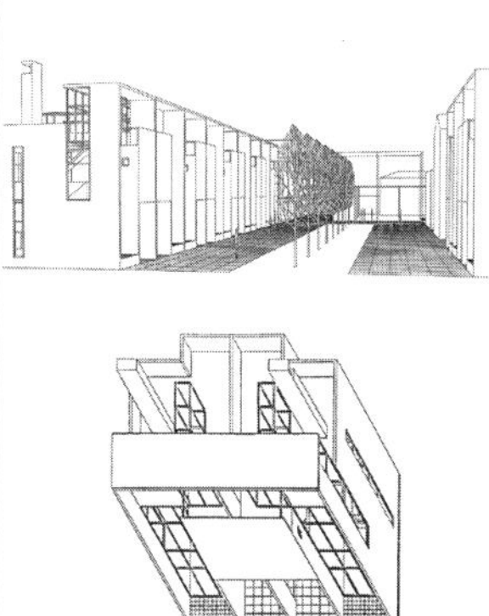	

Phase II 1991/92

	1	2	3	4
Objectives	Using elements to define abstract space to clarify concepts of space; two students work together - the importance of teamwork.	Transform abstract ideas into specific functions; sequence of spaces and clarification of public vs. private; organization of facades.	About structure and space.	Reconsider facade affected by structural and functional issues, and its formal logic.
Week	1	2	3	4
Lectures	Architecture Design 2.1 Construction 2.1	Architecture Design 2.2 Construction 2.2	Architecture Design 2.3 Construction 2.3	Architecture Design 2.4 Construction 2.4
Studio	Ex.1 Living unit Spatial Model 1:50 volumn formation	Ex.3 Living unit Plan with furniture 1:50 Structural plan modify and structural analysis 1:100	Ex.5 Living unit Structural model 1:50 Grouping of units	Ex.7 Living unit cavity wall construction section 1:50 Elevation structure 1:100
	Ex.2 Living unit Plan & section 1:100 Isometric 1:100	Ex.4 Living unit Elevation model 1:20	Ex.6 Living unit Plans of 1st and 2nd floor and section, with cavity wall construction 1:50	Ex.8 Living unit Front/back elevation 1:50 cavity or structrual wall
Process	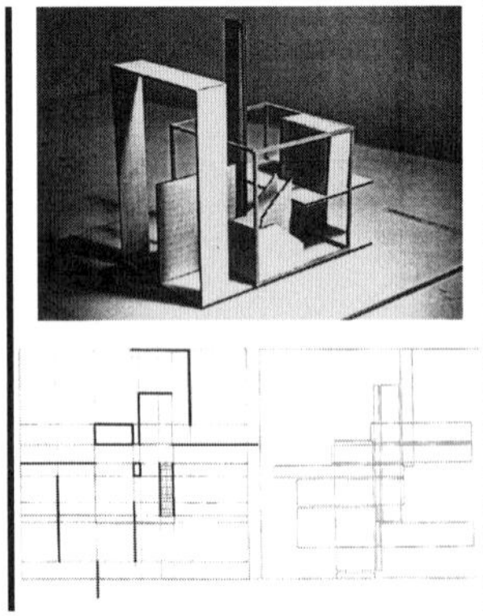	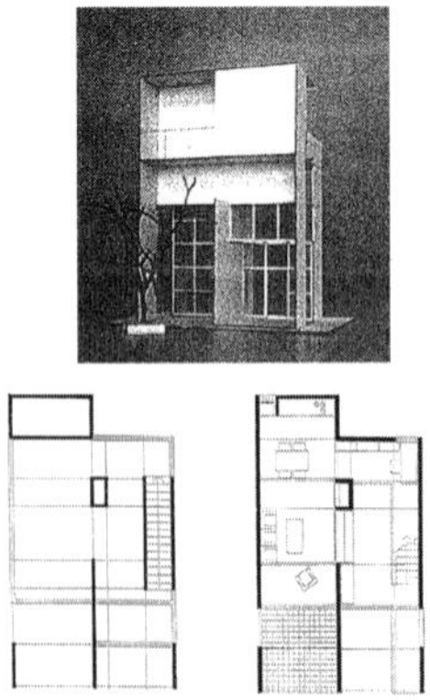	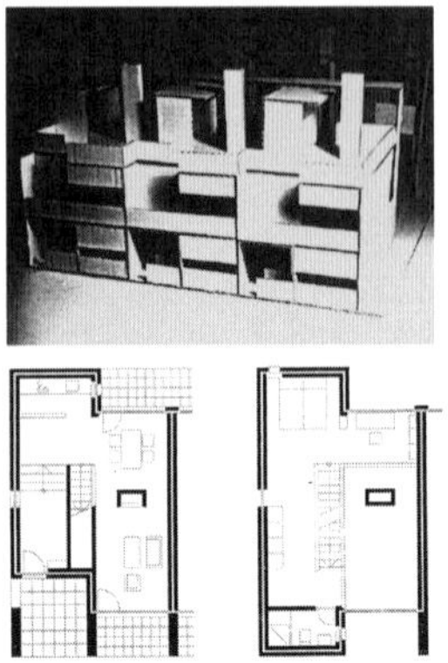	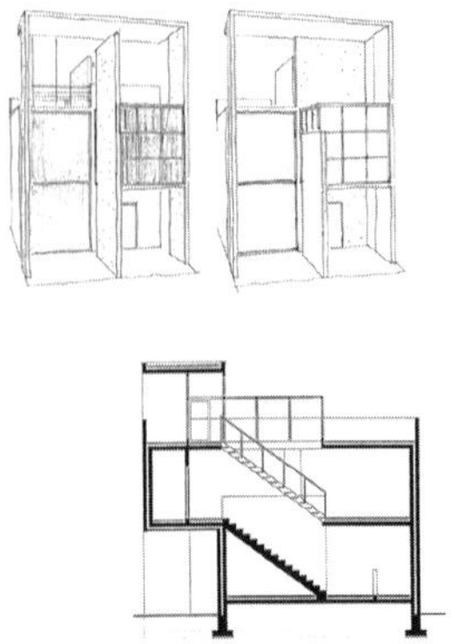

Objectives	To develop and represent the facades.	Use construction details to develop architectural form with large-scale drawings; investigate structure and materials; spatial organization of the site.	Perspectives and project presentation.	Students hear a summary of their progress including comments on group work.
Week	5	6	7	8
Lectures	Architecture Design 2.5 Construction2.5	Architecture Design 2.6 Construction 2.6	Architecture Design 2.7 Construction 2.7	
Studio	Ex.9 Living unit Rework plans and section 1:50 Ex.10 Living unit Rework elevation 1:50	Ex.11 Living unit Isometric detials 1:20 Ex.12 The site Structrual plan of Living units and market hall situation 1:200	Ex.13 Living unit Perspective/ Isometric drawing/ final submission Ex.14 Living unit Perspective/ Isometric drawing/ final submission	Review of Phase II
Process	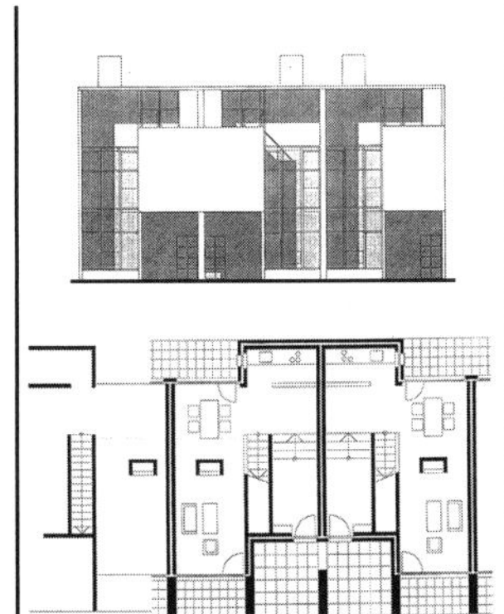	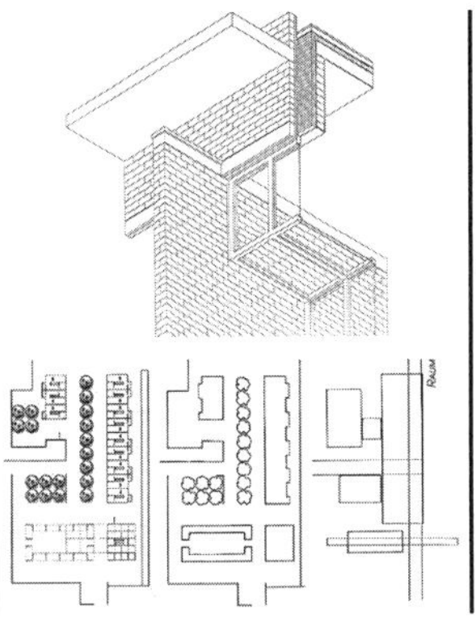	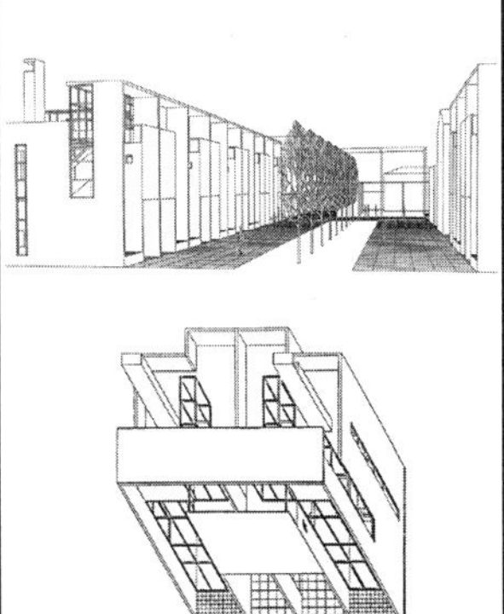	

第三阶

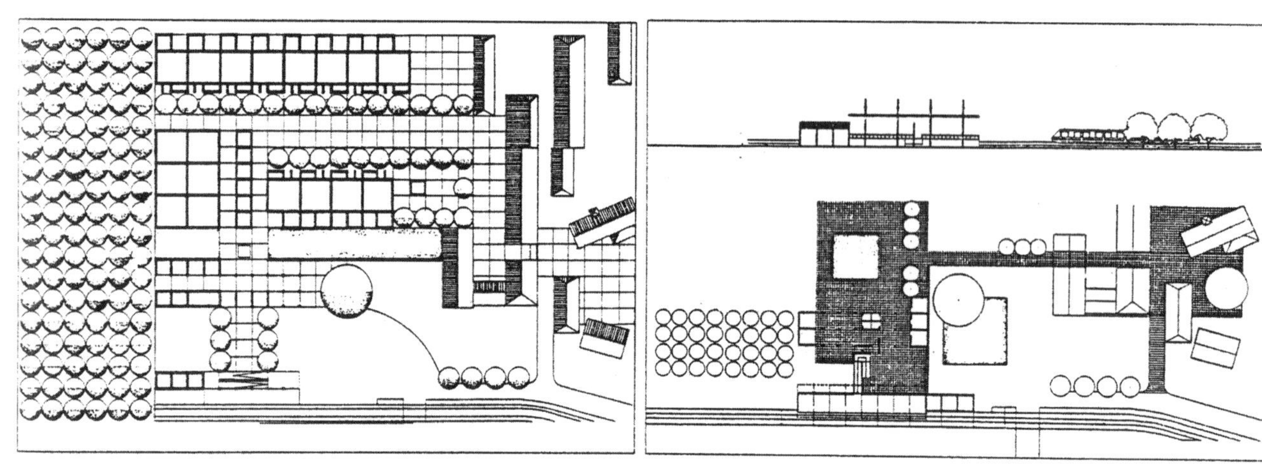

学生作业，第三阶，1986/87学年 | Students' work, Phase III, 1986/87

与前述第一阶一样，场地及前两阶的设计成果成为第三阶的出发点。第三阶是一个使用功能更复杂、空间尺度差异更大的设计。与前两阶比较可以发现它们形成了某种模式展露了教案中的变量与恒量。

难度的递增过程具有循环性，每次都包含分析问题、设计想法、设计深化和设计表达。循环模式巩固了学习的过程。在第一阶中勉强掌握的概念，在第二阶中就较能驾驭了，到了第三阶，学生作业就明显呈现出对建筑概念及设计惯例的理解。学生的理解程度可以通过更高要求的循环模式激发出来。

新任教师的学习曲线与此相似；在第一阶中保守的教学技巧、师生间小心翼翼的对话，在第二阶中为有计划有策略的指导所取代，到了第三阶，对教与学过程的愈加熟悉，以及对学生潜力的把握使教师可以在课程内拓展自己和学生对建筑形式的理解。教师的课堂经验越丰富，就越能够判断出构思、组织及呈现练习的最佳方式，并决定进行个体或小组辅导。教师通过课程设计中的经常性评估知道何时应该回应学生的问题，而何时应该让学生运用自己的能力去迎战问题。

Phase III

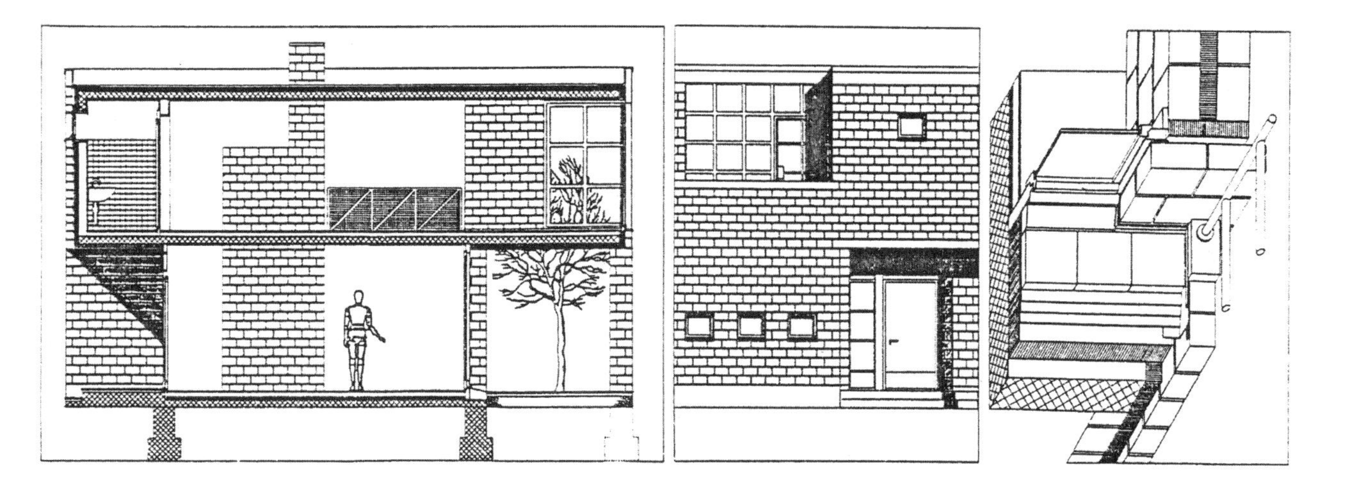

As in the previous phases, in Phase III, the site provided the common denominator. The first two phases served as existing conditions. In Phase III, a more diverse program requiring spaces of different heights and spans introduced a new set of questions and yet another level of complexity. If one compares the various phases, a certain pattern emerges, showing a specific relationship between the constant and the variables of the program.

The rising complexity is accompanied by the repetitive pattern of the phases, each involving problem analyses, design proposals, design development, and presentations. The repetitive mode reinforces the learning process. A notion barely grasped in Phase I can be used with some confidence in Phase II. And so by Phase III, reflected in the work of the students is a considerable level of understanding of architectural ideas and conventions. Students' understanding can be challenged and propelled by repetition as they progress through the levels with rising expectation.

A new teacher experiences a similar learning curve; conservative studio techniques and cautious interaction with students in Phase I become informed planning and strategic intervention geared to specific tendencies of the class and individual students in Phase II. In Phase III a growing familiarity with the program and the learning process, and an increasing perception of students' potential provide a teacher with the resources to challenge both him or herself and the students to expand their understanding of architectural form within the program. As a teacher increases his or her level of classroom experience, the ability to make judgments about events in the studio also increases—judgments about how best to conceive, structure and present exercises and work with the students as a group or individually. Constant reevaluation during the studio enables a teacher to know when to respond to students' questions and when to challenge students to utilize their own resources.

第三阶 1991/92学年

	1	2	3
目标	用体量模型建立空间和功能之间的关系；延续小组合作的方式以巩固高度的互动。	诠释及深化模型以使公共/私密、室内/室外、围合/开敞等关系清晰化；理解承重结构系统。	总平面组织关系对单元设计的影响。
周次	1	2	3
讲座	建筑设计 3.1 构造 3.1	建筑设计 3.2 构造 3.2	建筑设计 3.3 构造 3.3
课程设计	练习1 工作室 工作室单元纸圈模型 1:100	练习3 工作室 单元的结构模型 1:50	练习5 场地 总平面组织 1:200
	练习2 工作室 2～4个工作室单元群组的纸圈模型 1:100	练习4 工作室 4个单元（组）的结构平面 1:200	练习6 场地 总平面（含铺地及绿化）1:200
过程	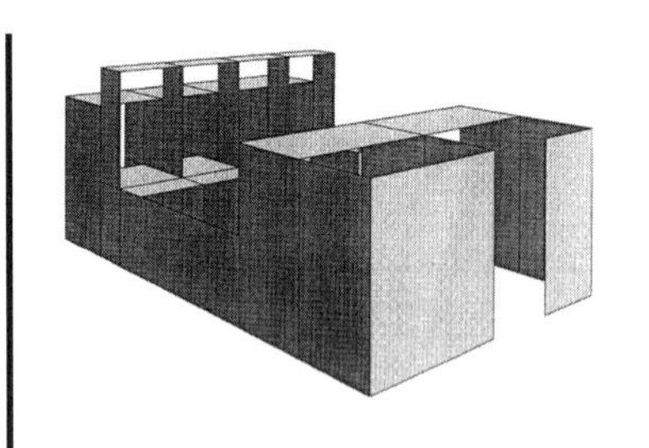	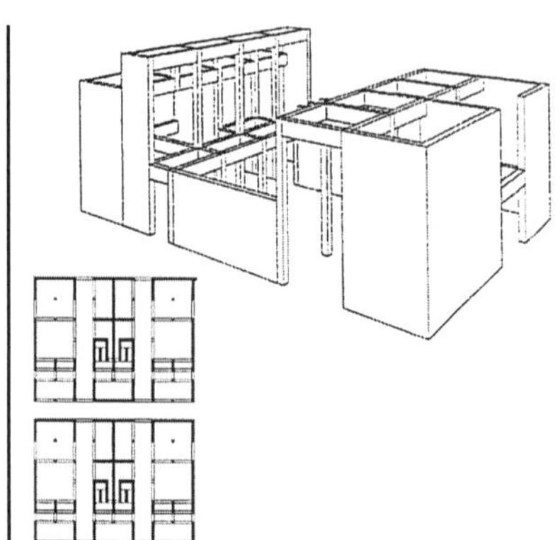	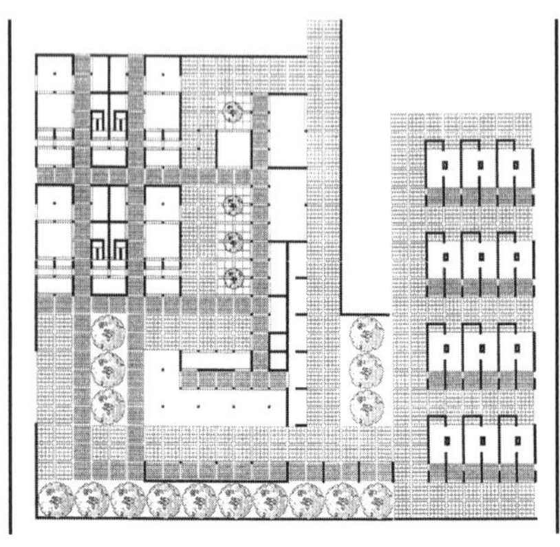

目标	通过立面及剖面图及大比例的节点研究建筑结构及材料。	用不同比例的图纸加深对细节程度与比例的选取之间关系的理解。	设计的整体表达：协调性，完整度及渲染效果。
周次	4	5	6
讲座	建筑设计 3.4 构造 3.4	建筑设计 3.5 构造 3.5	
课程设计	练习7 工作室 4～6个细部节点的轴测图 相邻单元立面 1:100 练习8 工作室 剖面 1:100 立面（含场地关系）1:100	练习9 工作室 剖面 1:50 立面 1:50 构造节点 1:20/1:10	练习10 工作室 透视与轴测图 第三阶评图
过程	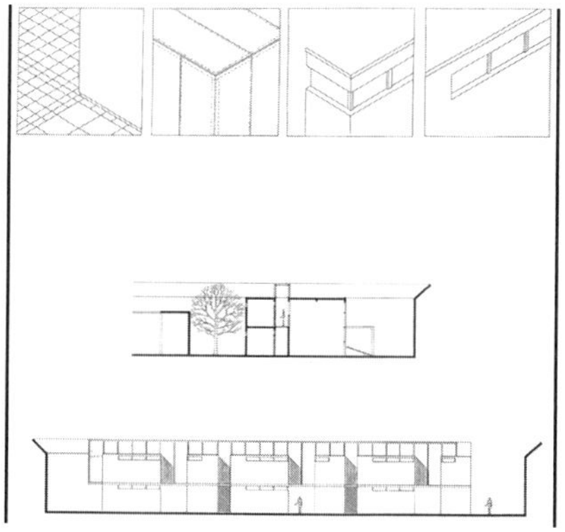	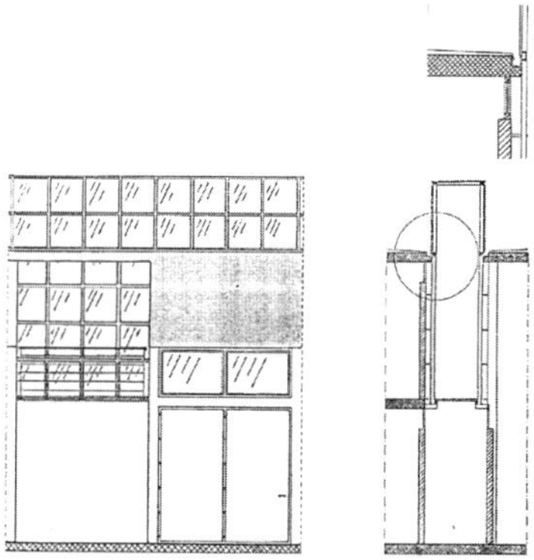	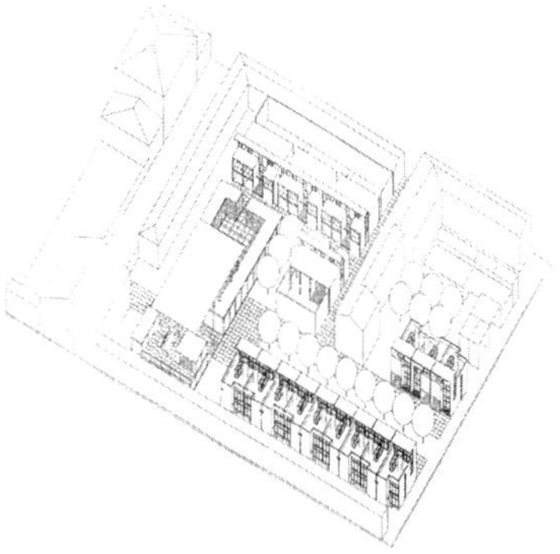

Phase III 1991/92

	1	2	3
Objectives	Use volumetric models to establish spatial and functional relationships; students continue to work in teams to maintain a high level of interaction.	Interpretation and development of models to clarify relationships: public vs. private, interior vs. exterior, closed vs. open; Understanding the structural system.	About the influence of site organization on the design of workshop unit.
Week	1	2	3
Lectures	Architecture Design 3.1 Architectural Technology 3.1	Architecture Design 3.2 Architectural Technology 3.2	Architecture Design 3.3 Architectural Technoogy 3.3
Studio	Ex.1 Workshop Workshop, one unit: loop model 1:100	Ex.3 Workshop Structural model of one unit 1:50	Ex.5 Site Site plan 1:200
	Ex.2 Workshop Workshop, 2 to 4 units: loop model 1:100	Ex.4 Workshop Structural plan of four units 1:200	Ex.6 Site Site plan with flooring and vegetation 1:200

Process

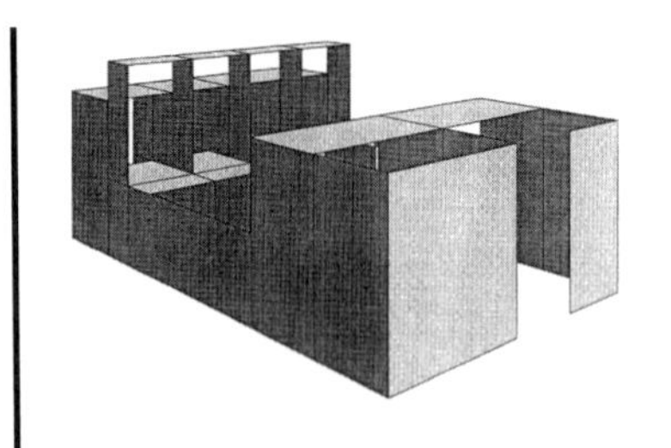

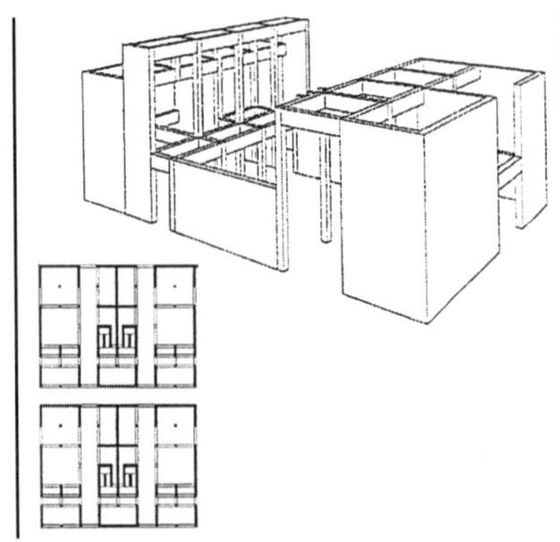

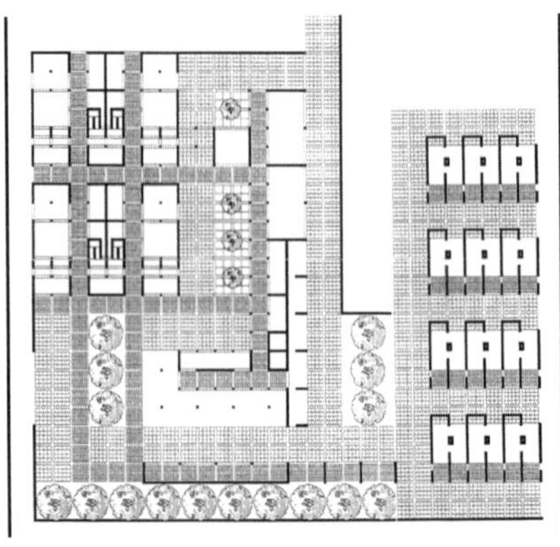

	4	5	6
Objectives	Use drawings and large-scale details of elevations and section to investigate building structure and material.	Use drawings in various scales to develop an understanding of how levels of detail and levels of scale work together.	Focus on the entire presentation: coordination, completion and rendering.
Week	4	5	6
Lectures	Architecture Design 3.4 Architectural Technology 3.4	Architecture Design 3.5 Architectural Technology 3.5	
Studio	Ex.7 Workshop 4 to 6 isometric details Elevation of 2 units 1:100 Ex.8 Workshop Section 1:100 Elevation with context 1:100	Ex.9 Workshop Section 1:50 Elevation section 1:50 Detail 1:20/1:10	Ex.10 Workshop Perspective, isometric drawings Review of Phase III
Process	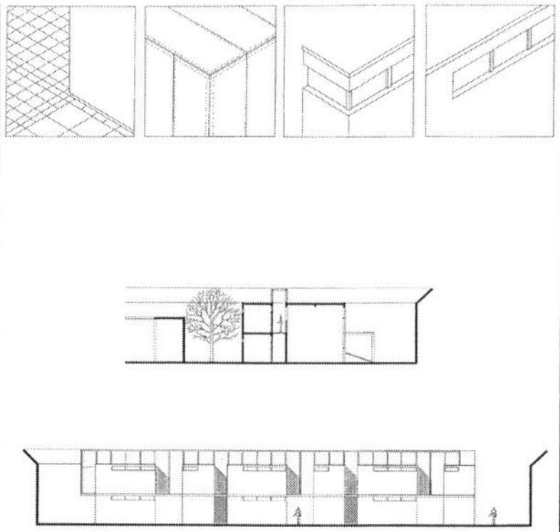	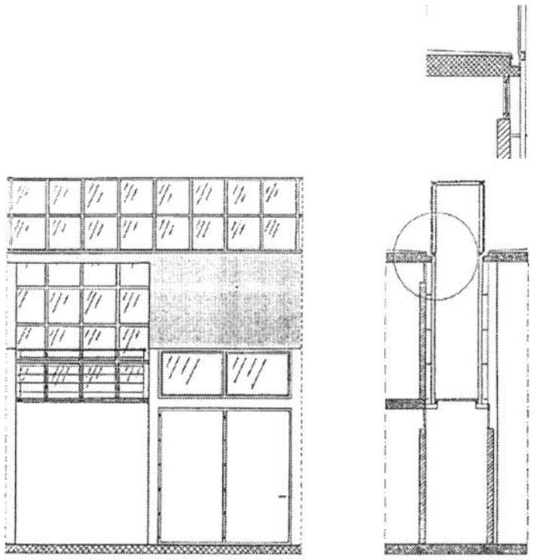	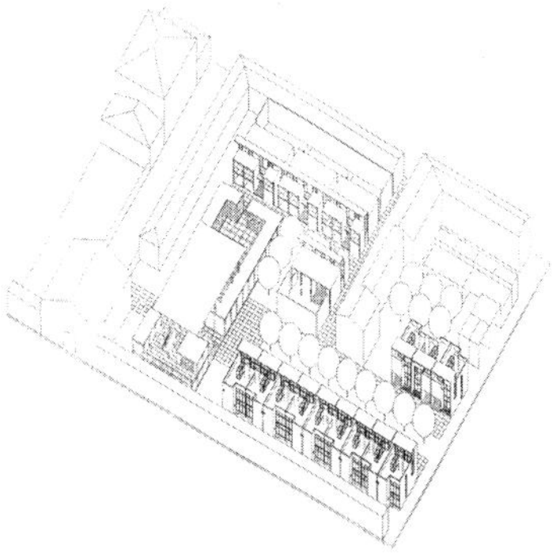

第四阶

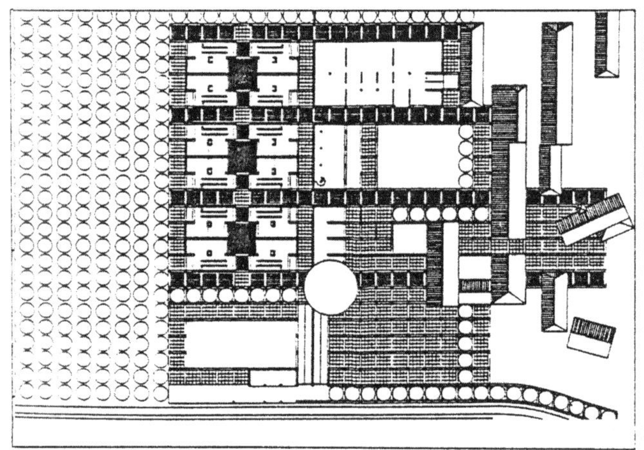

学生作业，第四阶，1986/87学年 | Students' work, Phase IV, 1986/87

通过前三个阶段的学习，学生熟悉了建筑形式的相关概念，以及设计过程的步骤和要点——他们将终身受用的专业方法。他们在设计过程中，通过循环的训练模式将设计过程消化理解，随着不断挑战更高复杂程度的设计问题而逐步提高。

在最后一阶，即第四阶，每周的练习模式变得自由，从而让学生证明自己的成就。布置第四阶的总题目，提供5块与前面三阶地块相邻的场地及若干组不同的功能需求。学生选择其中一块场地与功能，运用所学，加上自己的设计意图，证明自己的能力与理解。

第四阶的开放模式为教师带来新的契机，并非所有学生都选择在教室进行设计，因而教师有更多时间和留在课堂的学生进行讨论。焦点集中于文脉、建筑形式、构造及设计过程。主动性及设计能力较强的学生可以尽力发挥。一般的学生呈现了对建筑形式的理解，并尝试更进一步。能力较弱的学生则抓住最后的机会理解建筑和设计，尝试驾驭专业技能。最后一阶的关键是让每个学生尽力展示自己的所学。所有学生设计会在期末进行个别评图。

Phase IV

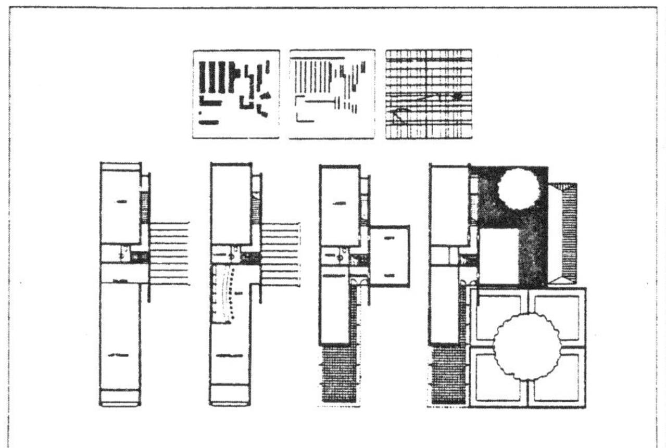

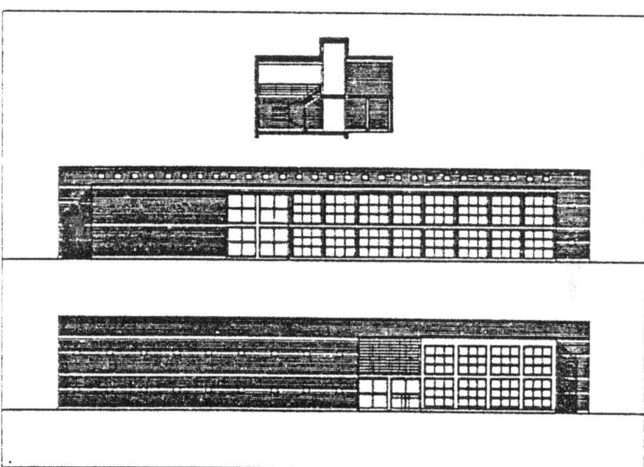

Through the learning process of Phases I-III, students become familiar with notions of architectural form and the steps and considerations of the design process—a process they will use throughout their careers. They have been immersed in the design process, and through structured repetition, they have assimilated the design process into their own understanding. They have challenged and furthered their understanding by working at increasingly higher levels of complexity.

In the fourth and final phase of the Basic Design Program, the structure of weekly exercises allows students greater freedom to demonstrate their individual achievements. Students are presented with five different sites adjacent to Phases I-III. Several different functions are proposed. Students select a site and a function. Using skills and abilities previously acquired, and their own initiative, students demonstrate their capabilities and levels of understanding.

The open structure of Phase IV brings new opportunities for a teacher. Not all students choose to work at the studio, allowing more time for discussion with students who do. The focus is on context, architectural form, construction and the design process itself. Motivated and advanced students have the opportunity to experiment and push their abilities as far as possible. Average students can demonstrate their basic understanding of architectural form and the design process, and endeavor to further their understanding. Lower-performing students have a final opportunity to grasp the notions of architecture and design, and to attempt to master skills and abilities required to pursue a career in the profession. It is important at the end of this phase that each student demonstrates—to the best of their abilities—a level of achievement in the Basic Design Program. All student work is reviewed on an individual basis at the end of the Phase.

第四阶 1995/96学年

目标	研究更大尺度的城市文脉，找到选择场地及功能的依据。	选取场地及功能；运用敏感性及判断来分辨关键设计问题及其对策。	运用问题及对策为一个特定的场地寻求解决办法。
周次	1	2	3
讲座	建筑设计 4.1 构造 4.1	建筑设计 4.2 构造 4.2	建筑设计 4.3 构造 4.3
课程设计	收集信息 教师：向学生阐明第四阶的性质，即展现在“基础设计”中习得的技能、能力及知识。	研讨会 学生倾向于用完整的体块结构作为城市策略，而非将建筑视为城市文脉中的物体；容易止步于小比例的城市研究，无视更为场所化的问题——街道正面、小型开放空间。 教师：重申在不同比例间切换工作从而解决不同层面问题的重要性；用图解说明设计问题。	研讨会 许多学生难以集中在城市图解的一个具体方面（一个场地），而倾向于解决更大范围内的笼统问题。 教师：研讨会将重点放在撒比奥内塔城的具体问题及一般城市策略上。
学生作业			

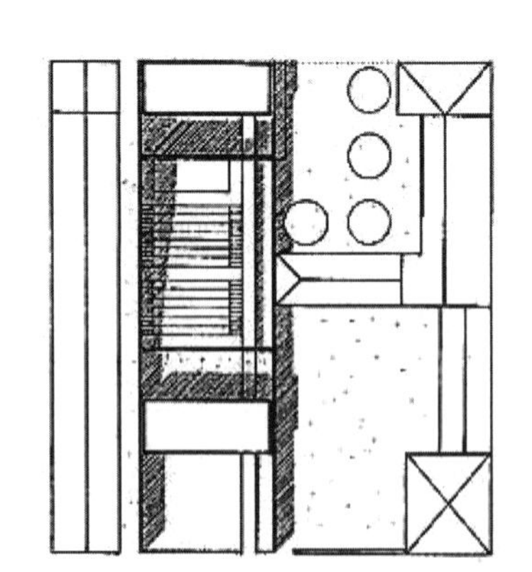

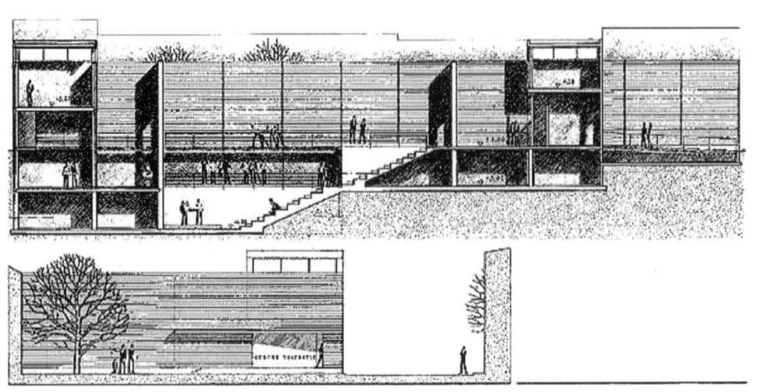

目标		
着眼于重要概念及品质的策略发展。	运用二维及三维手段表达第四阶的主要元素；在第一阶至第三阶中选出最能代表本年进展的一个。	从第一阶到第四阶的个人评图，发现进步与收获。

周次 讲座

4

建筑设计 4.4
构造 4.4

5

建筑设计 4.5
构造 4.5

6

课程设计

研讨会

大多数学生能够对场地及功能作出判断、表达观察所得、为场地寻求答案；少数学生能够保持建筑在更大尺度上与周边城市结构的关联。
教师：研讨会将重点放在具体的城市项目及其多个尺度与细节层面的发展。

深化设计/图纸表达

学生大都能专注于设计的具体问题，但其中一些不确定该如何更好地表达设计。
教师：在学期最后阶段激励学生并提供设计表达的建议。

期末评图

思路清晰的学生一般都能够在期末评图中做清晰的介绍。
通病在于无法选取合适的比例及绘图类型。以期末评图为契机总结学生的表现并为将来的学习提出建议。

学生作业

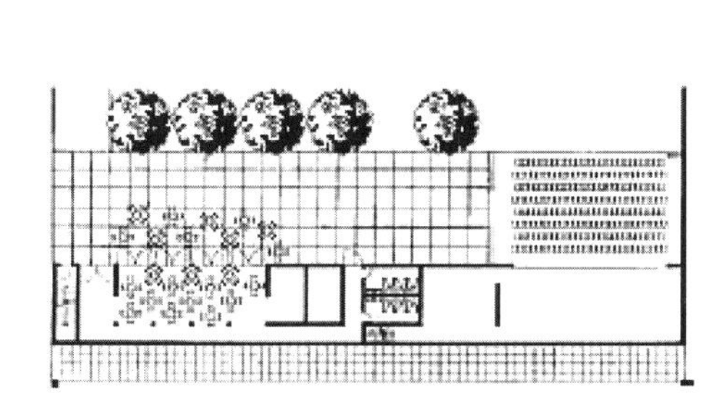

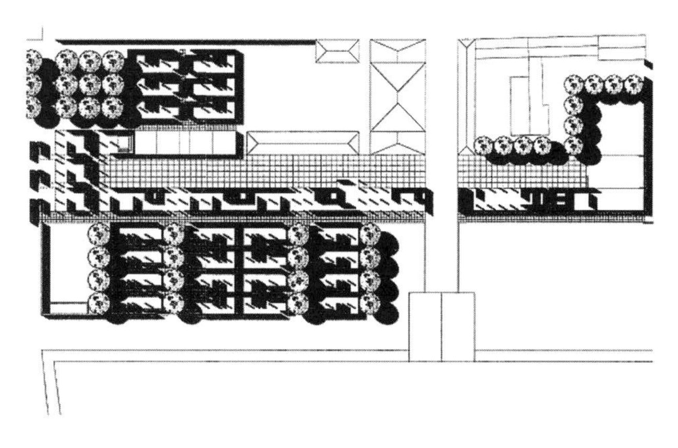

Phase IV 1995/96

Objectives	Analysis of larger urban context to provide basis for selection of site and function	Selection of site and function; Use of sensibilities and judgement to clarify important design issues and appropriate strategies	Applying issues and strategies to find a solution for a specific site.
Week	**1**	**2**	**3**
Lectures	Architecture Design 4.1 Construction 4.1	Architecture Design 4.2 Construction 4.2	Architecture Design 4.3 Construction 4.3
Studio	Survey/Collecting information Students are quick to define unbuilt space as "public space", regardless of existing or non-existing urban order. Clarify for students the nature of Phase IV as representative of skills, abilities and knowledge gained in the Basic Design Program.	Seminar Students tend to use complete block structures as the urban strategy rather than regard buildings as objects embedded in the urban context. Students tend to stop at small-scale urban analyses, ignoring more localized issues—street front, smaller open spaces. Reiteration of the importance of working at different scales to solve different problems; use of diagrams to clarify design issues.	Seminar Many students have difficulty focusing on a specific aspect (one site) of their urban diagrams; they tend to solve problems at a general level for a larger area. Seminars on Tuesdays and Wednesdays focused on issues specific to Sabbionetta and general urban strategies.

Students' work

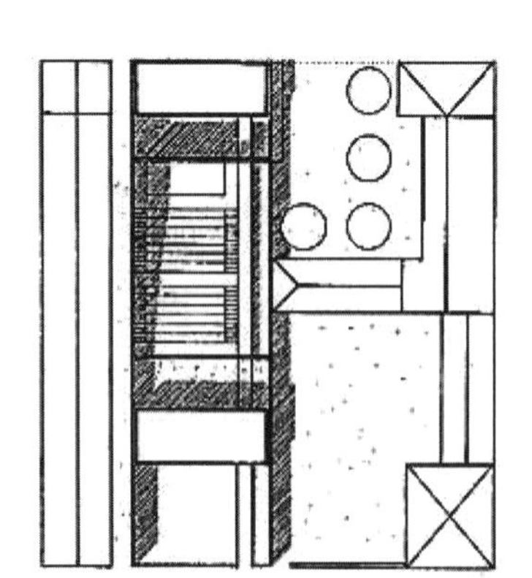

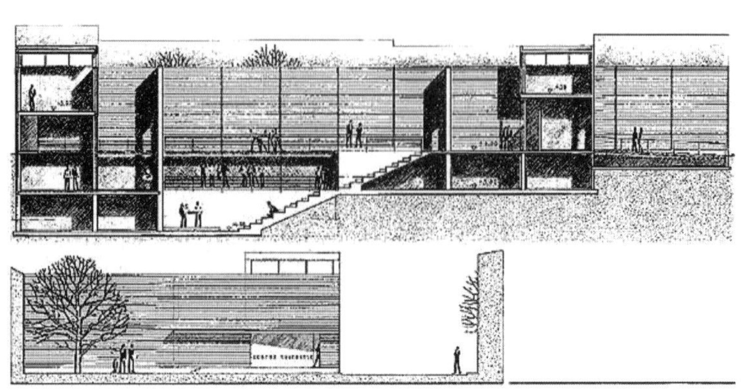

Objectives

Develop strategies concentrating on important ideas and qualities.

Use 2 and 3-D techniques to express major elements of work for Phase IV; select work for Phase I-III to best represent progress through the year.

Review students' work Individually for Phase I-IV to see progress and achievements.

Week

4

Lectures

Architecture Design 4.4
Construction 4.4

Studio

Seminar

Most students have been able to make judgements and express observations about site and function and are moving towards a solution for the site itself; fewer students are able to maintain a strong relationship between buildings and the surrounding urban structure at a larger scale. Seminars focused on specific urban projects and their development at various levels of scale and detail.

Week

5

Lectures

Architecture Design 4.5
Construction 4.5

Studio

Representation

Many students are well focused on specifics for their projects but some are unsure how best to represent their work.
Encouragements and presentation advice for students during final stages of semester work.

Week

6

Studio

Final Review

Students with a clear understanding of their work are, in general, able to portray this clarity in their final presentations.
Students are generally weak in the selection of an appropriate scale and drawing type.
Using final reviews as an opportunity to sum up students' performance and provide advice for further studies.

Students' work

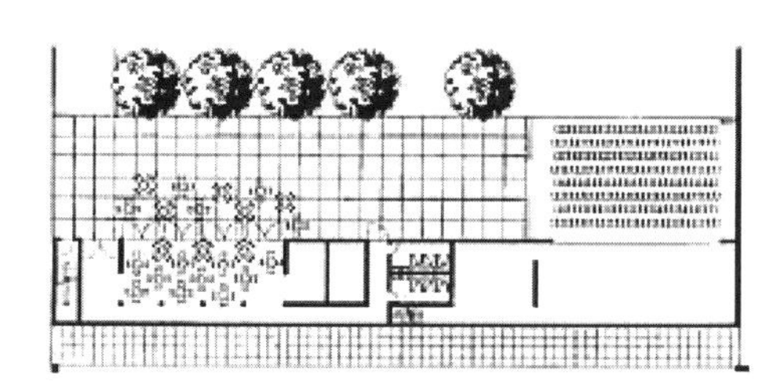

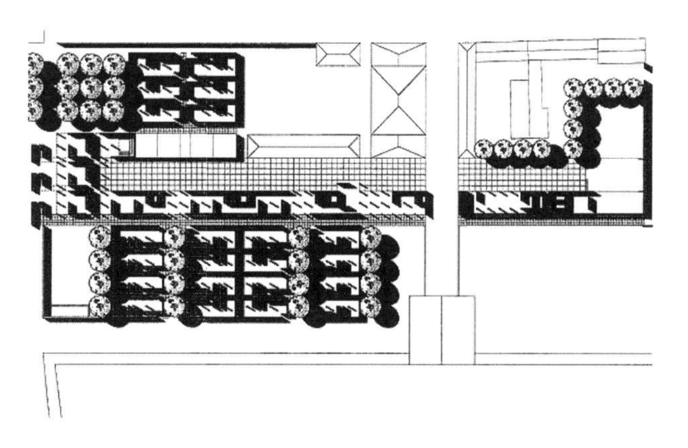

2 演化 | Evolution

为了保持教案的新鲜感，避免教学团队产生厌倦，每年的主题和教案编排会有调整和变化。课程设计通常包含若干建筑类型，但总会包含一个住宅群。题目设置基于两个考量：其一，建筑类型需具有空间形态（跨度、高度、氛围等）上的多样性，并且为学生所熟悉。其二，主题要与场地相关。如1987/88学年选地于苏黎世湖畔，便以划艇俱乐部为主题，而1990/91学年的题目则是古城撒比奥内塔的修复研究所。

20世纪90年代初，设计题目的难度出现了一次大的提升，原先第一阶中单层的入口设计改为两层的市场设计，并附加屋顶体量。同时加强了城市层面问题的训练，具有复合空间的公共建筑课题被提前到第三阶，以便在课程的最后阶段探讨城市肌理的修补，从更宏观的尺度考量建筑问题。在结构和构造方面，钢筋混凝土框架结构和钢结构逐步取代了木结构和砖混结构。教程发展后期，在前三个阶段中逐一强调三大设计要素的模式变得清晰。1994/95学年，一直设置在第二阶中的居住单元设计被提前到第一阶，用以讨论活动与空间功能分区的关系；市场设计放在第二阶，讨论不同结构跨度的处理问题；第三阶是工作坊设计，讨论建筑体量与周边环境的关系。由此，形成了一种新的逻辑关系，即从单一体量，到重复性空间组织，再到差异性空间组织。

建筑形式是教学过程所关注的。建筑形式的三大要素中，组织建筑形式时探讨及使用的不是“文脉主义”，而是作为边界条件的文脉（体量和空间）。环境参照的首要特质是其物理形态。每一个环境都具有二维或三维的形态特征。形态同时决定了比

Different themes and programs are developed each year in order to maintain the life and energy of the exercise. The theme for each year's Architecture Design is a compound that includes several types of buildings, but always involves a residential area. The setting of the theme satisfies two requirements: it involves a diversity of building types that differ in spatial morphology (span, height, atmosphere, etc.); they are more or less familiar to the students. Another concern is that the theme has to connect with the site, e.g. the brief in 1987/88 was a rowing club by the Zurich Lake which in 1990/91 was a restoration institute in the historical town of Sabbioneta.

At the beginning of the nineties, the level of complexity of the program was lifted. A two-story market hall with additional roof structure replaced the preliminary project of a single-floor entrance. At the same time, more emphasis was places on urban issues. The design project in the fourth phase was shifted to the third, whereas the last phase deals with the connection of urban fabrics to their context. On the construction level, steel as a material for construction became more common in addition to wood and brick. In the later years, the pattern in which each of the first three phases stresses one of the determinants of architectural form while relating to the other two became clearer. In 1994/95, the themes of Phase I and II were exchanged—the design of a living unit preceded that of the entrance. Thereby, the first phase related architectural space to the activities it accommodated; the second phase dealt with large and small spanned spaces; the third explored the transition space with the courtyard and the volumetric organization regarding the context. The variation brought in another logical sequence, departing from operation within a single volume, to that with a group of similar volumes, and finally, to the configuration of a variety of volumes and courtyards.

The concern of the actual teaching/learning program is architectural form. While the form is understood as the result of the interaction of function/space, construction/material, site/place, "contextualism" is not the key. Nevertheless, the context as a boundary condition is explored and used in the organization of architectural form (volumes and spaces). The first quality of a reference environment is its physical form. A form in 2 or 3 dimensions characterizes each

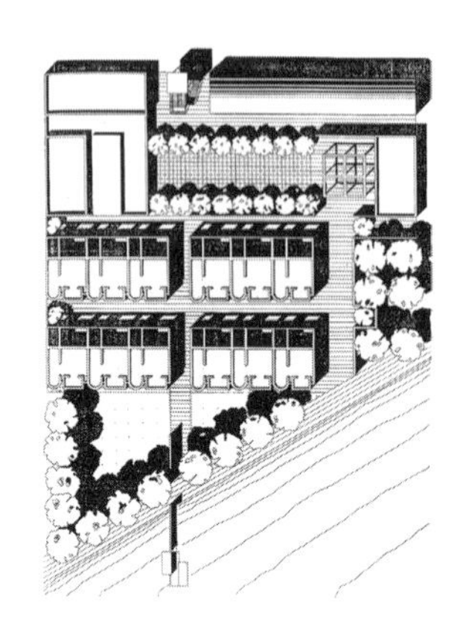

总平面（苏黎世湖畔）， 学生作业，1985/86学年
| Site plan (by Zurich lake), student's work, 1985/86

例、尺寸、尺度，构成关系，引出问题。因此文脉可以训练对实体与空间关系，以及整体、部分及构件关系的考量。总体而言，设计场地的选择以平地为主，偶尔具有一些微小的高差。为获得一定的建筑密度，以3m为模数，场地尺寸的通常设定为60m宽、84m长。

起先，选取的场地相对独立于其周边环境，如1985/86、1987/88学年择址于苏黎世湖畔。其后，随着对城市层面问题的关注，场地的选取与克莱默教席对无名建筑及村落形态的研究紧密联系起来。1986/87学年及1988/89学年的场地位于提契诺地区的村庄圣母皮亚诺。早年克莱默曾带领学生进行详尽的建筑测绘及历史演化调研。从1989/90学年起，意大利北部古城撒比奥内塔被选为场地，开启了文脉的新旧对话[1]。

撒比奥内塔之重要性源于它是意大利唯一的“理想城市”。它是由曼图亚城的贡扎加家族所建立，体现了文艺复兴时期意大利城市的所有特点，是一个完美的“学习”环境。

在课程之外，还以位于中国、坦桑尼亚及苏丹的场地为环境参照进行住宅集群设计研究。项目往往从当地的建筑类型、技术及影像出发，寻求“统一中的多样性”。

[1] 在此之前，克莱默指导的几位来自哈佛设计研究生院的交换生将其作为课程设计的场地，基于更早的调研资料构建了三维计算机模型。

[1] A few years earlier, field research had been conducted in this town. Later, five exchange students from Harvard University built a computer model of Sabbioneta when they chose the town for their design projects.

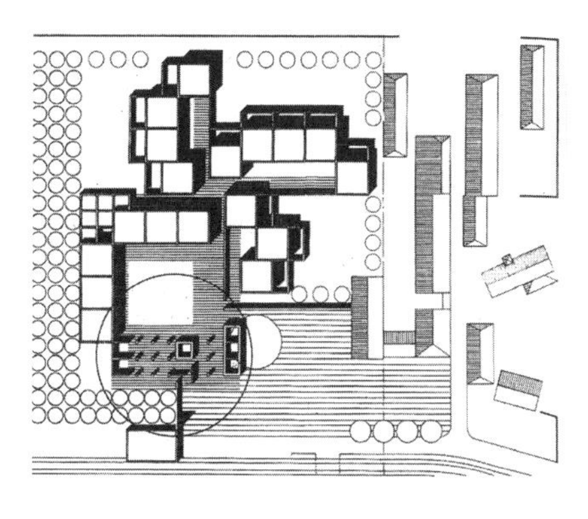

总平面（提契诺区）， 学生作业，1986/87学年 | Site plan (Ticino area), student's work, 1986/87

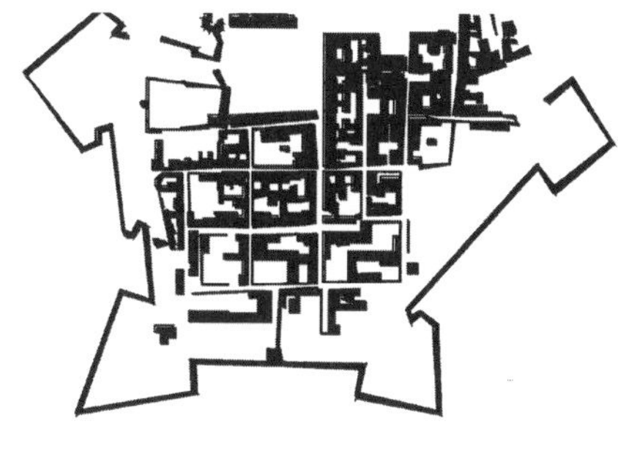

撒比奥内塔城，意大利 | Sabbioneta, Italy

同里，中国 | Tongli, China

environment. The form is also a vehicle to determine proportions, measures, dimensions, to establish relationships and to generate architectural problems. Thus, the context is an instrument to train students in the notions of volumetric and spatial relationships: the whole, the parts, and the elements. The site for the program is generally flat, occasionally with small variations in height. The size of the site is defined to accommodate the intended amount of buildings, normally measuring around 60mx84m.

In the earlier years, the site was more or less isolated from its larger context. As the issue of the urban began to gain more attention, in 1986/87 and 1988/89, the site was chosen in Madonna del Piano, one of the towns that had been studied by Kramel's Chair before. The historical town of Sabbioneta was introduced in 1989/90, initiating the discussion of contextual relevance when confronting old substances and the insertion of the new [1].

Sabbioneta has been important since it is the only "ideal" city in Italy. It was constructed by the Gonzaga family of Mantua and represents all the properties of an Italian city of the Renaissance. Sabbioneta is a perfect "learning" environment.

In addition to the curriculum, residential cluster design studies have been conducted on sites in China, Tanzania and Sudan. In all these projects, the points of departure have been existing typologies, technologies, images and the search for "diversity within unity".

历年教案汇总 | Summary of Teaching Plans

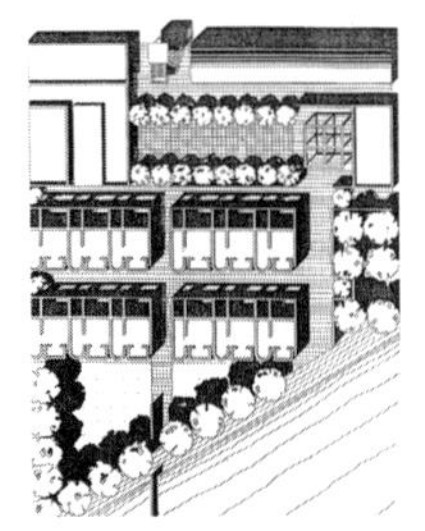
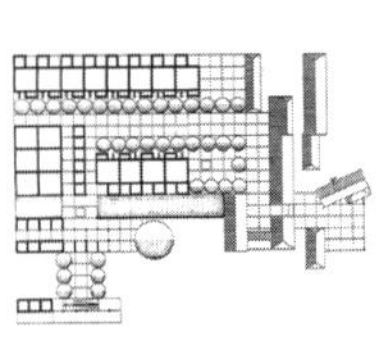
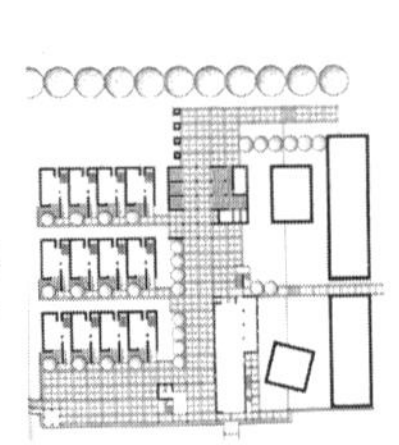
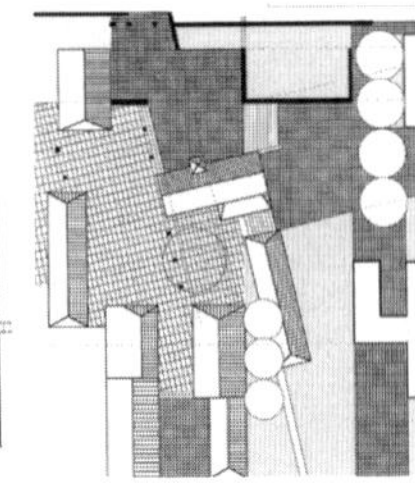
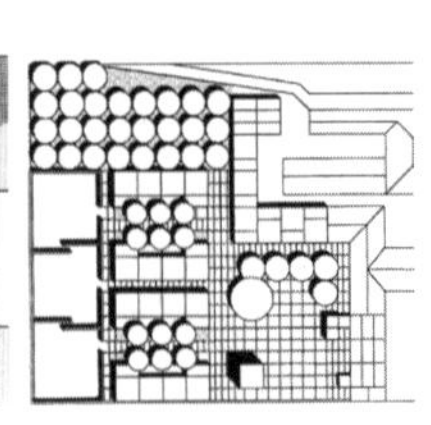

学年 | Year	1985/86	1986/87	1987/88	1988/89	1989/90
主题 | Theme	飞行俱乐部 Aviation Camp	设计研究所 Design Institute	划艇俱乐部 Rowing Club	村落更新 Village Renewal	音乐学校 Music School
场地 | Site	苏黎世湖畔 Zurich Lake	提契诺地区村庄 Ticino Village	苏黎世湖畔 Zurich Lake	圣母皮亚诺村 Madonna del Piano	撒比奥内塔地块1 Sabbioneta plot 1
第一阶 | Phase I	问询处 Information pavilion	入口 Entrance	入口 Entrance	入口 Entrance	入口 Entrance
第二阶 | Phase II	居住单元 Living units	居住单元 Living units	居住单元 Living units	居住单元 Living units	居住单元 Living units
第三阶 | Phase III	控制塔 Control tower	火车站 Railway station	瞭望塔 Watchtower	户外过渡空间 In-between space	居住单元的构造深化 Living units (construction details)
第四阶 | Phase IV	俱乐部 Club house	校舍 School building	俱乐部 Club	社区集会厅 Community hall	校舍 School building

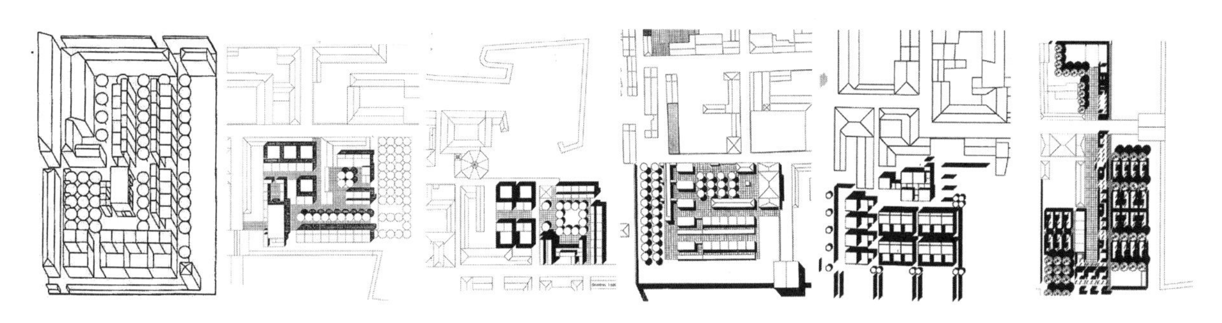

1990/91	**1991/92**	**1992/93**	**1993/94**	**1994/95**	**1995/96**
建筑修缮研究所 Restoration Institute	培训中心 Training Centre	培训中心 Training Centre	流离失所者营地 Displaced Persons Camp	--	--
撒比奥内塔地块 2 Sabbioneta plot 2	撒比奥内塔地块3 Sabbioneta plot 3	撒比奥内塔地块 4 Sabbioneta plot 4	撒比奥内塔地块 5 Sabbioneta plot 5	撒比奥内塔地块 1 Sabbioneta plot 1	撒比奥内塔地块 5 Sabbioneta plot 5
入口 Entrance	市场 Market hall	市场 Market hall	社区中心 Community centre	居住单元 Living units	居住单元 Living units
居住单元 Living units	居住单元 Living units	居住单元 Living units	居住单元 Living units	入口 Entrance	社区中心 Communal building
居住单元（构造） Living units (construction)	工作室 Workshops	工作室 Workshops	工作室 Workshops	工作室 Workshops	工作室 Workshop
工作坊 Atelier for technician	小型公共建筑 Tower/Gate/Theater		自定题目 Free theme	自定题目 Free theme	自定题目 Free theme

教案选例 | Examples

1985/86学年 I Academic Year 1985/86

飞行俱乐部 I Aviation Camp

苏黎世湖畔 I Zurich Lake

1 问询处 I Information pavilion

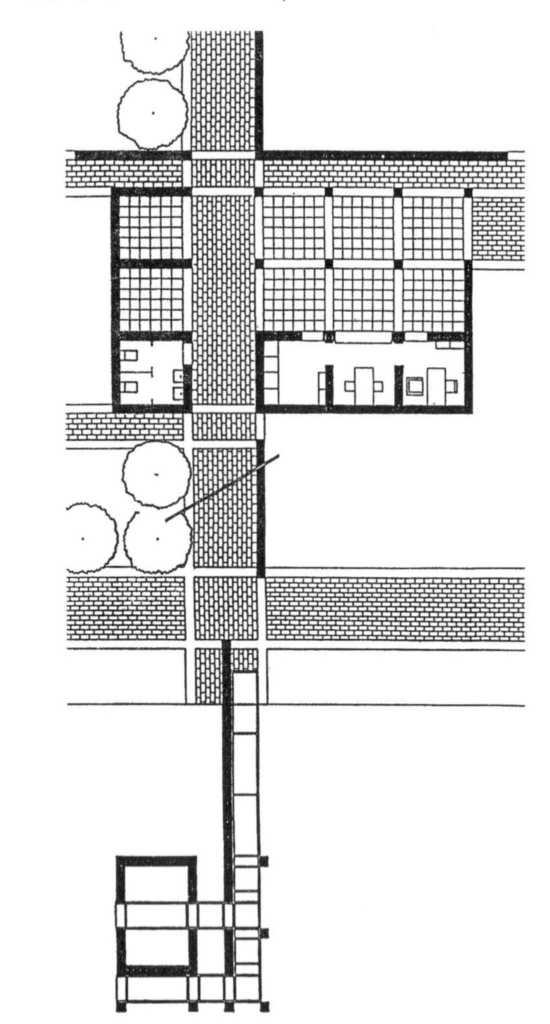

2 居住单元 I Living units

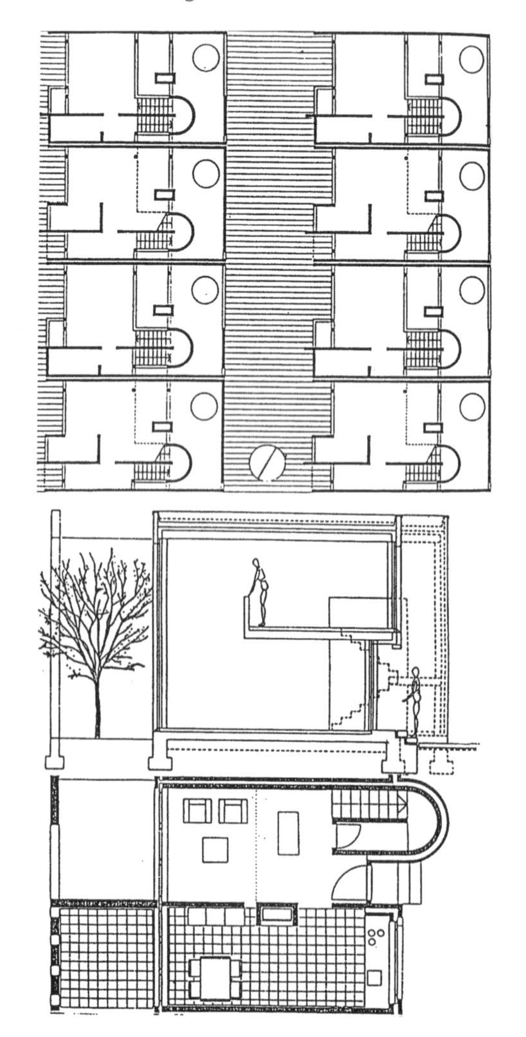

3 控制塔 | Control tower

4 俱乐部 | Club house

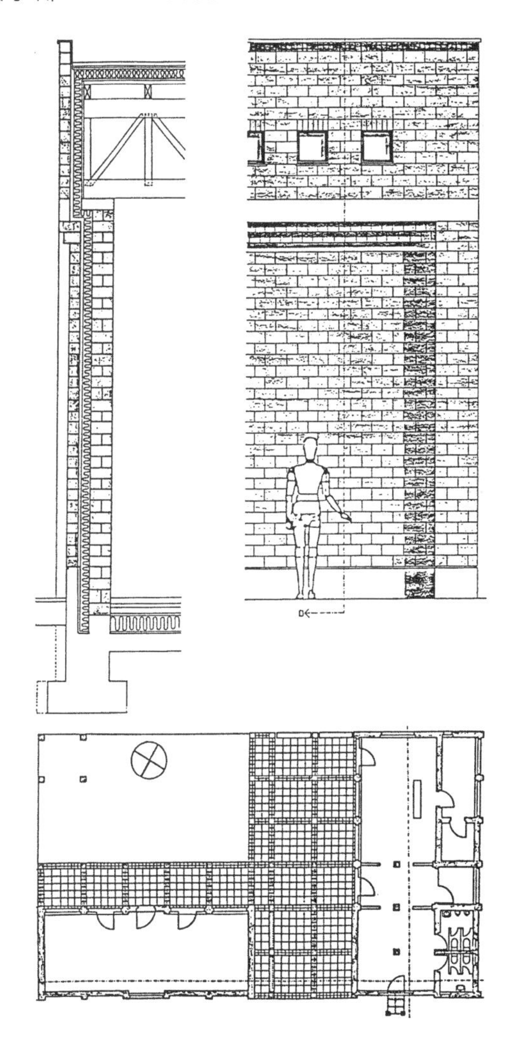

1988/89学年 I Academic Year 1988/89
村落更新 I Village Renewal
圣母皮亚诺村 I Madonna del Piano

1 入口 I Entrance

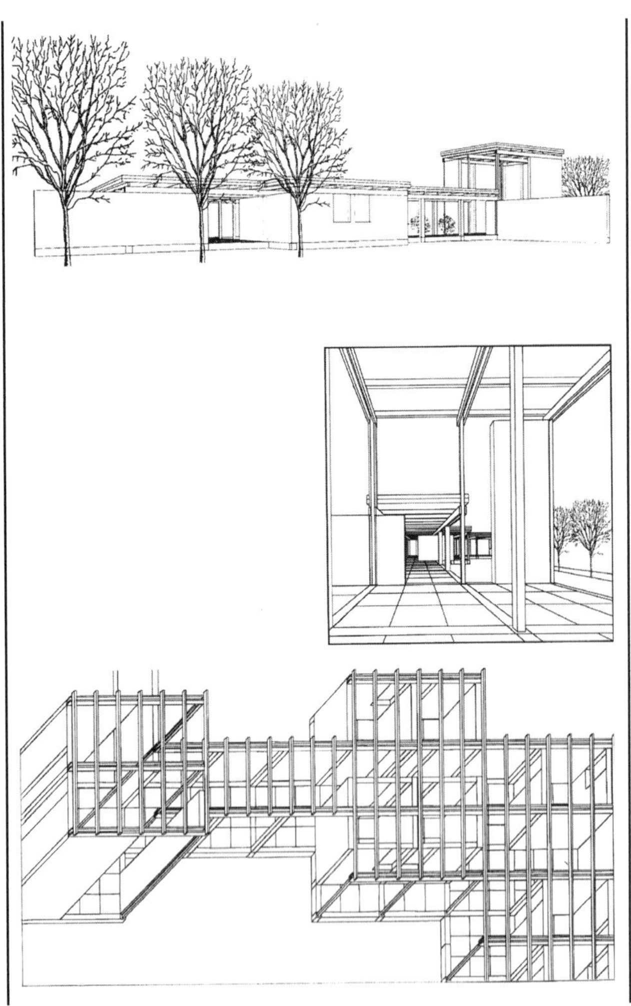

2 居住单元 I Living units

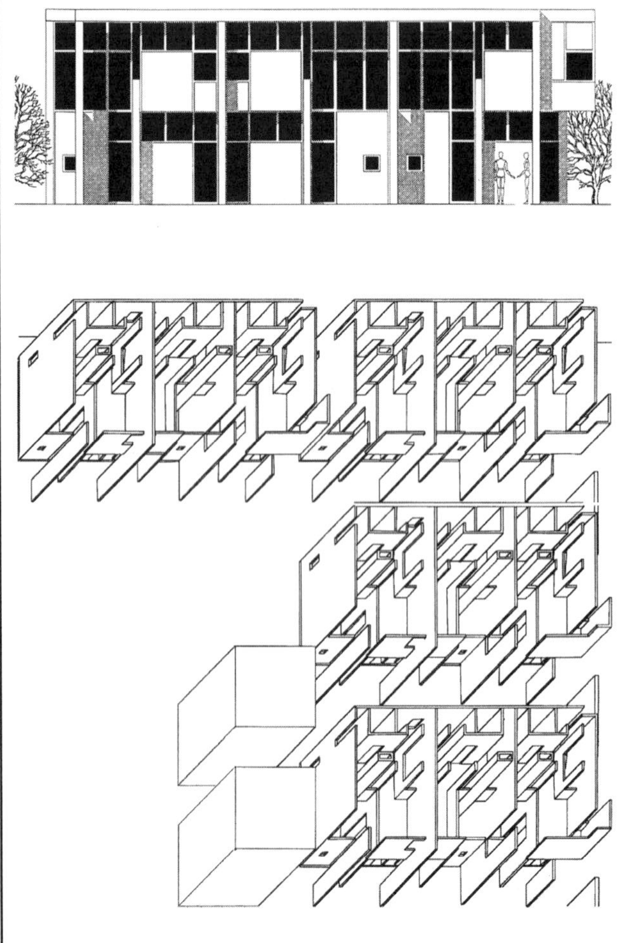

3 户外过渡空间 | In-between space

4 社区集会厅 | Community hall

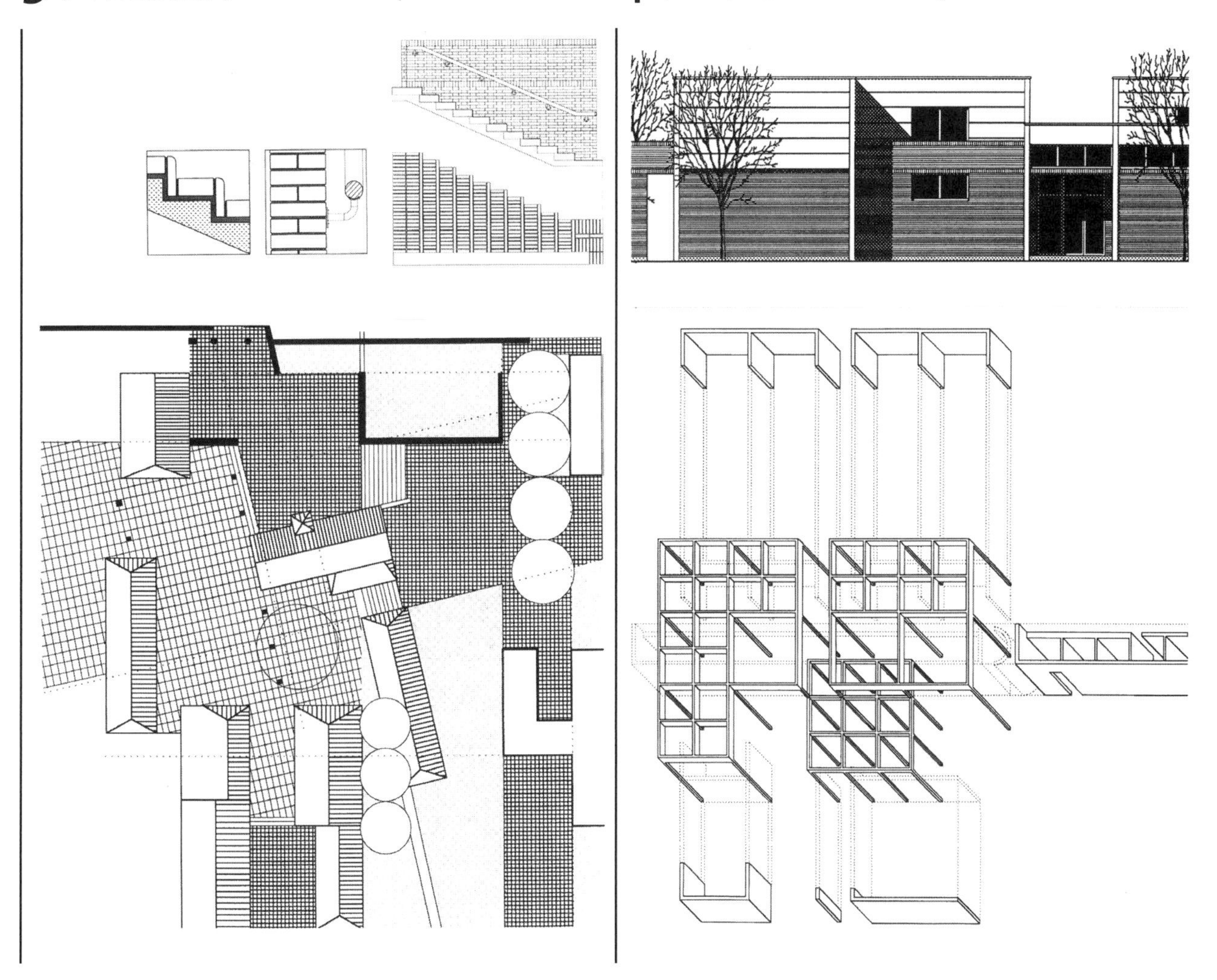

1990/91学年 I Academic Year 1990/91
建筑修缮研究所 I Restoration Institute
撒比奥内塔地块2 I Sabbioneta plot 2

1 入口 I Entrance

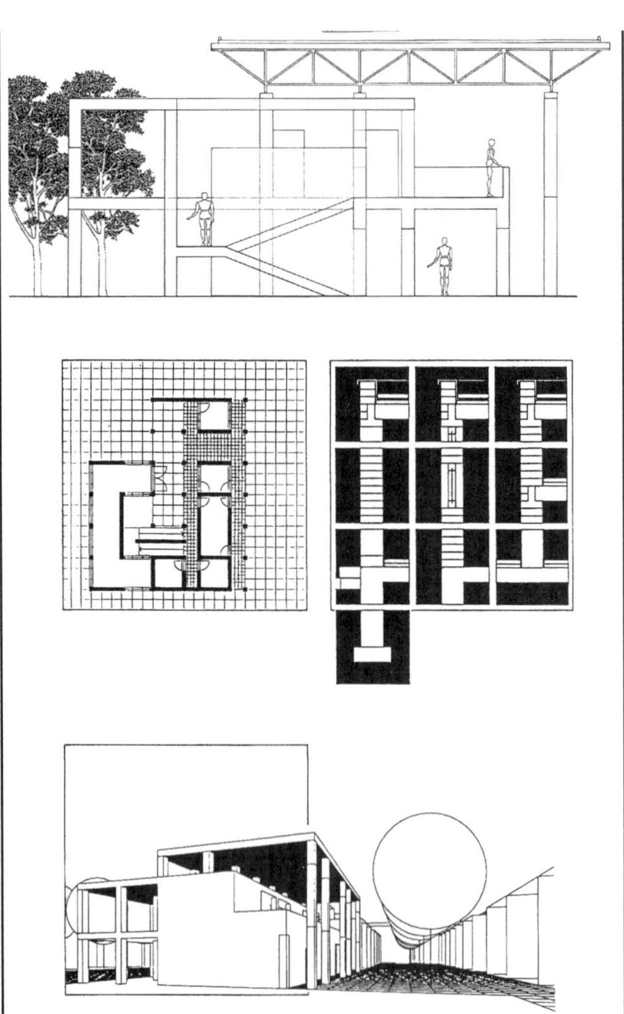

2 居住单元 I Living units

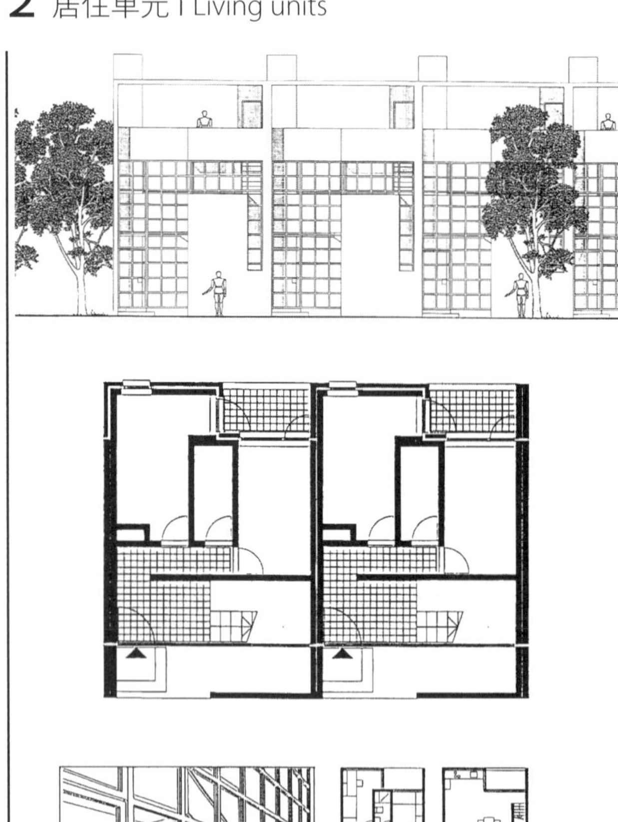

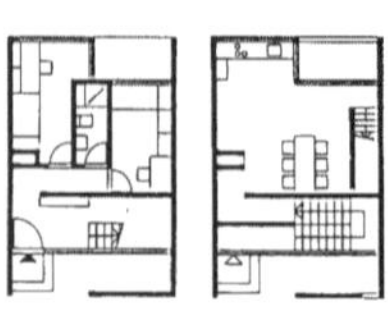

3 居住单元（构造）| Construction detail

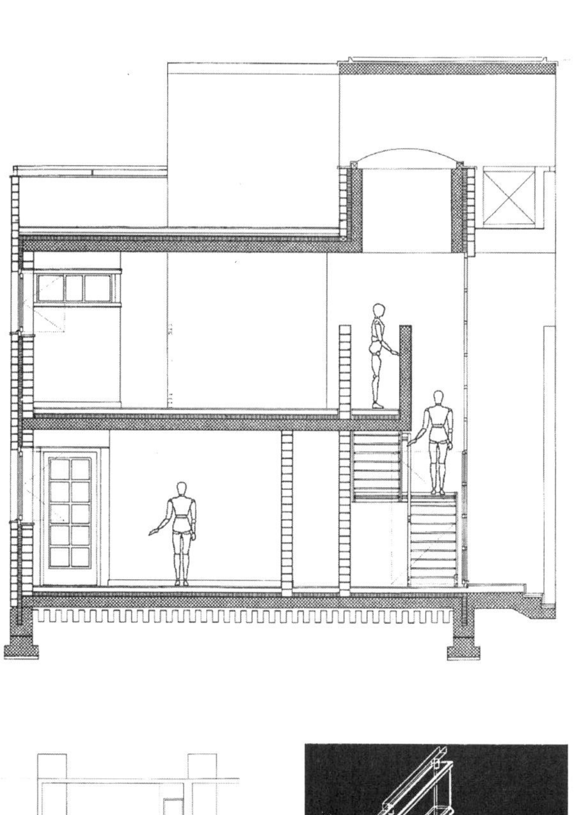

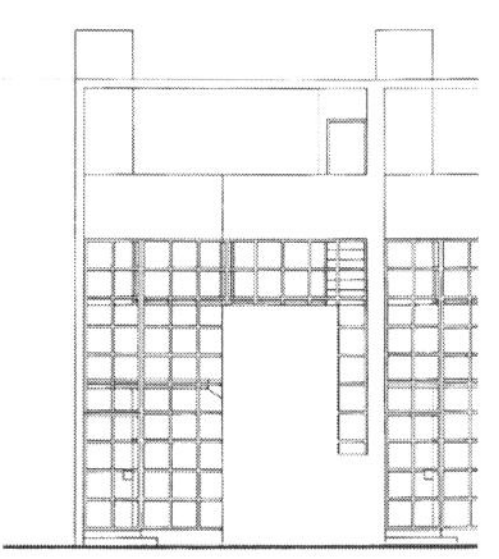

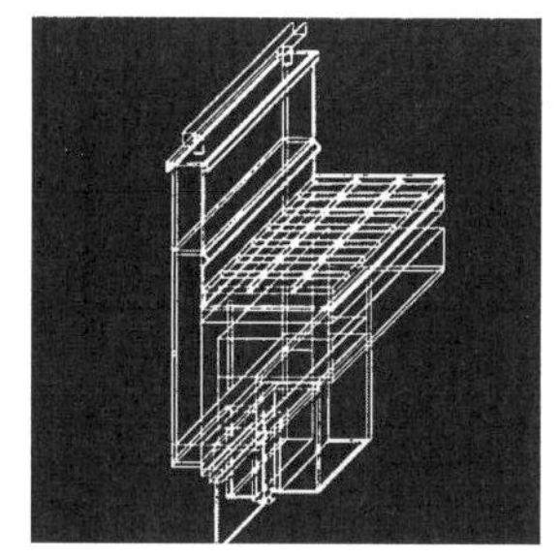

4 工作坊 | Atelier for technician

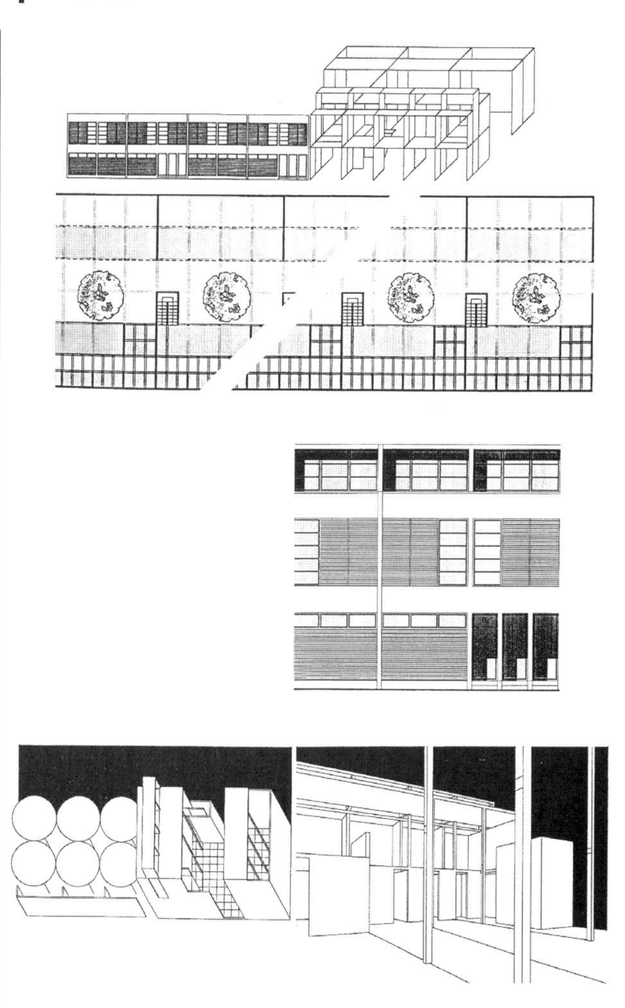

1993/94学年 I Academic Year 1993/94
流离失所者营地 I Displaced Persons' Camp
撒比奥内塔地块5 I Sabbioneta plot 5

1 社区中心 I Community centre

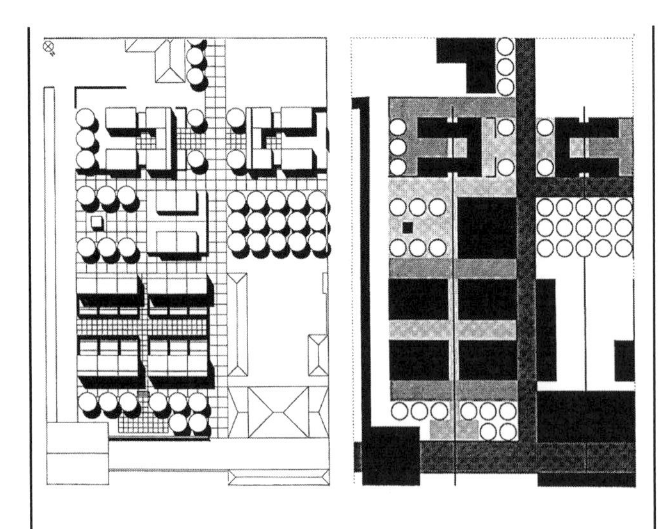

2 居住单元 I Living units

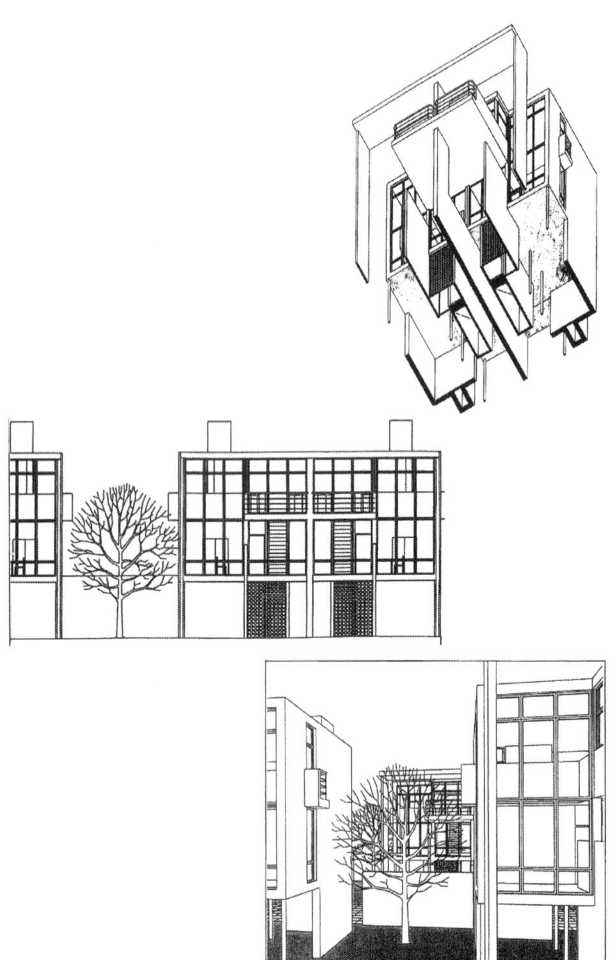

3 工作室 I Workshops

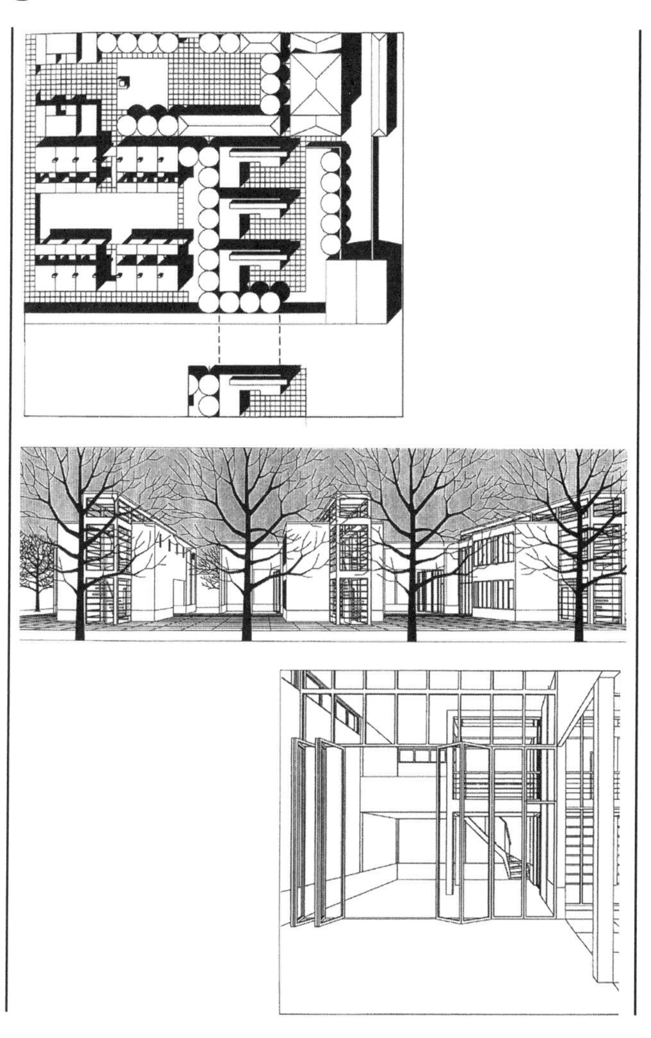

4 自定题目 I Free theme

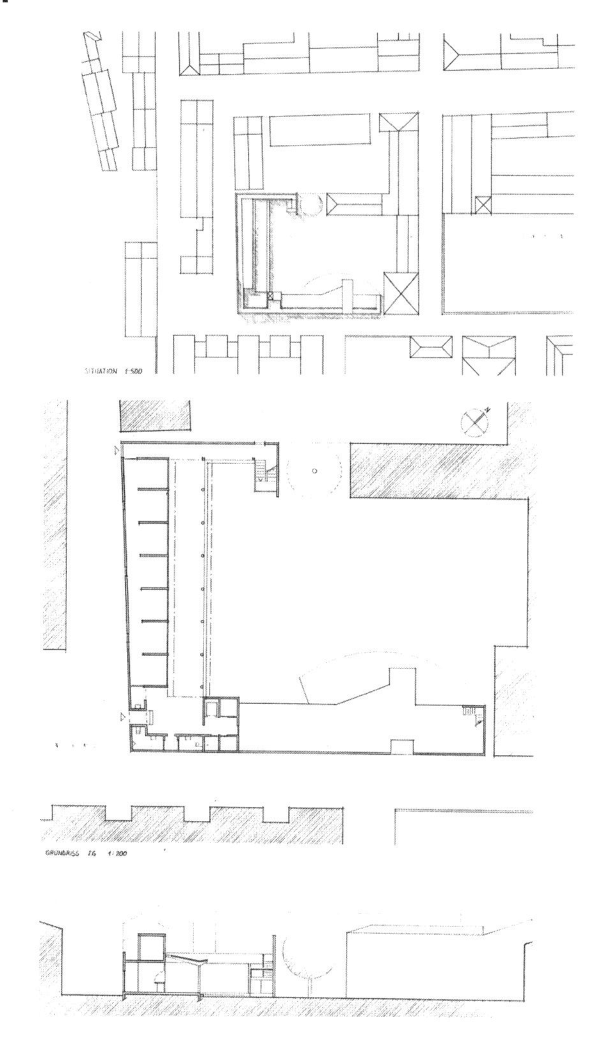

3 实例 | Cases

以下基于不同时期、主题和场地选取了3个例子，前2个例子呈现了相似教程结构所能获得的丰富性与多样性。为呈现更为完整的设计过程，我们选用了助教的课程试做文本。第3个例子并未出现在ETH的课程中，而是相同设计方法在中国文脉中的一次尝试。3个例子均由参加“南京交流”的东南大学青年教师完成。

Three examples are selected for the variation of time, theme and site. The first two show the richness and diversity derived from similar program structures. In order to present a more complete design process, we have selected the documentation of the test-runs done by the teaching assistants. The third example was an attempt to apply the same design approach in the Chinese context. All three examples were completed by the young teachers from Southeast University who had participated in the "Nanjing Exchange".

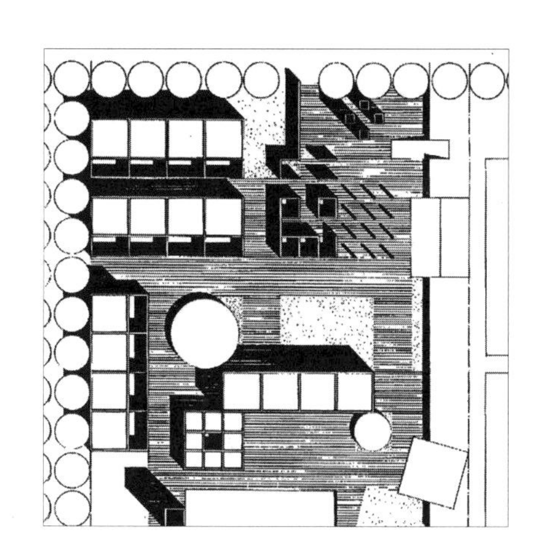

1987/88 划艇俱乐部，顾大庆 | 1987/88 Rowing Club, Gu Daqing

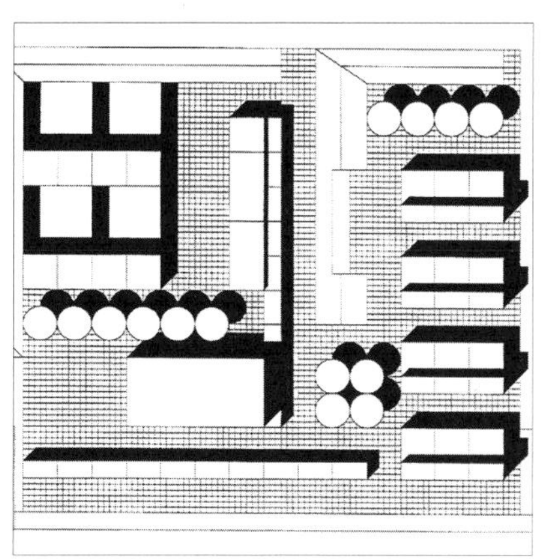

1991/92 游客中心，张雷 | 1991/92 Tourist Centre, Zhang Lei

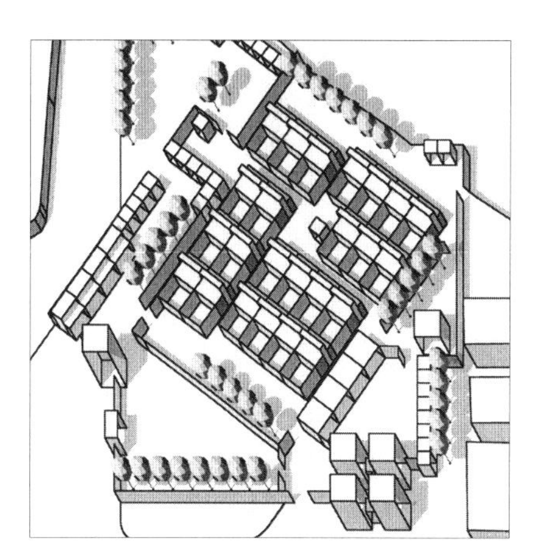

1998/99 旅馆，张彤，鲍莉 | 1998/99 Hotel Complex, Zhang Tong, Bao Li

1987/88

划艇俱乐部，苏黎世湖

Rowing Club, Zurich Lake

顾大庆
GU Daqing

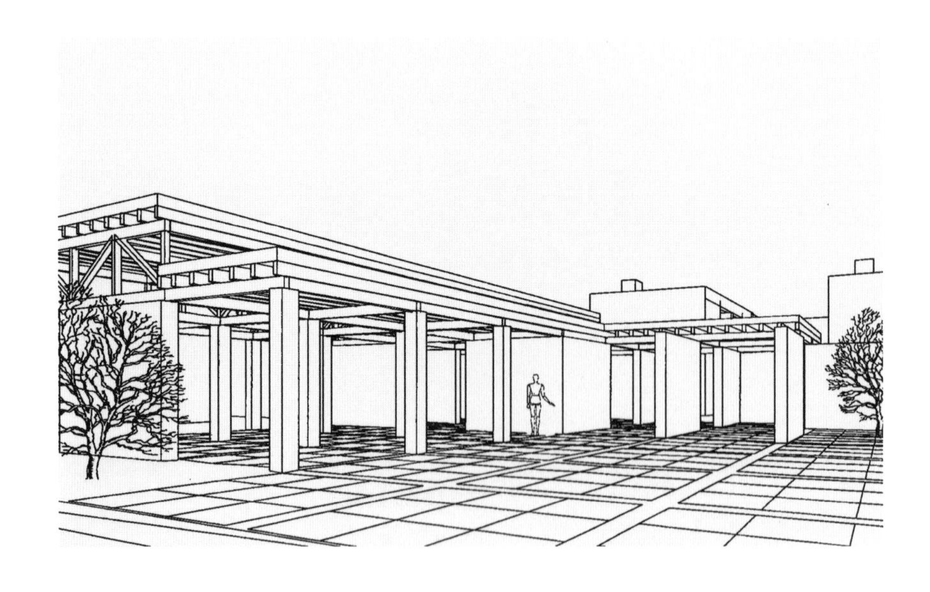

1987/88学年“基础设计”的主题是一个虚拟场地上的划艇俱乐部，包含了4个功能性部分，对应4个设计阶段。每一阶具有特定的设计问题与教学目标，但通过场地彼此联系，这种联系不仅是功能上的，而且使独立的设计练习上升到城市设计层面。例如，俱乐部的入口、度假屋、划艇比赛的瞭望台和瞭望塔，以及划艇存放仓库及工作间。

最初，练习完全是由手工模型和手绘图完成的。后来在学习电脑3D建模及绘图后补绘了部分轴测图、透视图等，并非学生作业的要求。

The program for the academic year 1987-88 was a canoeing club design that contains four different functional parts corresponding to four design phases based on a hypothetical situation. Each phase had its specific problems and purposes but were interrelated with others. This interrelation not only functioned as links between phases but also brought exercises to the urban design level. For instance, an entrance for the club, a settlement for the people on holiday there, a stand and an observing tower for this kind of sport, and a boathouse and workstation were needed. Initially, the exercises were completed entirely by hand-made models and hand drawings. Later, some axonometric drawings, perspective drawings, etc. were completed after learning computer 3D modelling and drawing, which was not required for the students' assignments.

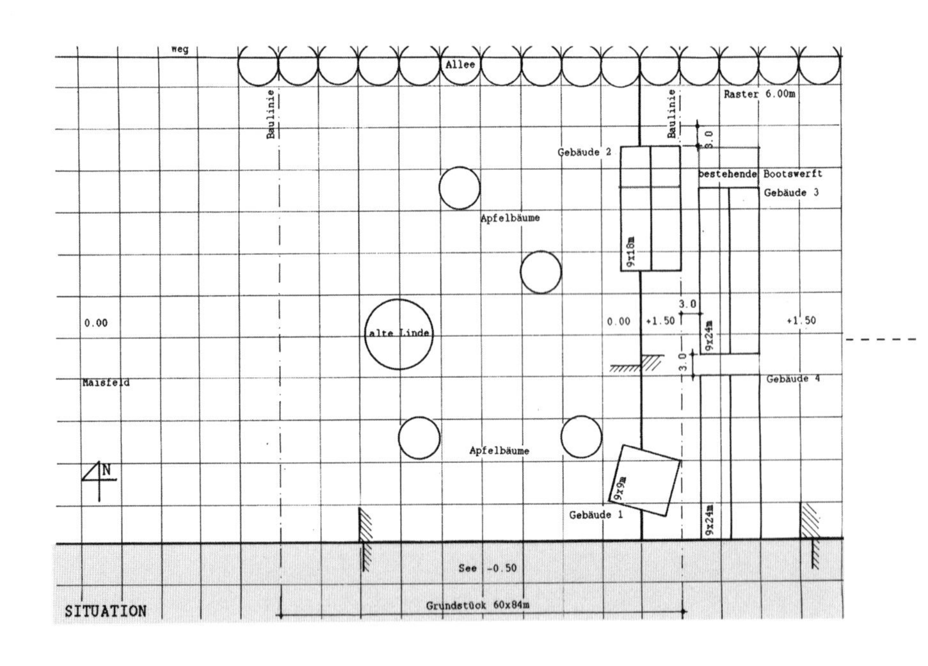
Weg
Allee
Baulinie
Baulinie
Raster 6.00m
3.0
Gebäude 2
bestehende Bootswerft
Gebäude 3
Apfelbäume
9x18m
3.0
0.00
+1.50
+1.50
0.00
alte Linde
9x24m
3.0
Maisfeld
Gebäude 4
Apfelbäume
N
9x9m
Gebäude 1
9x24m
See -0.50
SITUATION
Grundstück 60x84m

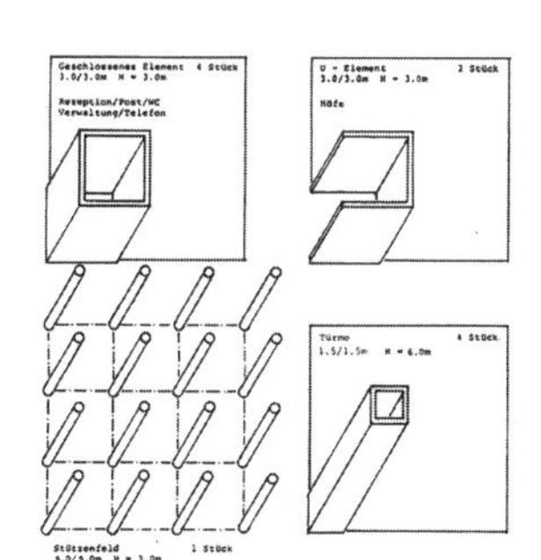

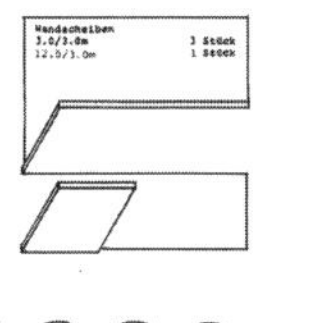

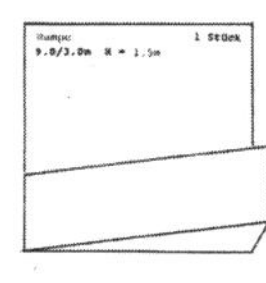

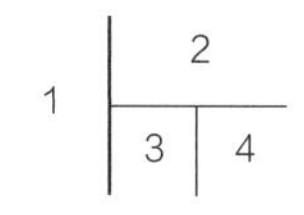

1 经网格模数处理的场地 2 空间定义模块 3 用“纸圈”模型组织场地 4 初步的总平面

1. Modularised site with grids 2. Space-defining modules 3. Site organization with the "paper tube" models 4. Preliminary site plan

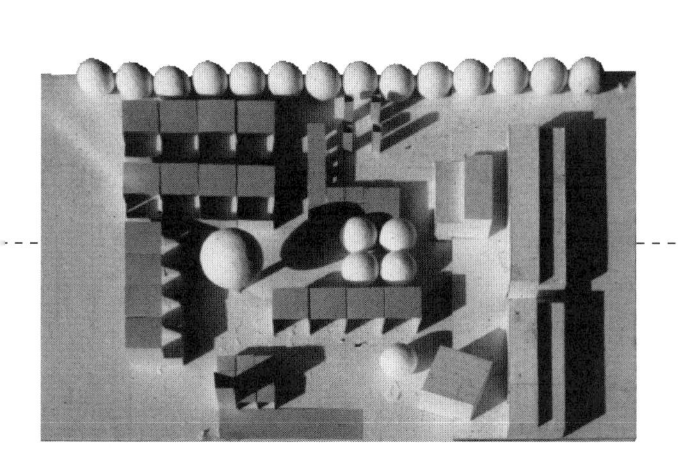

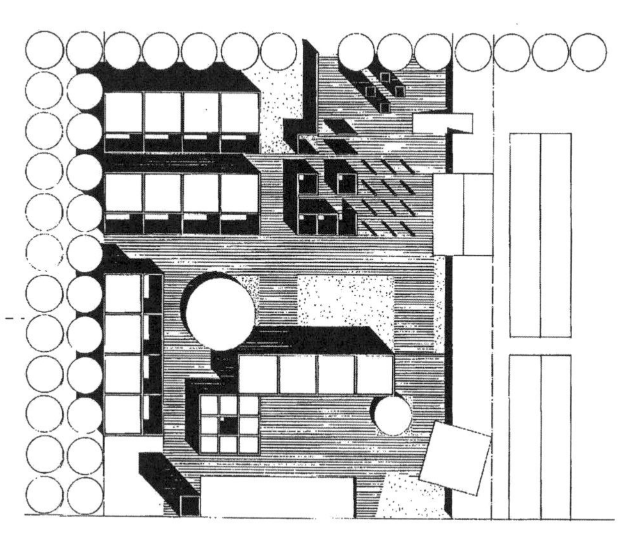

第一阶包含两个部分，首先是整个场地的总平面组织。

从制作1:200的模型开始，场地包含4组不同功能的建筑，因此需要设定功能分区、与原有建筑之间的关系并组织建筑单体。1:200的总平面引入了若干概念：硬质、软质地面；空间与体量；图一底关系；作为空间定义元素的栽植。

Two tasks were involved in phase one. The first was to build up a site plan for the whole project.

Students started by making a 1:200 model. The site plan contained four groups of buildings corresponding to the four phases such that it was necessary to differentiate functional zones, to recognize the relation between old and new, and to arrange elements in a proper way. By drawing the 1:200 site plan, a group of concepts were introduced: hard and soft surfaces, space and volume, figure-ground relationship, and planting to create space-defining elements.

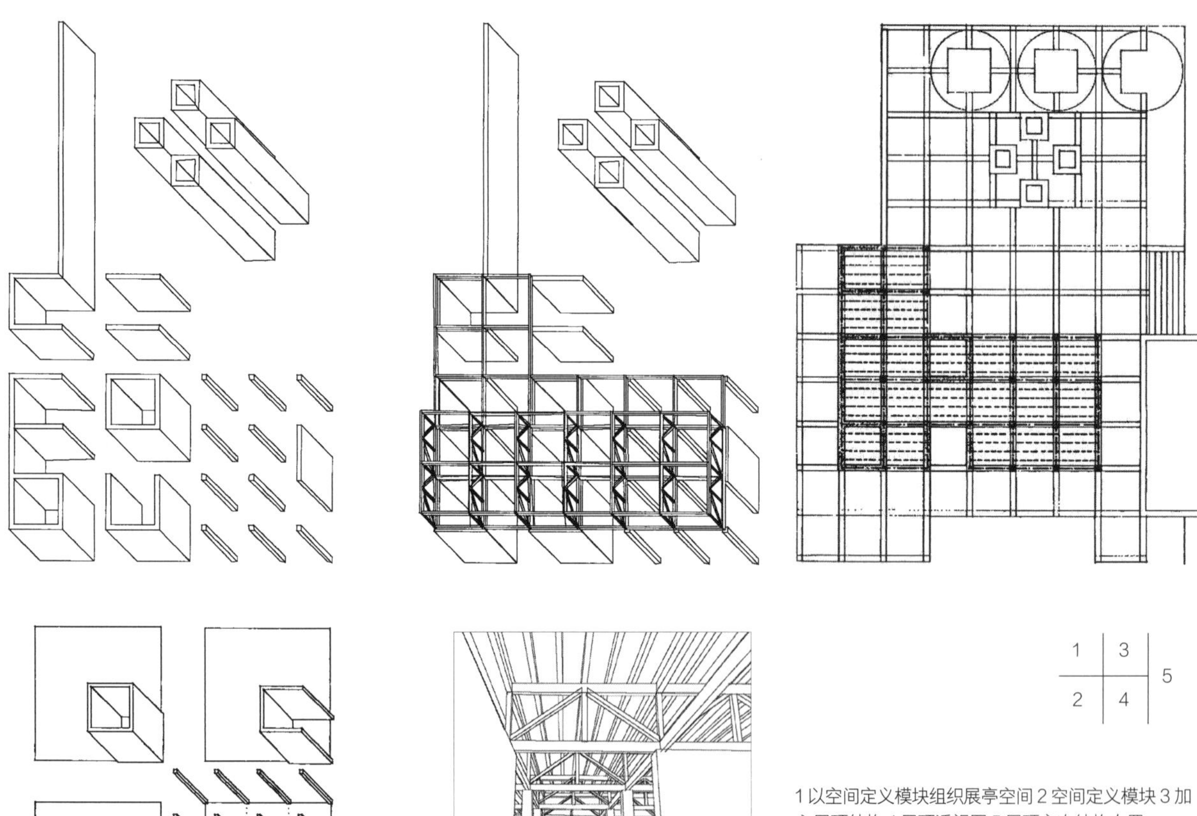

1 以空间定义模块组织展亭空间 2 空间定义模块 3 加入屋顶结构 4 屋顶透视图 5 屋顶主次结构布置

1. Space organization of the pavilion with given modules 2. Space-defining modules 3. Adding roof structure 4. Perspective of the roof 5. Structural layout of the roof

第一阶的第二部分是设计作为场地入口的展览亭，用以陈列三艘划艇，设置访客询问处及管理办公室。

1:100的模型呈现了从空间定义到结构系统的发展。简单的空间定义元素提供了恰到好处的知识量。制作模型的目的之一是形成对主/次结构及两者协同作用的结构系统的意识。

The second part of Phase I was to design an exhibition pavilion, which was also the entrance of the entire site. It housed the arrival information desk and an administrative office; three boats were also exhibited therein.

The 1:100 entrance model showed the development from space definition to structural system. Here, the simplified elements gave students the proper depth of knowledge without distractions. The awareness of structural systems in light of the primary and secondary structural elements, and the logical way they work together was one of the goals in the process of model making.

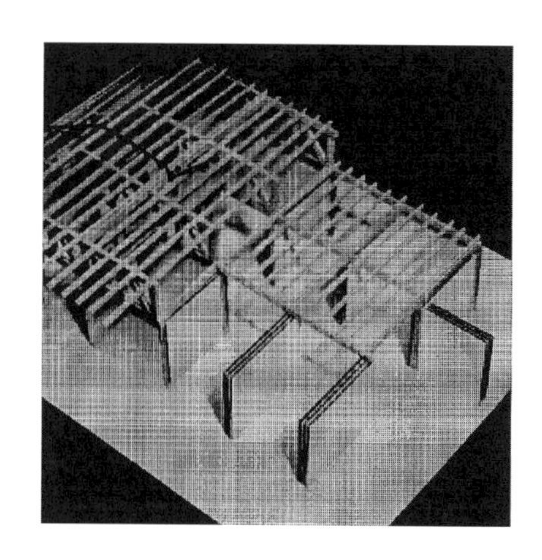

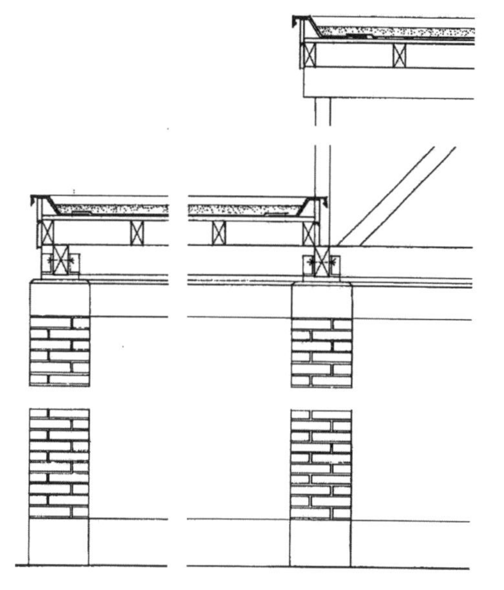

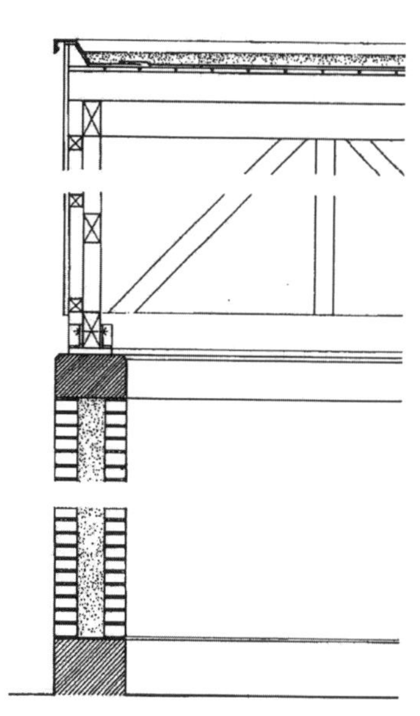

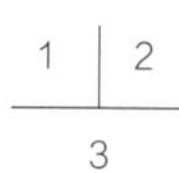

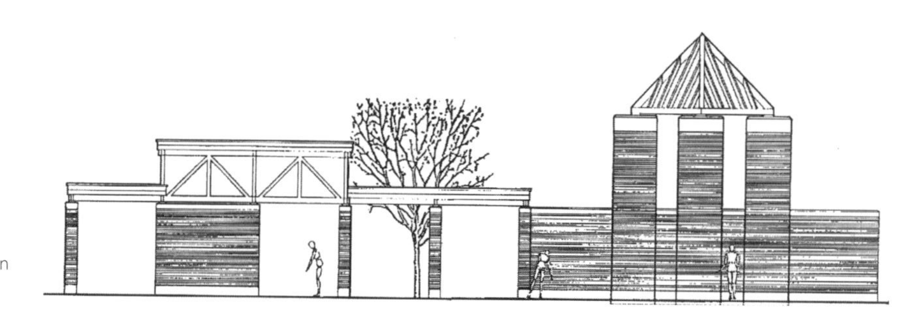

1展亭模型（除去屋面板）2关键节点剖面3立面
1. Pavilion model without roofing 2. Desposition sections 3. Elevation

建造层面的技术问题集中在节点、屋顶及基础，涉及了木材、砖等最普通的建筑材料。材料的表达及形式的推敲与结构系统、材料及构造细部紧密结合。

第一阶的最后一步是将入口设计放回到场地中，激发了场地设计的新想法。虽然对页的透视图是由计算机生成，但学生亦掌握了用尺规求透视的方法。

Concerning the process of construction, the technological problems focused on conjunction, roof detail, and foundations, involving simple materials, such as wood and brick. The materialization and articulation of the visual forms were closely related to the question of the structural system, materials and construction details.

The final step of Phase I was to insert the entrance design back into the situation, which gave rise to some new ideas about the whole site plan. Although the perspective drawings on the facing page were done on computers, the students already knew how to draw perspectives using rulers step by step.

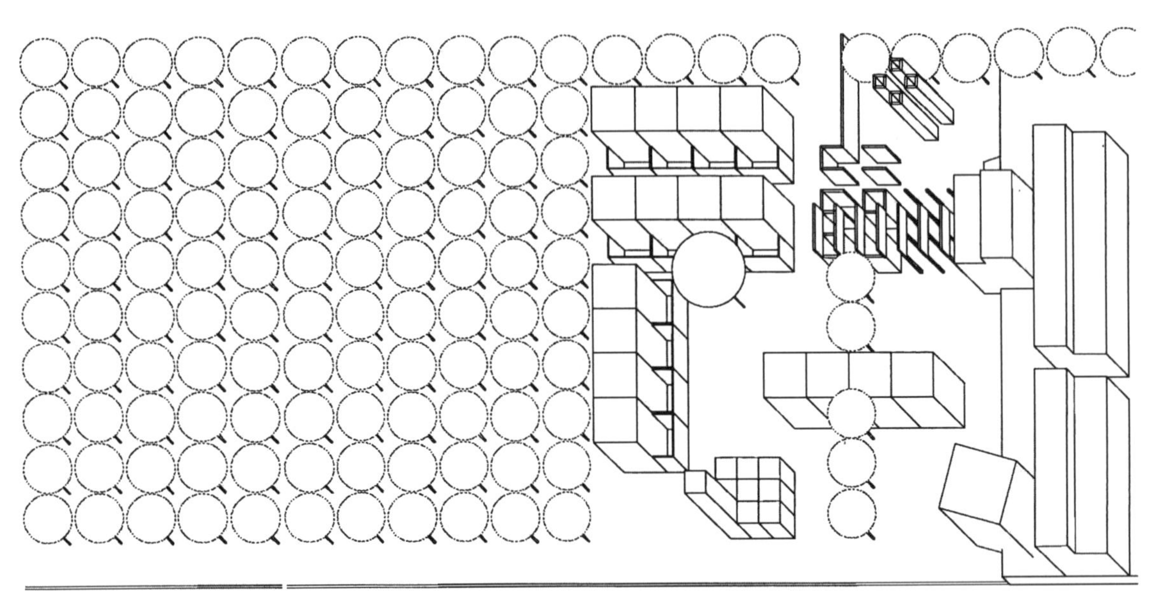

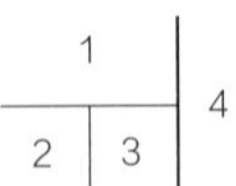

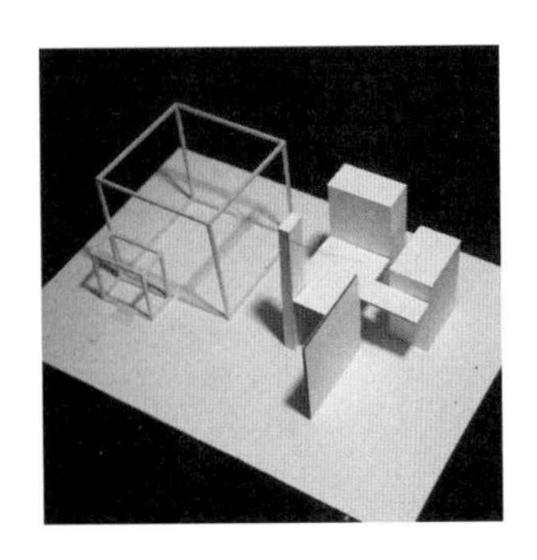

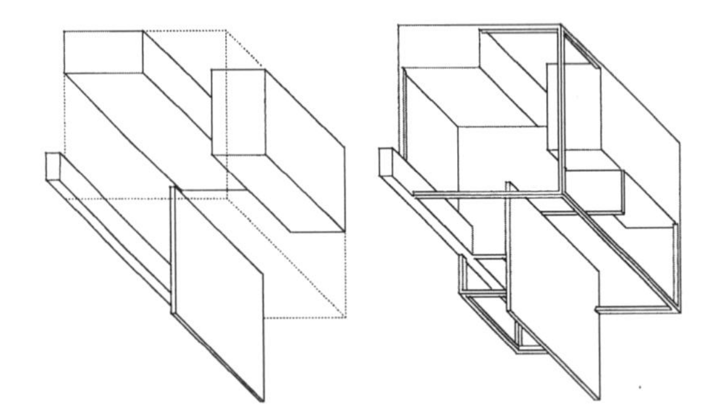

第二阶，设计12个居住单元，每个单元可容纳2名度假者。首先设计一个单元，然后以复制或镜像的方式在场地上尝试多种组合形成组群。

In Phase II, the students were to design twelve living units, with each accommodating two holiday residents. The structure was built up in such a way that the students first had to create one unit and then manipulate it in various ways to form a settlement in the situation.

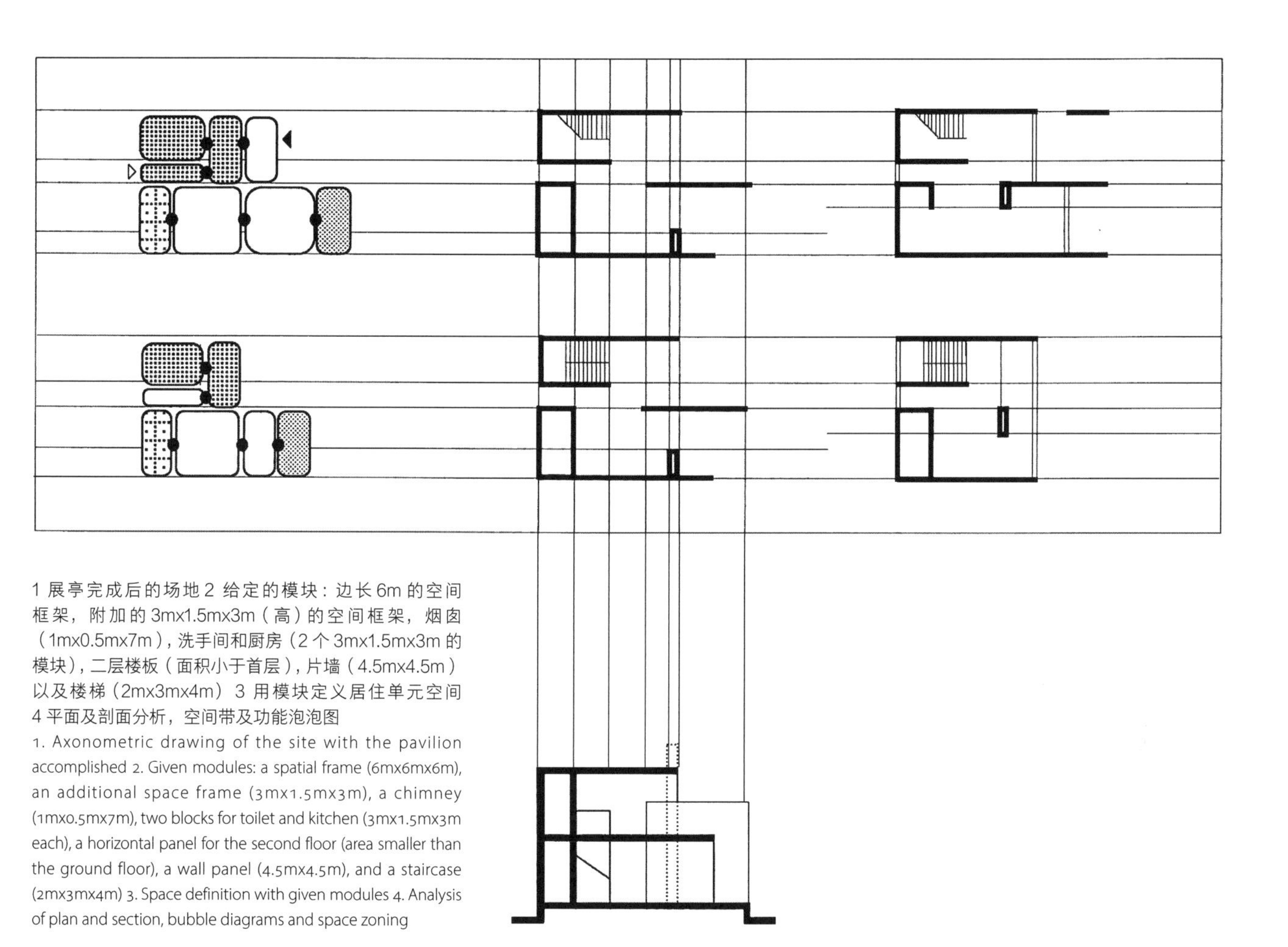

1 展亭完成后的场地 2 给定的模块：边长 6m 的空间框架，附加的 3mx1.5mx3m（高）的空间框架，烟囱（1mx0.5mx7m），洗手间和厨房（2 个 3mx1.5mx3m 的模块），二层楼板（面积小于首层），片墙（4.5mx4.5m）以及楼梯（2mx3mx4m） 3 用模块定义居住单元空间 4 平面及剖面分析，空间带及功能泡泡图

1. Axonometric drawing of the site with the pavilion accomplished 2. Given modules: a spatial frame (6mx6mx6m), an additional space frame (3mx1.5mx3m), a chimney (1mx0.5mx7m), two blocks for toilet and kitchen (3mx1.5mx3m each), a horizontal panel for the second floor (area smaller than the ground floor), a wall panel (4.5mx4.5m), and a staircase (2mx3mx4m) 3. Space definition with given modules 4. Analysis of plan and section, bubble diagrams and space zoning

给出若干限制条件——空间定义元素及空间构架。通过要素的组织围合出“之间”的空间。这不只是操作抽象的线、面、体，还关于空间的居住品质。

本页图解尝试将实体模型中的设计概念投射到平面分析图中。从而在设计过程中引入泡泡图及空间结构平面图式，有助于设计概念的生成。

A set of limitations (space-defining elements and space-generating structure/frame) was given. The “givens” had to be organized, and space as an “in-between” had to be generated. It concerned not only the manipulation of abstract lines, planes, volumes and spaces but also the quality of life.

The analysis of created objects was an attempt to translate design concepts from 3D models to 2D drawings. Here, the techniques of bubble diagrams and structural plans were introduced into the design process in order to formulate the design concept.

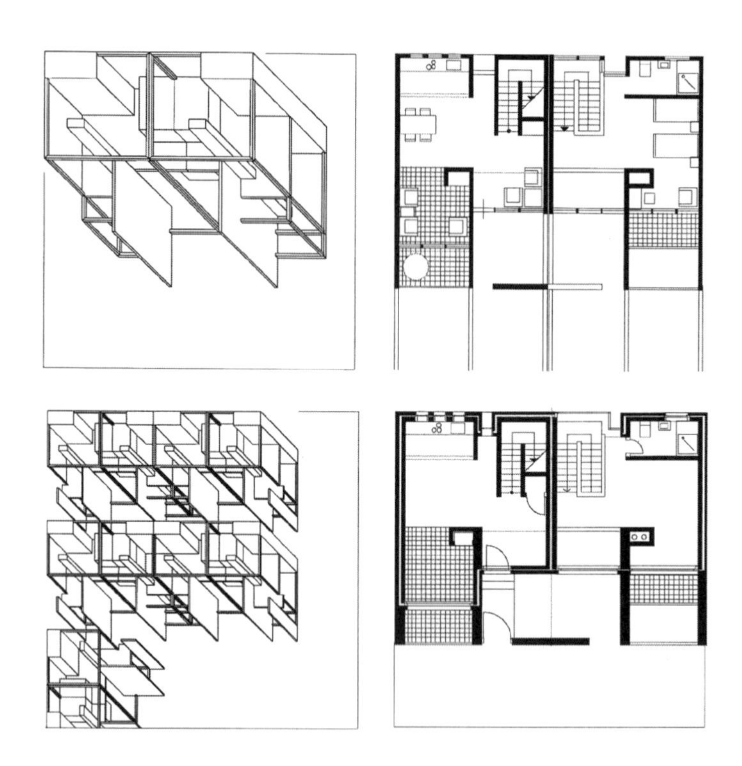

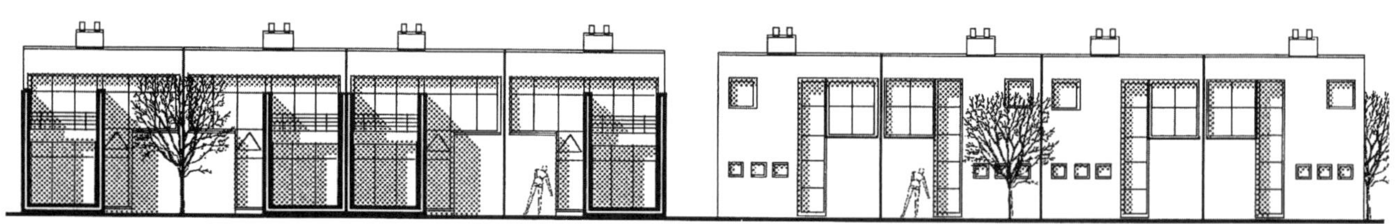

1:50平面图是设计过程的自然推进，家具的摆放及室内空间的推敲使空间与功能更为明确。因为外围护引入了保温层构造，平面图绘制必须在原有外墙的外部再加上一层墙体，同时考虑墙体之间的连接。

4个单元的连续立面呈现了单体和组合间的互动；继而在环境层面讨论从私密向公共空间的过渡。

As a natural consequence of development, a 1:50 plan with furniture layout and interior space rendered the problems of space and function more apparent. Another question was how to build up a floor plan with the given elements. With the consideration of insulation, an additional wall had to be added outside the previous wall, and the special conjunctions were considered.

The facade of four units was used to explain the interaction between the definition of the unit and the possibilities of grouping. Attention then turned to the transition from private space to public space at the environmental level.

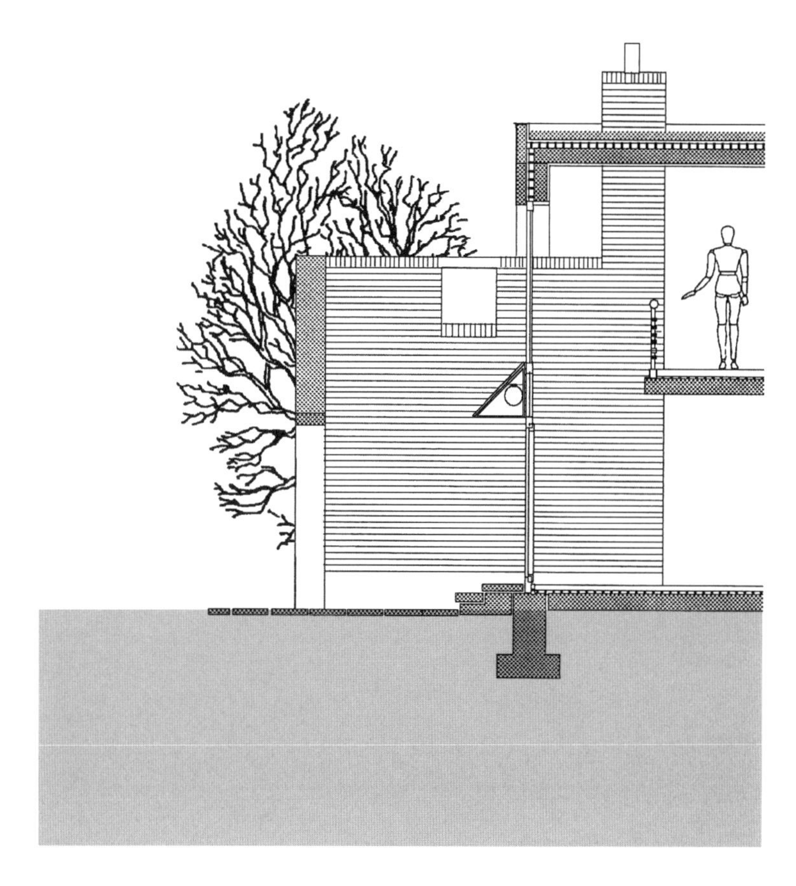

1 | 2 | 4 | 5
3

1 居住单元的组合 2 平面的发展 3 联排住宅立面 4 立面研究模型 5 住宅主要空间的剖面
1. Grouping of living units 2. Development of the plan 3. Elevations of units in a row 4. Facade model 5. Section through main space

1:20的立面模型表达了1.5m的实际进深范围，以表达立面与内部空间的关系。之后依据模型绘制剖面和立面图。

The 1:20 facade model measuring 1.5m in depth was used as a means to explore the representation of facade in terms of interior space. Subsequently, corresponding to the model, the section and facade drawings were assigned.

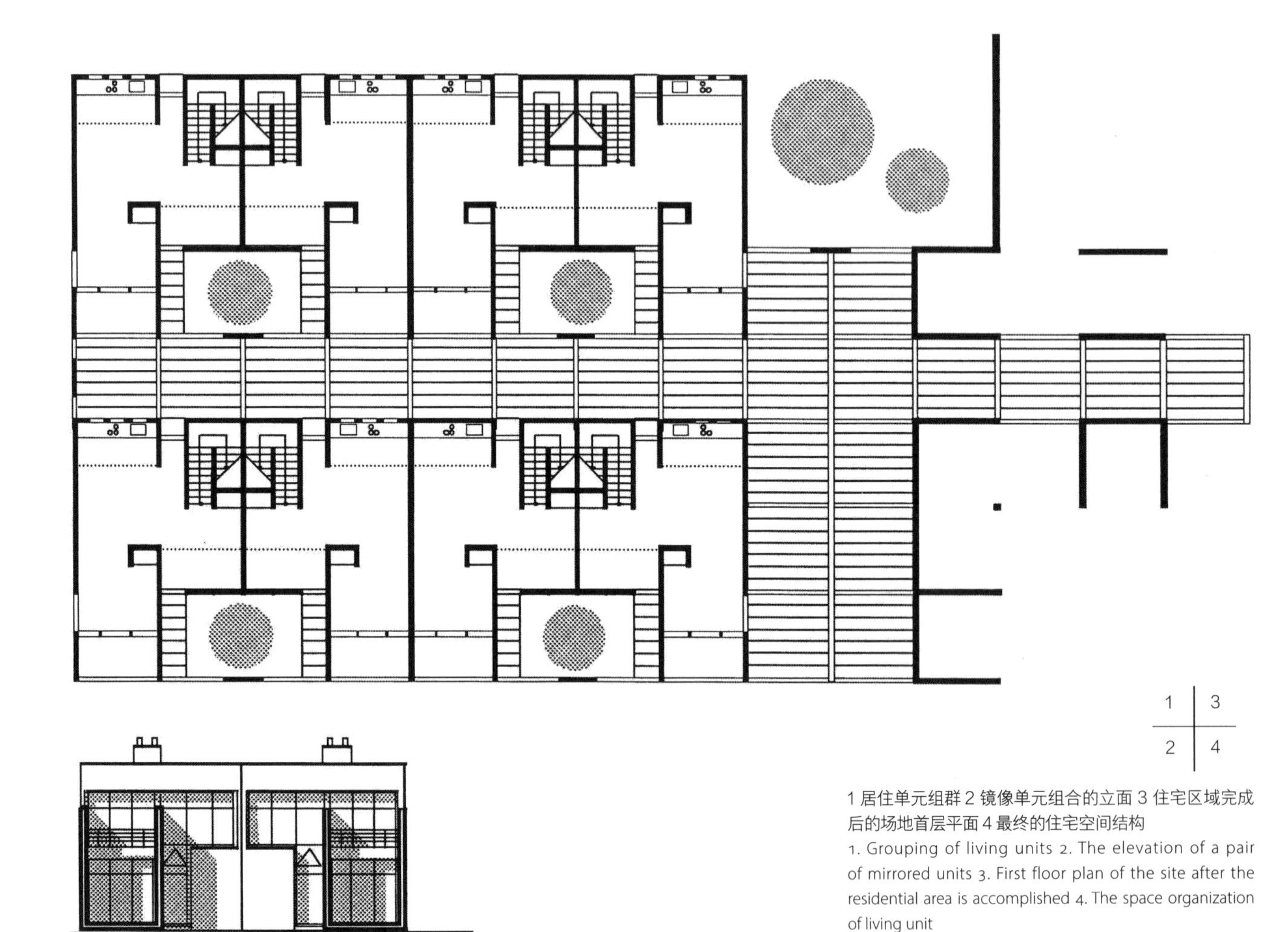

1 居住单元组群 2 镜像单元组合的立面 3 住宅区域完成后的场地首层平面 4 最终的住宅空间结构

1. Grouping of living units 2. The elevation of a pair of mirrored units 3. First floor plan of the site after the residential area is accomplished 4. The space organization of living unit

单元的组织与单元的组群相关，重点在群组或行列的边缘或尽端状态。

The organization of units was related to the grouping of the units, with an emphasis on the articulation of the end conditions.

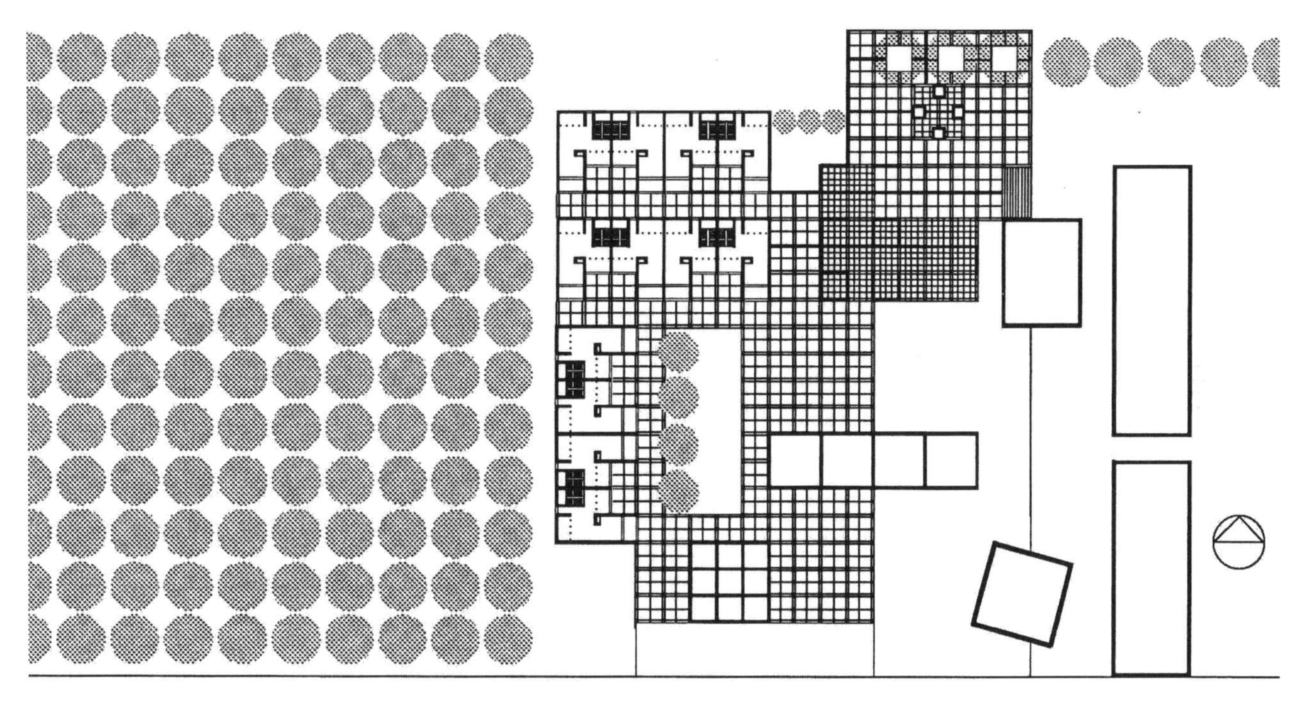

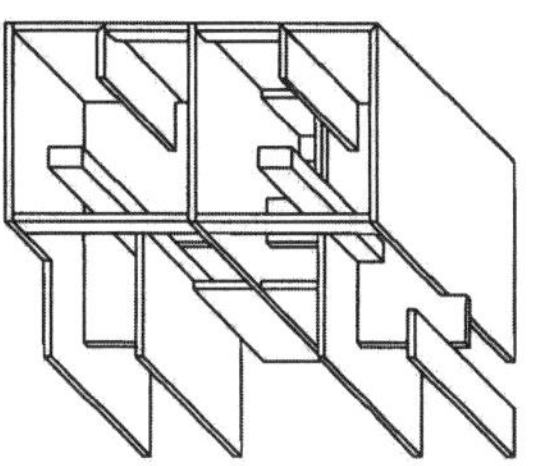

重新绘制整个场地的首层平面图有助验证设计的整体合理性。

Obviously, to redraw the whole situation served as a tool to examine the overall feasibility of the proposal.

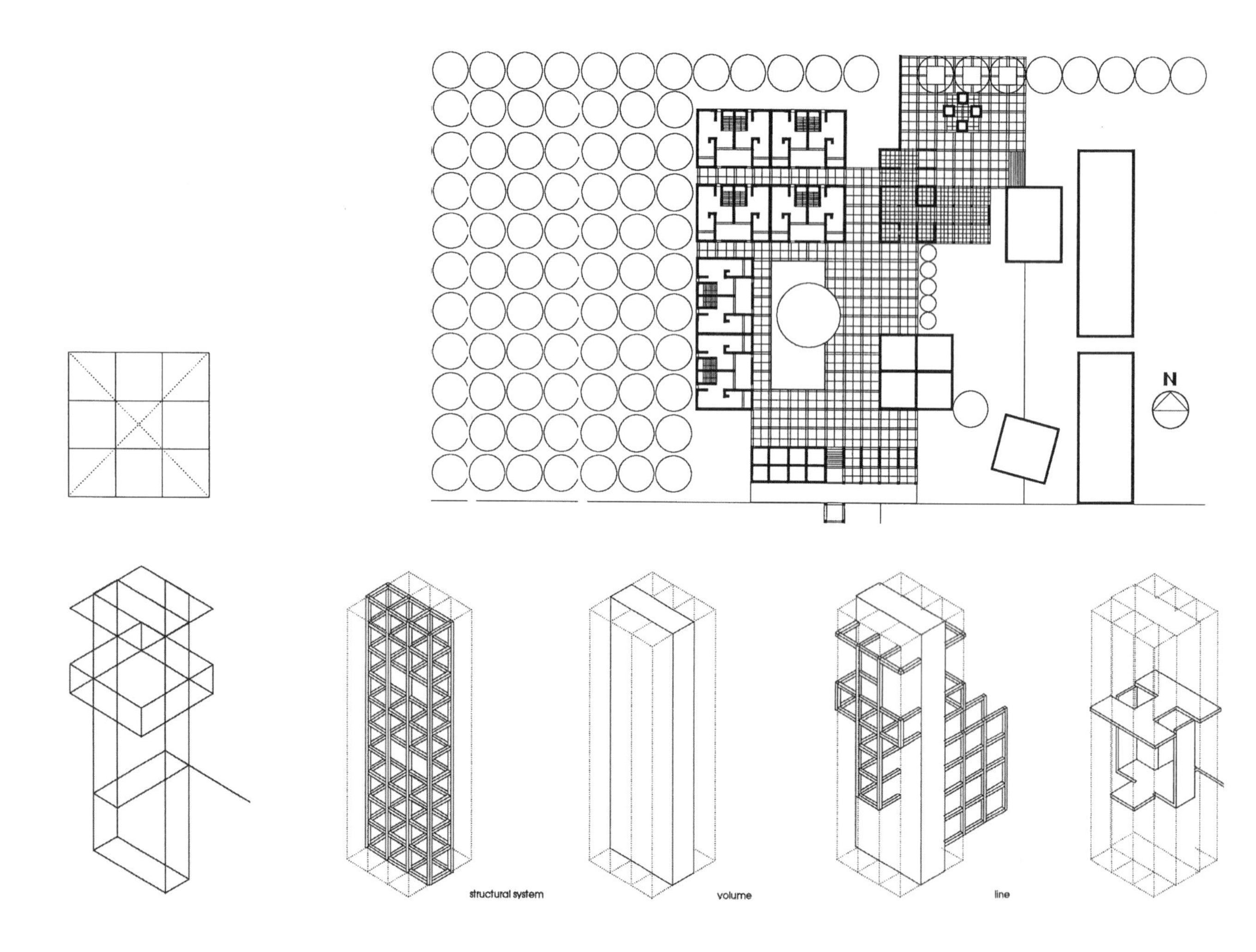

在相对较短的第三阶，设计一个木结构的瞭望塔，以探索建造技术在设计找形过程中的角色。因此必须重新组织总平面。

基于抽象的几何结构尝试了多个结构系统。设计概念的发展过程与建造过程相关。系列轴测图显示了结构生成的过程。

In the relatively short Phase III, the students were to design an observation tower in wood construction. This functioned as one of the means to explore the role of technology in the form-giving process. The new assignment for the observation tower and stand necessitated the reorganization of the situation.

Based on basic geometry, various proposals were considered for the observation tower in terms of structural systems. The process of developing the design concept was related to the process of construction. The sequence of isometric drawings reflect the generation of the structure.

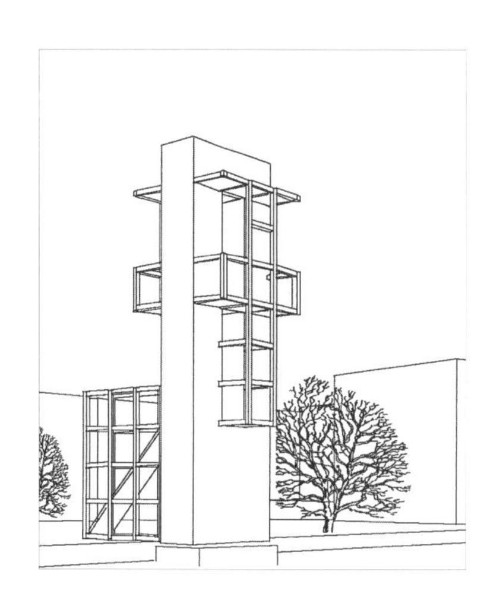

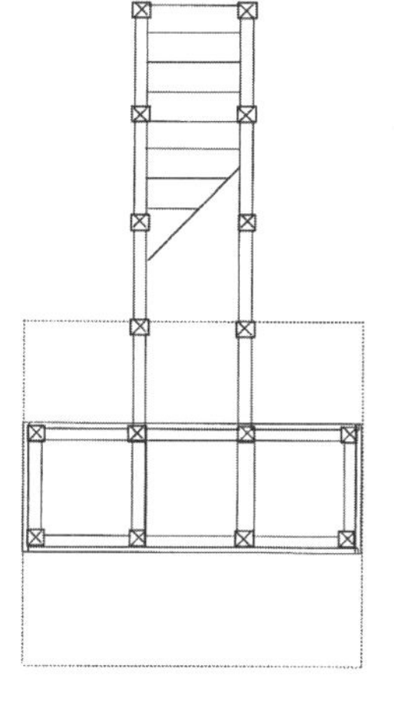

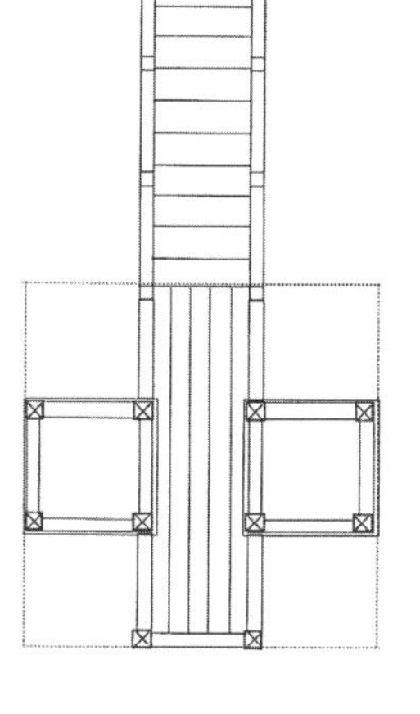

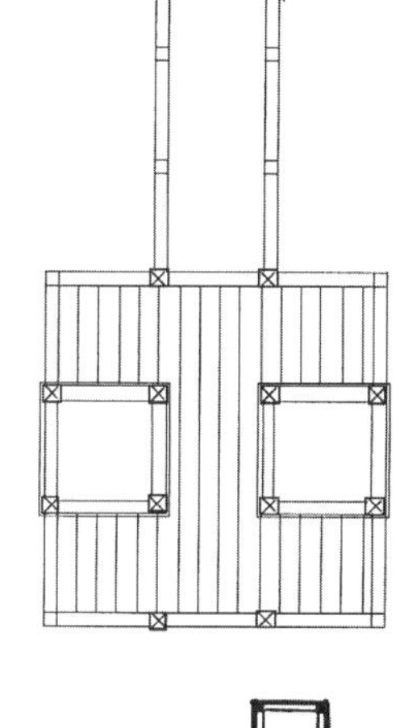

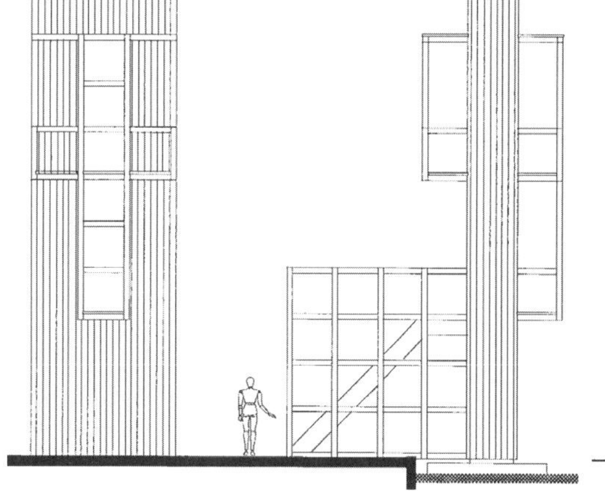

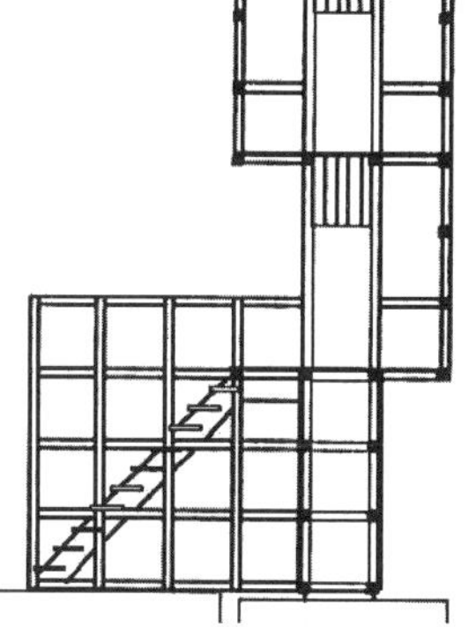

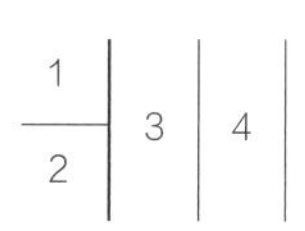

1 根据场地现状调整看台及瞭望塔的平面布局 2 瞭望塔的发展过程 3 框架与实体 4 瞭望塔平、立、剖面

1. Adjustment of the stand and watch tower according to the new site plan 2. Development of the watch tower 3. Frame and solid 4. Plans, section, and elevations of the watch tower

平面、剖面和立面展示了设计的特点。选取的结构系统和建筑材料决定了建筑形式。

The floor plan, section, and facade were drawn to show the characteristics of the design. The structural system and materials chosen determined the derived form.

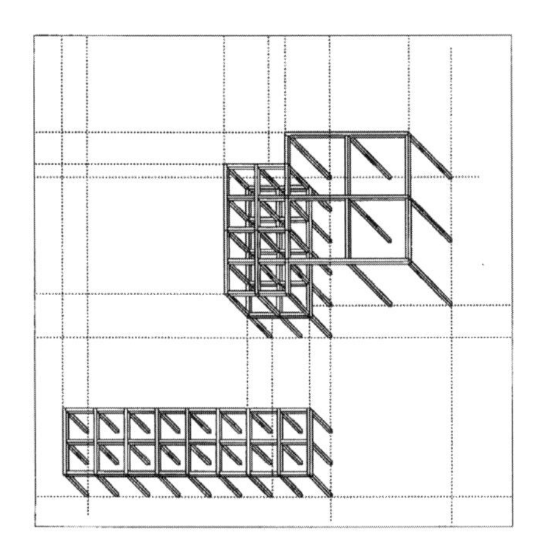
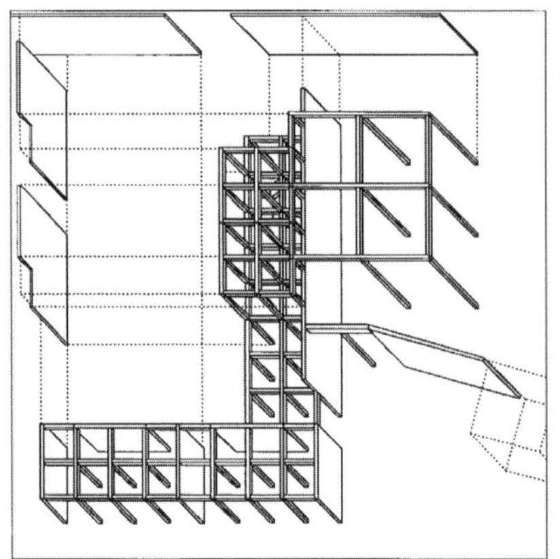
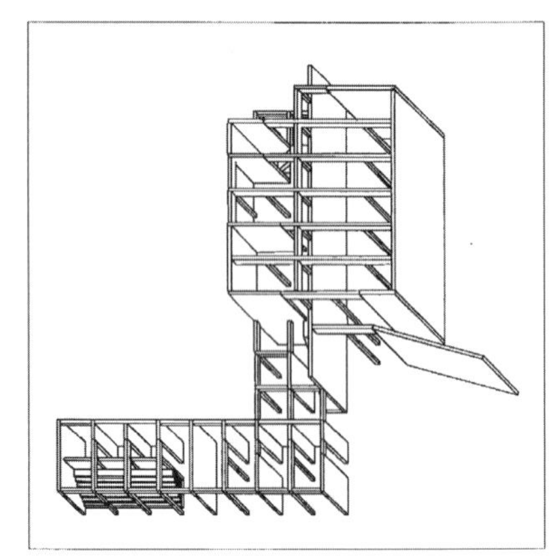
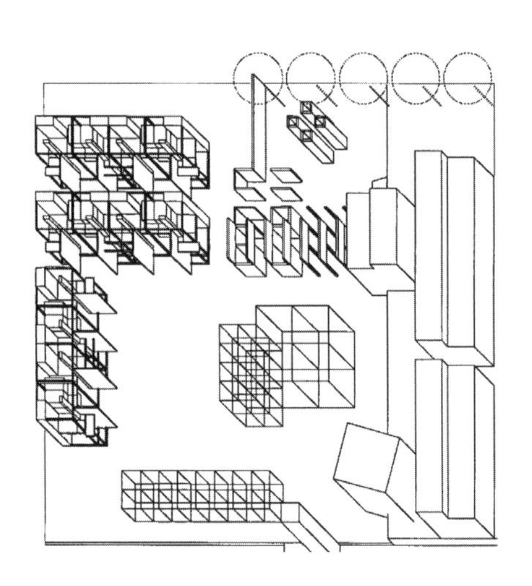
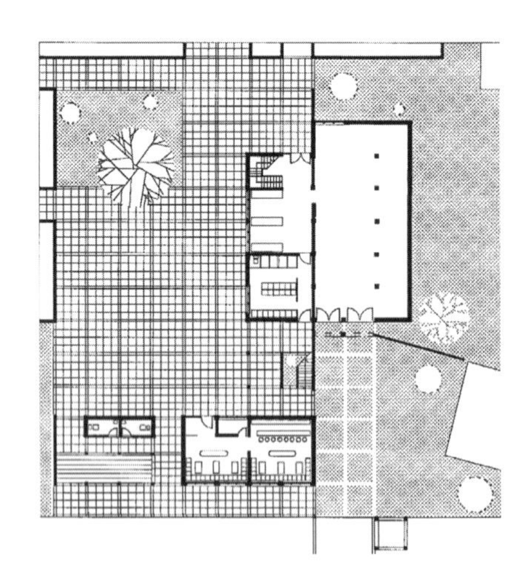

在第四阶，设计可供运动员存放及修理划艇、带衣帽间和办公室的划艇俱乐部。问题变得更为复杂，场地对建筑形式的影响更大。设计重点在于从功能图解出发，通过结构框架转化为空间组织。首先在场地关系中思考体块图解，然后根据图解设立混凝土的结构框架。

In the 4th phase, finally, the boathouse was designed to accommodate cloakrooms and administrative offices, as well as for athletes to store and repair boats. The problems tended to be more complex; the site was more significant in the determination of architectural form. The emphasis was placed on the transformation from a functional diagram to spatial organization through a structural frame. The first step was to examine the block diagram in the context. Then, the structural frame as a concrete skeleton was built up according to the diagram.

1	4
2 3	5

1 俱乐部设计发展过程：从结构框架到空间围合 2 在现有总平面中调整俱乐部的建筑体量 3 俱乐部首层平面 4 俱乐部建筑群空间结构模型 5 划艇俱乐部与联排住宅的立面关系

1. Development of the club: from structural frame to space definition 2. Adjustment of massing in the new context 3. First floor plan of the club 4. Spatial structure model of the club 5. Elevation showing the relation between the club and townhouse

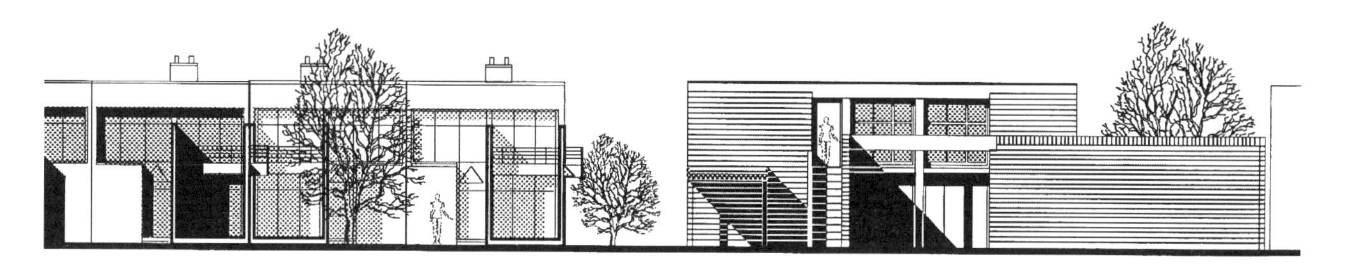

推敲结构框架的切分强调了建筑使用需求与场地的相互关系。借助三维模型可以看清建筑在场地中生成的过程。

概念一旦形成，就可以运用不同尺度的平面表达来推敲功能、建造、场地的问题。

划艇俱乐部的立面在形式上与前面几阶的设计关联也很重要。

The investigation of subdivisions in the structural frame stressed the interrelationship between the building program and its context. One can easily see the object generated in the context by taking advantage of a 3D program.

Once the concept was developed, the representation of floor plans in different scales dealt with the problems of function, construction, and context.

In addition to the elevation themselves, the formal relation with other designed buildings was also quite important.

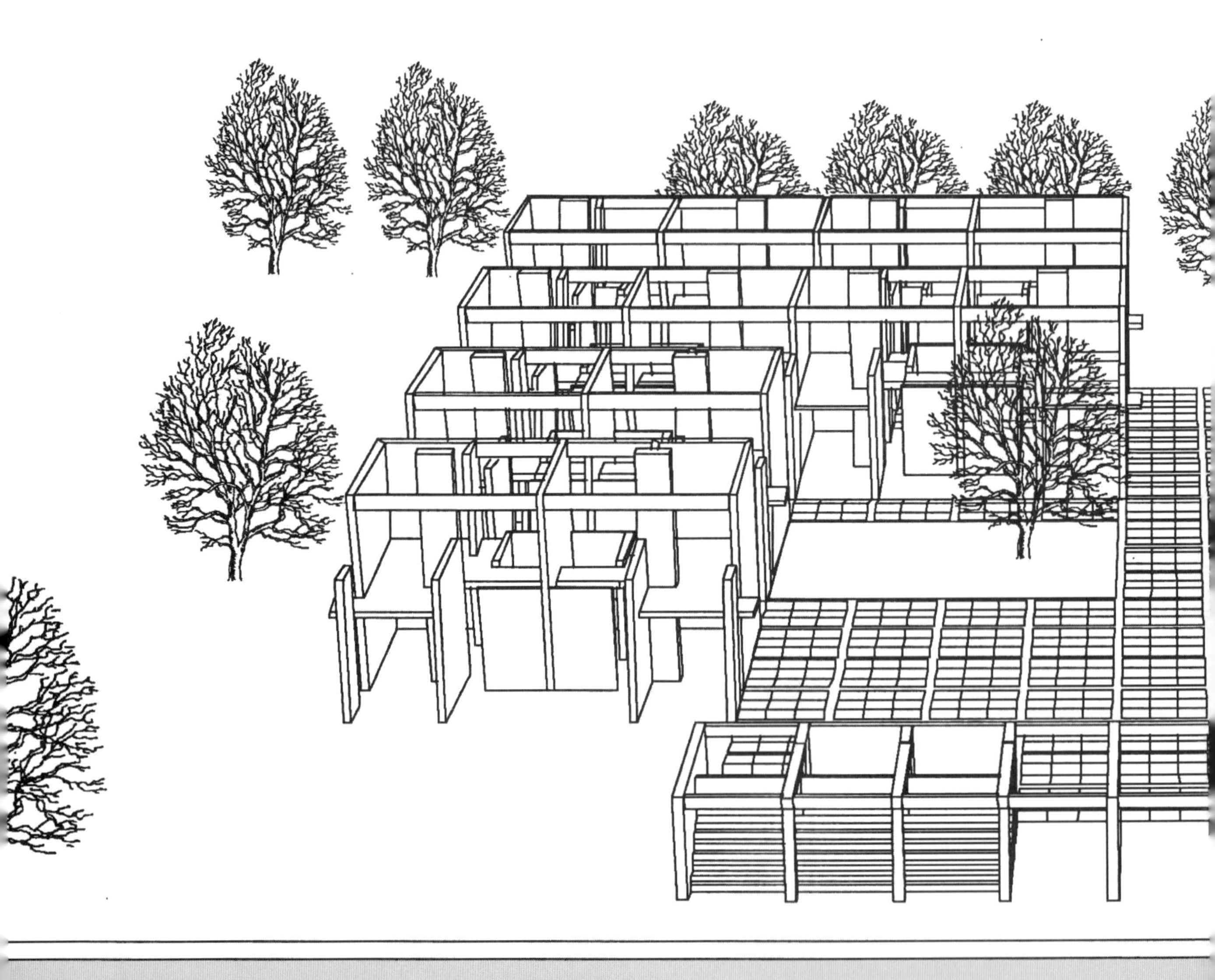

划艇俱乐部建筑群鸟瞰 | A bird's eye view of the rowing club complex

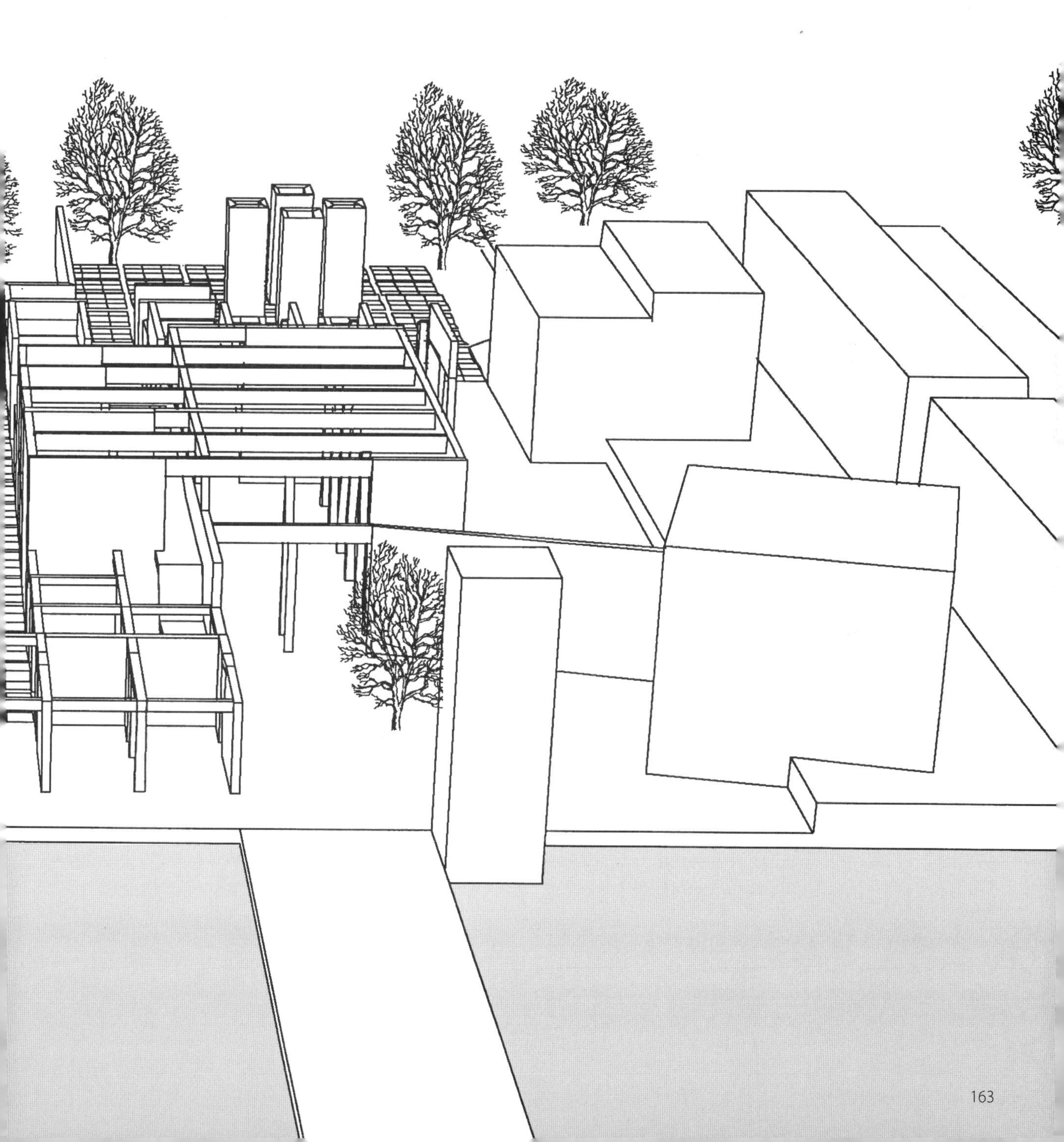

1991/92

游客中心，撒比奥内塔

Tourist Centre, Sabbioneta

张雷
ZHANG Lei

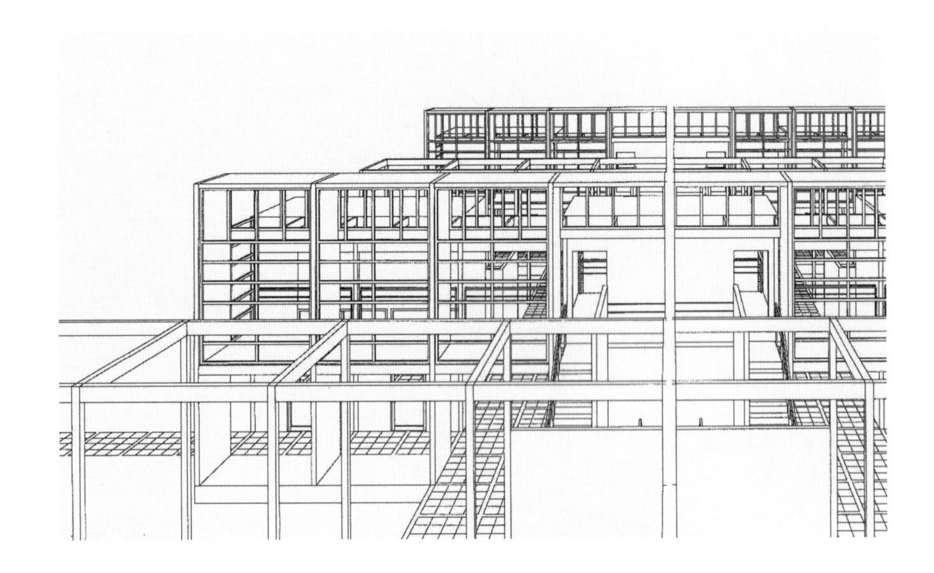

1991/92学年的“基础设计”的主题是位于意大利古城撒比奥内塔边缘的游客中心。与往年一样，共分为4个设计题目：第一阶为市场，第二阶为居住单元，第三阶为工作坊，第四阶为开放题目，学生与教师讨论后自行决定（本书只展示前三阶）。

场地位于文艺复兴时期伦巴第地区的理想城撒比奥内塔。其城市系统长久以来并未扩充，一直保持了原貌，因而提供了审视城市的“整体”与“部分”的机会，以及教学与观察的场所。

For the academic year of 1991/92, the building program and theme for the first-year design exercises was a new urban development for tourism on the outskirts of Sabbioneta.

Similar to the basic design course in the previous year, there were altogether four phases (this book only demonstrates the first three): a market hall (Phase I), the housing units (Phase II), studio-workshops (Phase III), and a free theme settled on by the students and their teachers after deliberation (Phase IV).

The general theme for the subjects was built on Sabbioneta, an ideal city of the Renaissance in Lombardy. Remaining untouched throughout the years, the city system has not been extended since its genesis. Thus, Sabbioneta serves as a training and observation ground, providing the opportunity to examine both the “whole” and the elements of a city.

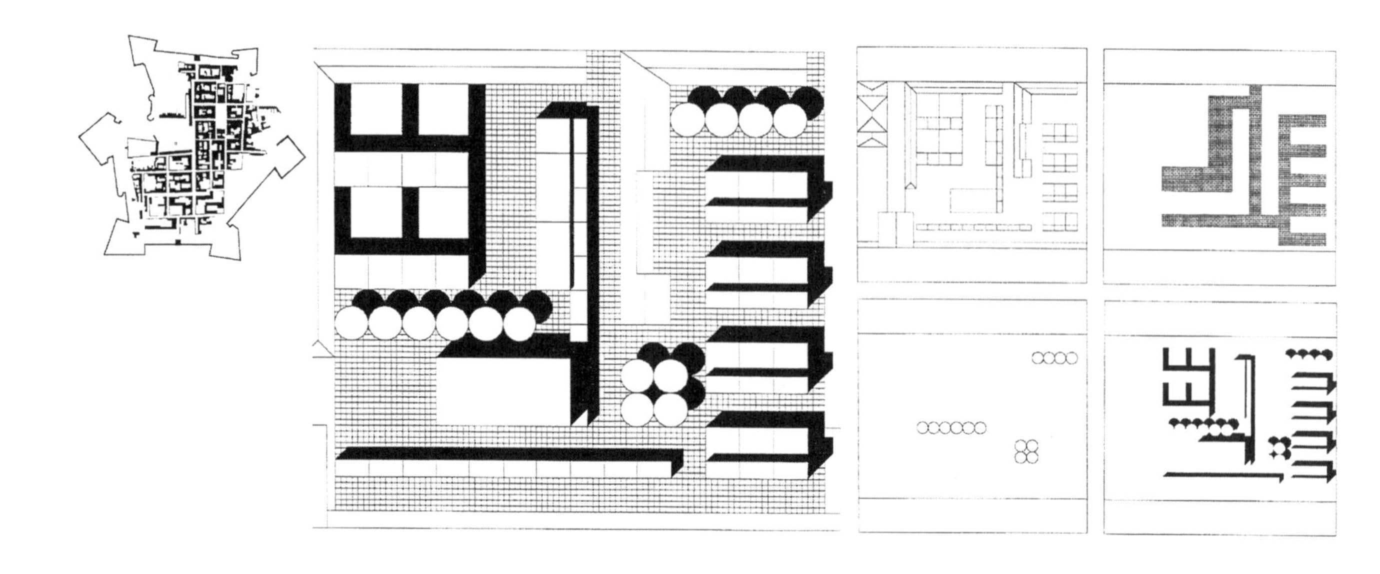

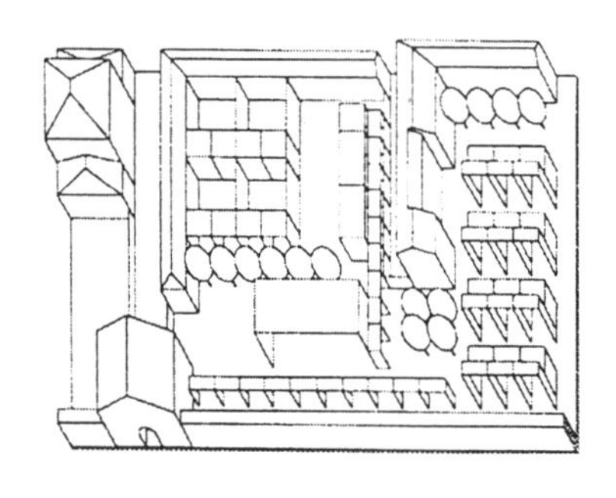

1 撒比奥内塔城图底关系图 2 场地总平面组织及分析 3 场地组织的“纸圈”模型 4 用符号表示不同的功能进行平面组织 5 根据功能选择不同的围合方式

1. Figure-and-ground relationship of Sabbioneta 2. Site plan and analysis 3. Site organization with the "paper tube" models 4. Space organization with squares marked with different symbols 5. Space definition regarding function

第一阶由场地模型开始，以屋顶的构造结束。在场地的基础上给出了若干体量及使用需求，作者做出最初的总平面设想，它可被解析为体量、铺地、绿化等层级。由此引入部分与整体的互动。

Phase I starts with a site model and ends with a roof detail. Based on the site, a number of given volumes and a general program, the author interprets the site and the program and then form a first hypothesis. This page illustrates how an “architectural solution” can be decomposed into layers such as volume, surfaces and planting. Following this, the notion of the parts interacting with the whole is introduced.

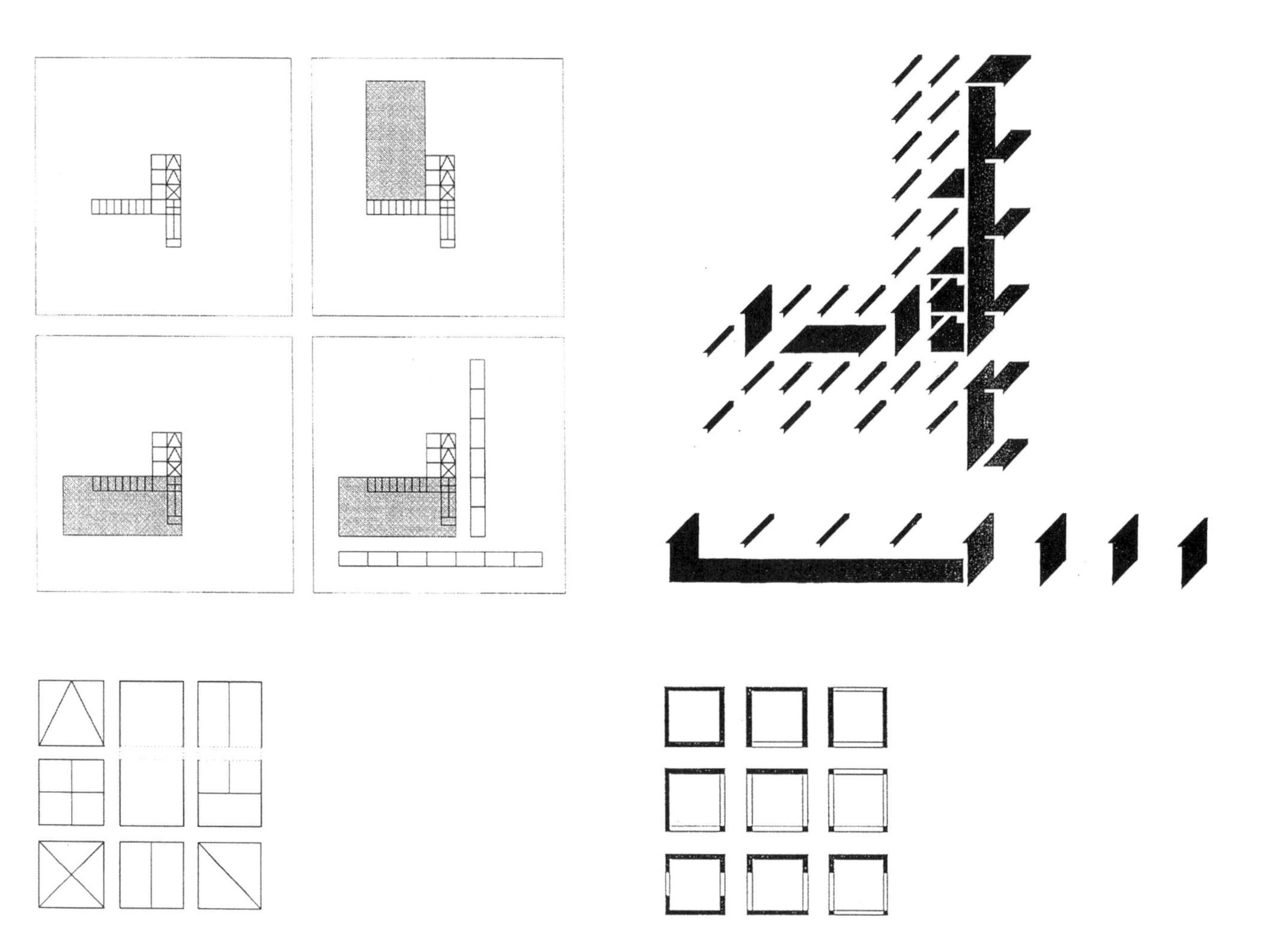

不同的使用功能以标符示意，将功能元素转化为“棋子”。在棋盘（即总平面）上把玩这些模块，以便快速地认识不同组合的内在逻辑。探索空间的组织、方向性及其品质。

运用空间基本类型将几何模块转化为三维空间并形成建筑平面，引入建筑基本单元的概念，说明了“数量”与“质量”的区别。

Different functions had been transformed into signs, turning each functional element into a “playing piece”. Placing the pieces on a game board (the site) enabled quick recognition of the internal logic of each combination. The organization of space, the notion of structured movement through space, as well as the qualities of space were explored.

The geometrically defined organization of elements was transformed into organized space, and a plan was created through the introduction of a typology of spaces. With it, we tried to introduce the differentiation between quantity and quality as well as the notion of BAU as an objective reference for our operations.

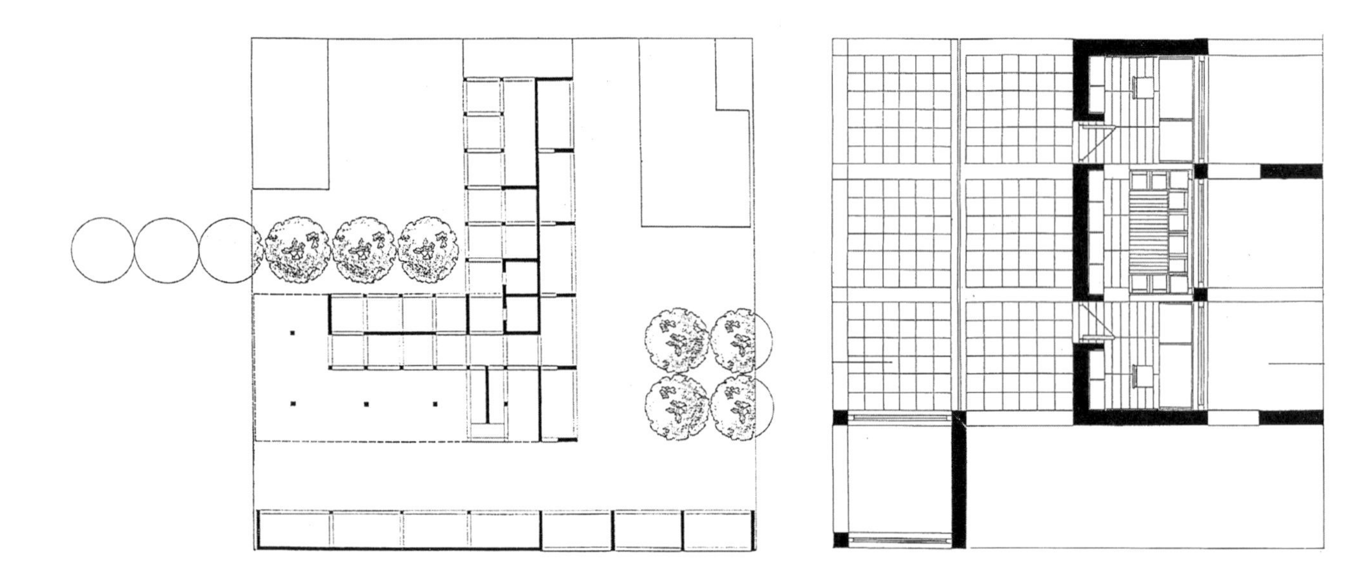

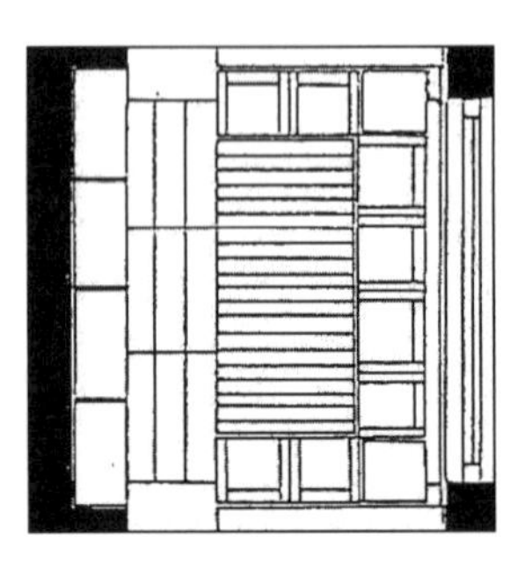

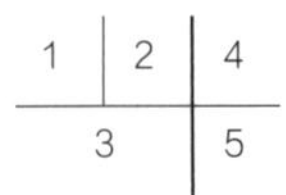

1 市场首层平面 2 平面局部家具布置及铺地 3 一个空间单元的家具布置 4 市场立面及剖面（局部）5 市场立面
1. First floor plan 2. Part of the floor plan showing pavement and furnishing 3. Furnishing of one BAU 4. Part of the elevation and section 5. Elevation

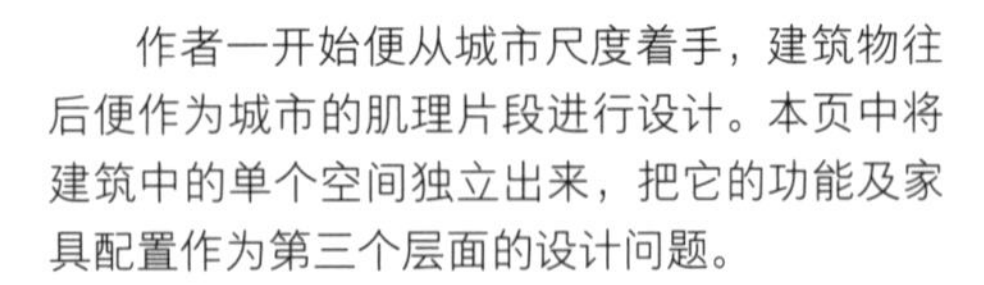

作者一开始便从城市尺度着手，建筑物往后便作为城市的肌理片段进行设计。本页中将建筑中的单个空间独立出来，把它的功能及家具配置作为第三个层面的设计问题。

From the very beginning, the author encounters the urban scale as a point of departure. The object is then developed as part of the urban fabric.
This page points at an attempt to introduce the individual space, the 'room', with its function and furnishings serving as a third level of consideration.

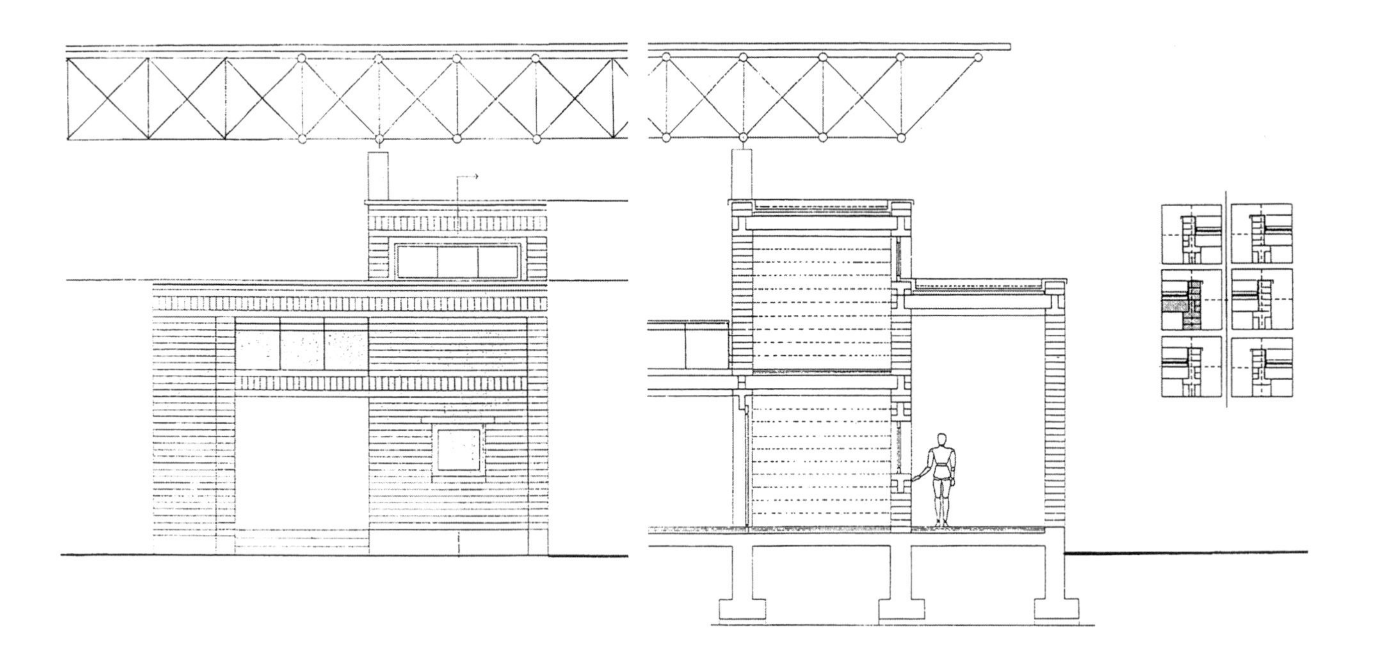

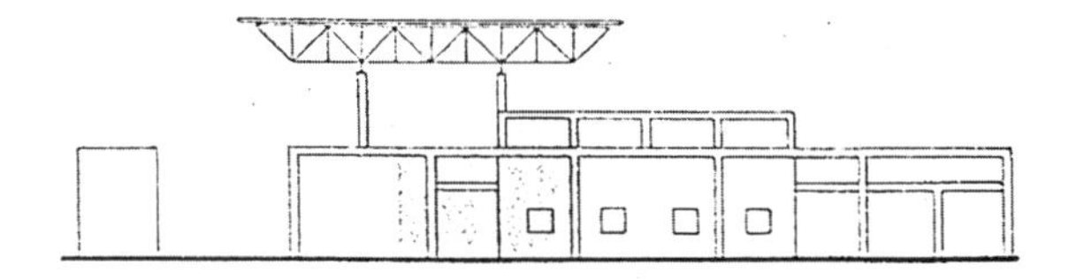

对构思与建造间相互关系的探寻贯穿整个学年，第一阶就涉及建造的逻辑及材料的质感。

与其他设计步骤中的简单限制一样，给定条件不是为了让学生能够马上解决问题，而是理解在建造过程中需要考虑的问题。“屋顶”形式决定了建筑的形态，因此用模型进行了多种尝试。

The question of how conception and construction are interrelated is explored throughout the year. Phase I already involves the logic of construction and the sensual qualities of the material.

As in other design steps that constitute Phase I, these exercises also introduce simple limitations that do not provide instant solutions but enable students to understand some of the problems involved in the process of construction.

Since the "roof" of the market hall determines the figurative qualities of the organization of elements, various positions had to be investigated (in a model).

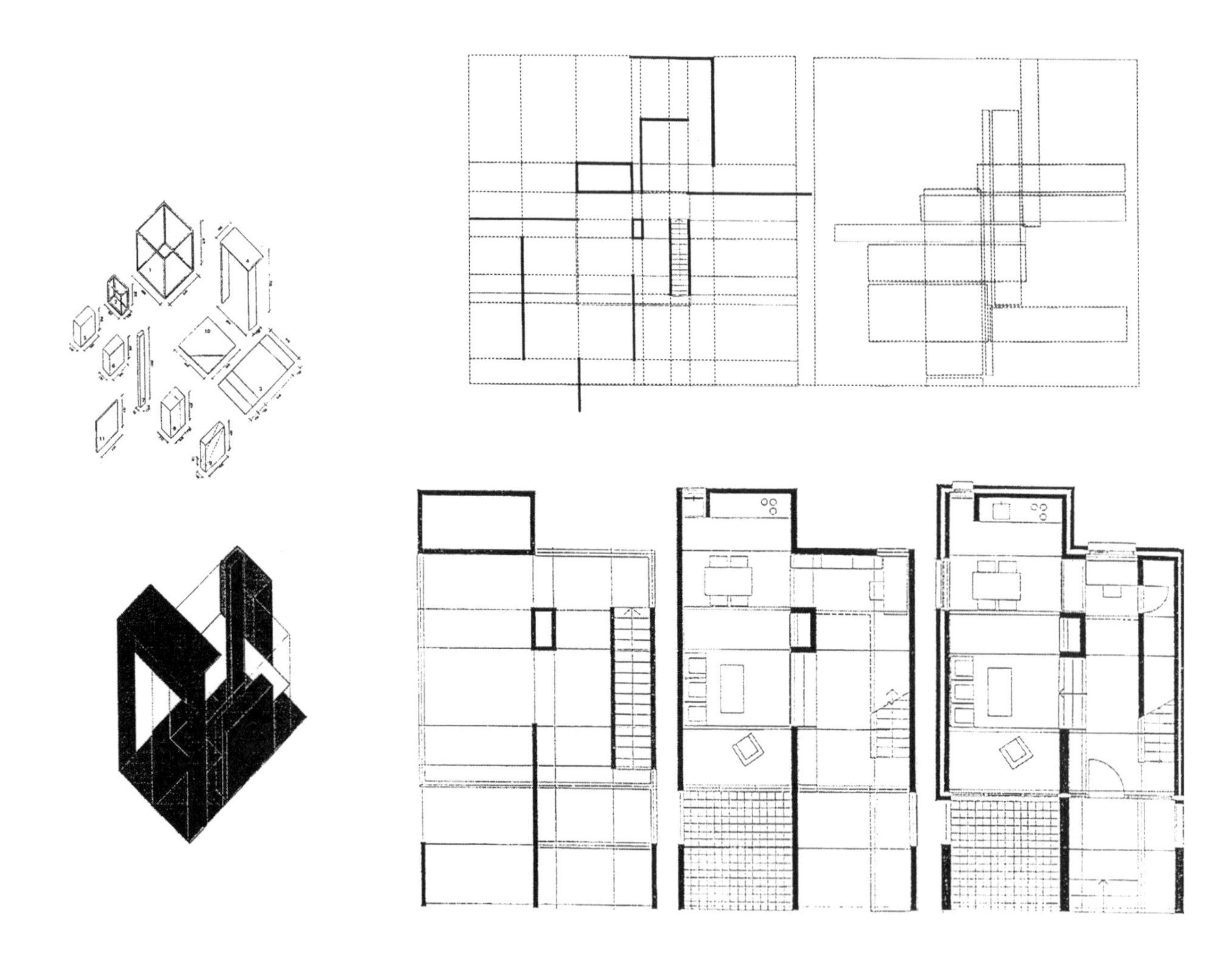

总平面的最初设想以及市场设计成为第二阶的既有条件。在此基础上，居住单元设计需要对建筑空间更细致的考量，组织更复杂的使用需求，进一步理解形式。

类似于第一阶，给出一系列空间操作的组件。以图—底关系研究建筑空间的形成；以空间带来揭示组件的对位关系。

In Phase II, this site plan, together with the new context of the developed market hall, became the starting reference point. Within this new context, the housing units had to expand the notion of architectural space, a more complex program had to be organized and the understanding of form had to be further developed.

As in Phase I, a set of limitations was given, part of which could be seen as a kit of parts. The generation and organization of architectural space was the issue. Figure-ground considerations entered the debate. The spatial zonings reveal the alignment of the parts.

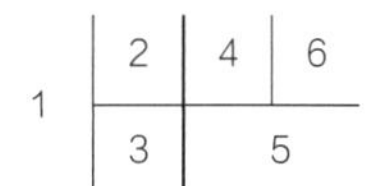

1 居住单元设计的给定模块及组合 2 空间带分析 3 平面的发展 4 居住单元与城墙的剖面关系 5 居住单元的联排组合 6 联排住宅的平面与正面
1. Modules for the living unit and one possible composition 2. Analysis of spatial zones 3. Development of the plan 4. Relation between living unit and city wall in section 5. Grouping of living units in townhouse mode 6. Plan and facade of the townhouse

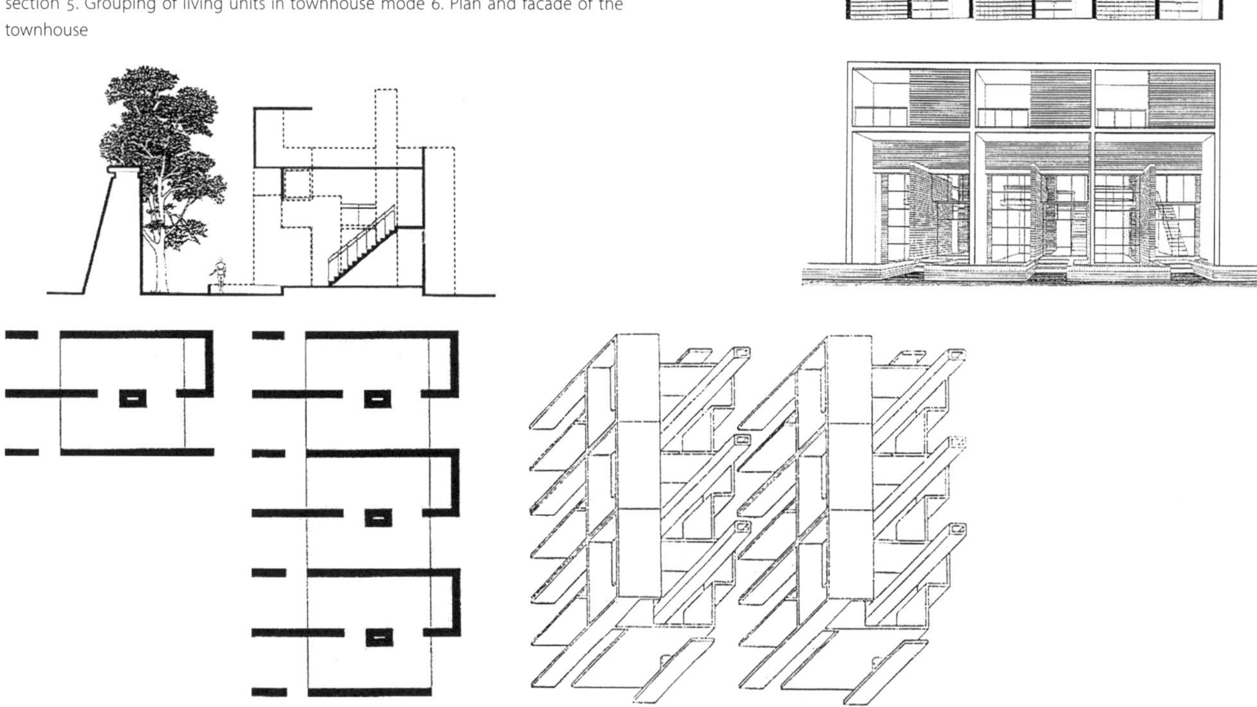

第二步，组合居住单元，形成群组或行列。根据功能需求（居住条件、家具等）深化单元设计，在构造方面引入空心墙体及保温隔热的问题。单元立面设计考虑虚实关系，引入开口的概念（开窗对应于空间结构，而不是在墙面上挖洞）。呈现建筑立面与立面图的对比，以及空间组织与建造的互动关系。

Deduced from step I, by step II, the house becomes a unit in a larger agglomeration (the group, the row, etc.)
Furthermore, the unit is developed functionally (e.g. living conditions, furniture) and in terms of construction (e.g. the cavity wall and the resulting thermal issues).
In the development of the object, solid and void, a typology of openings is introduced (structural opening, as opposed to the punched hole). Consequently, the differentiation between elevation and facade is debated. The interaction between the spatial organization and construction is introduced.

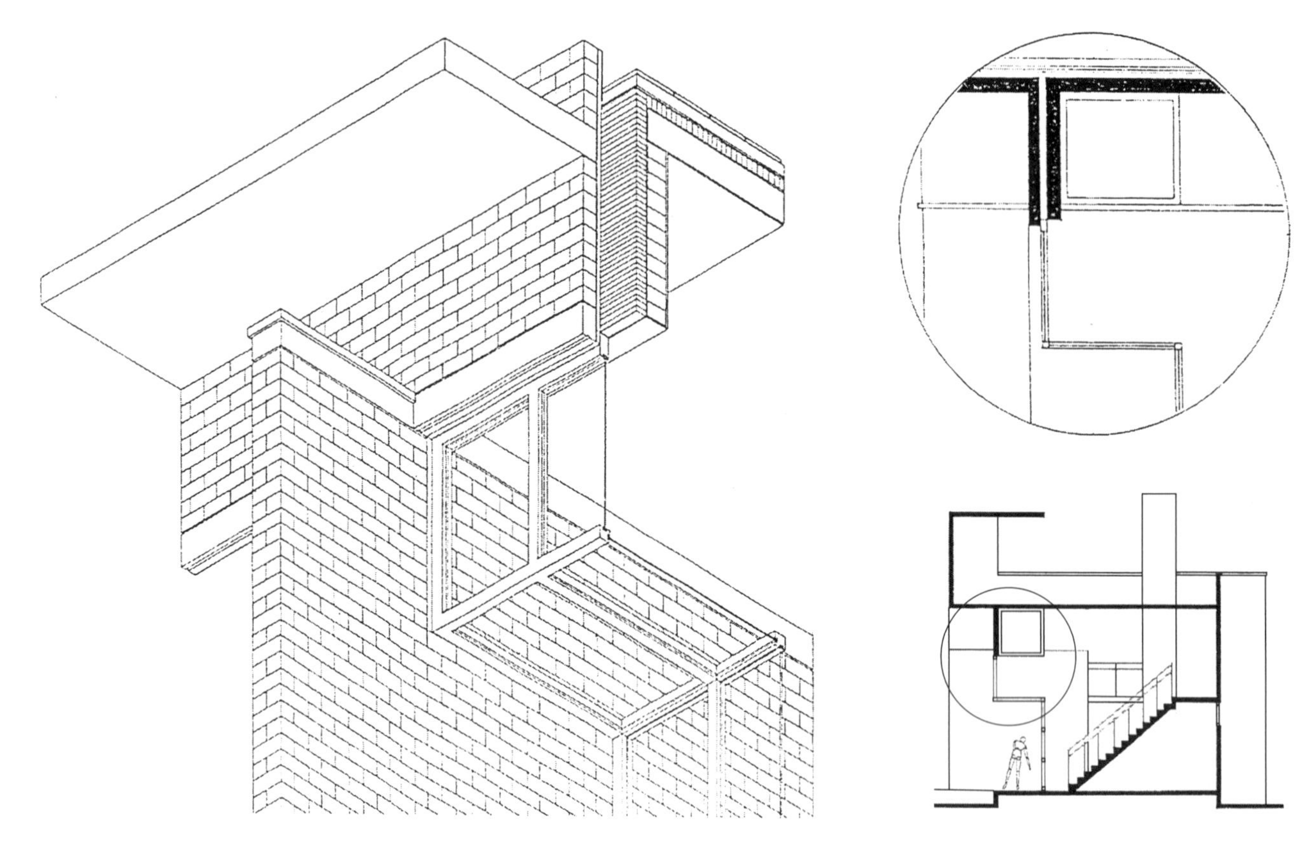

1 剖切构造轴测图 2 居住单元剖面及构造轴测图位置示意 3 外墙剖面 4 关键位置剖面 5 楼板与外墙交接部位、室内地面与基础部位细部 6 居住单元剖面

1. Cross section axonometric drawing 2. Section of the liveing unit indicating the location of cross section axonometric drawing 3. Section of the elevation 4. Disposition section 5. Detail drawings of floor, exterior wall and foundation 6. Section of the living unit

不同精度的剖面表达。剖面表达难于平面、立面，因此借助了实体模型。剖切构造轴测图有效地表达了建筑围护要素之间的关系。

The section has been used on various levels of resolution. Since the section is one of the most complex forms of representation, it was introduced with the models. The technique of using cross section drawings as a means of visualizing the building envelope and testing its performance has proved to be very effective.

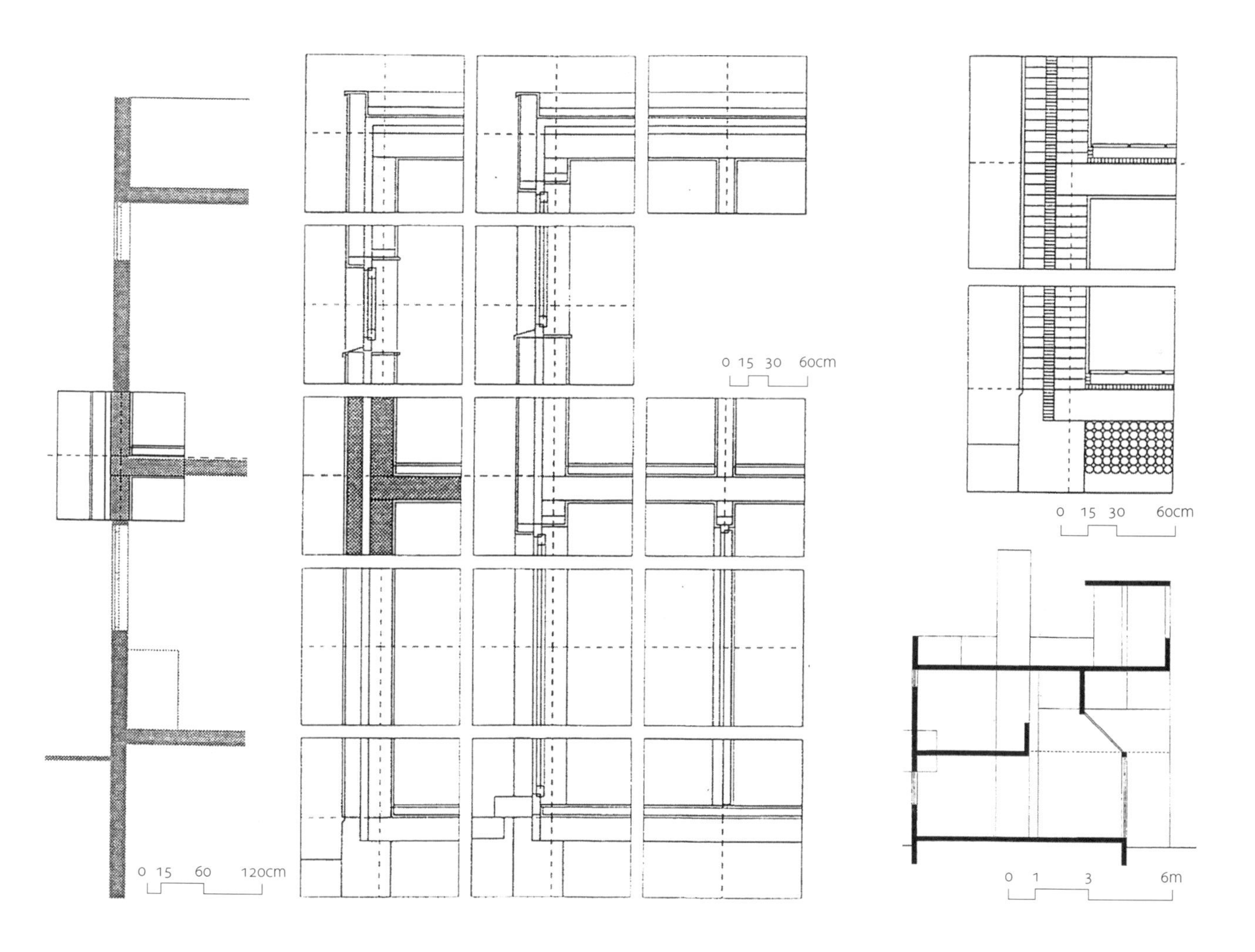

在我们的教学中，理解“特定的比例对应于特定的设计问题（对问题的感知），并指向不同层面的信息”是基本的要求。本页图展示的正是这个原则在剖面上的表达。一开始，墙体、楼板和屋面只是粗实线，而在1:20的图中则表达出了物理功能的分层，1:10的图纸则细致表达了材料及构件属性。

The understanding that a level of scale means also a level of problems (the perception of problems) which leads to various levels of information is basic to our teaching. This page shows how it applies to the section. While, at first, walls, floors and roofs are only thick lines, on the scale of 1:20, the walls consist of layers (of different functionality) that then take on material and product attributes in 1:10.

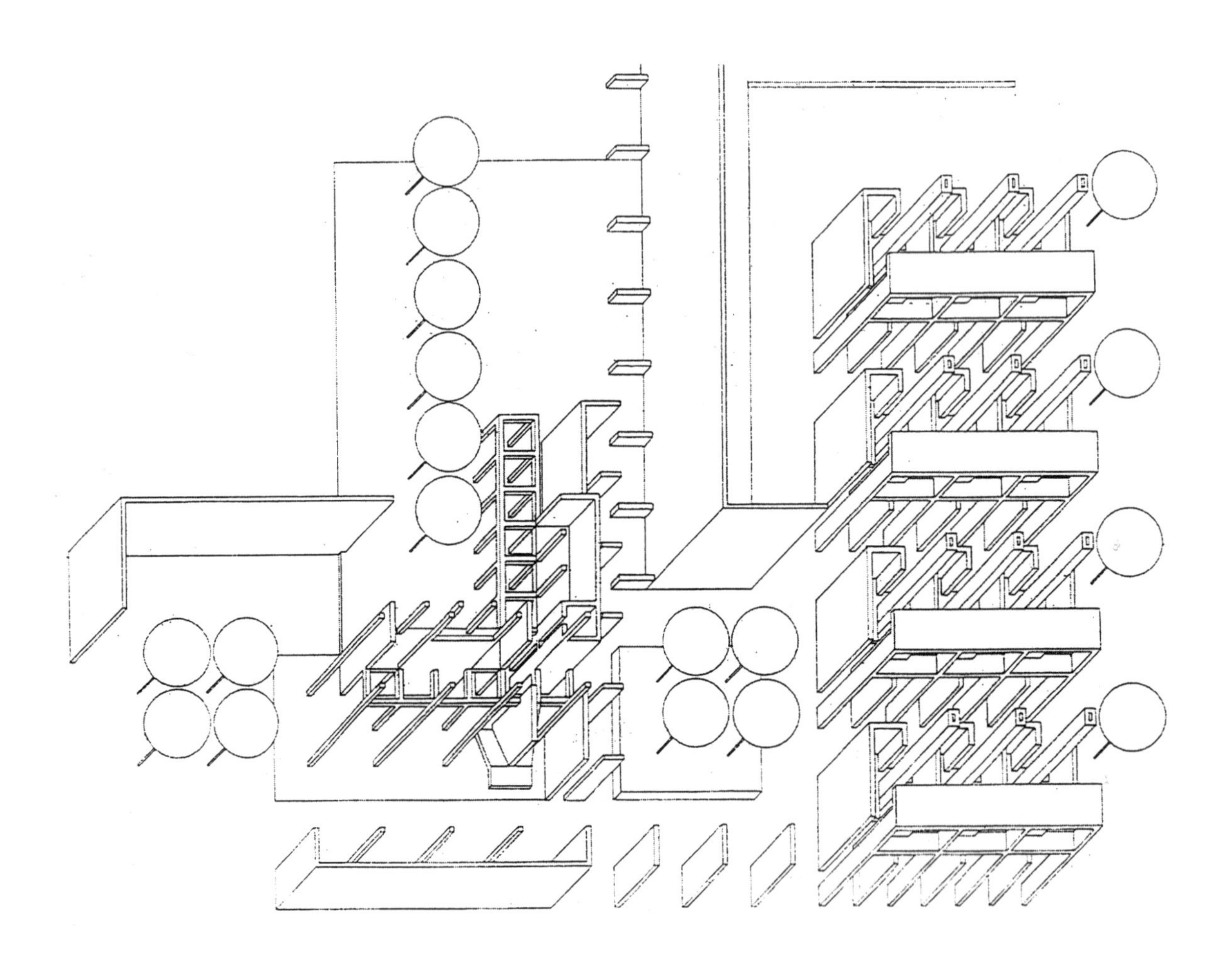

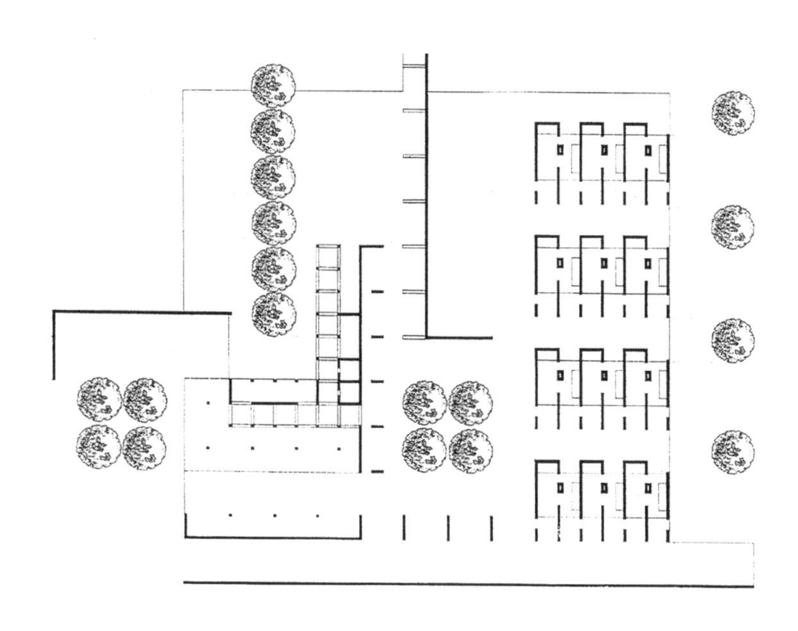

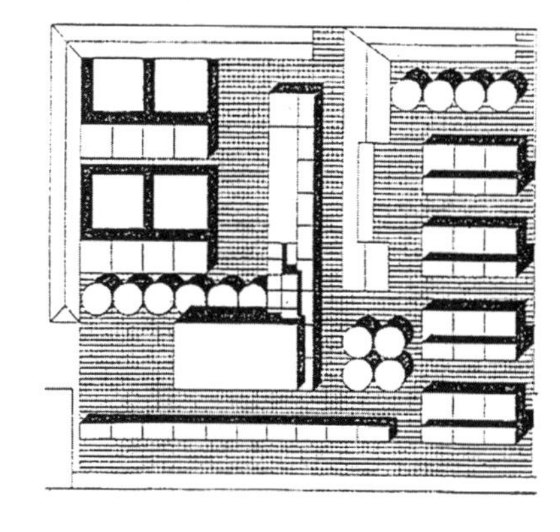

1 | 2 | 3

1 市场与住宅建筑群空间结构轴测图 2 空间结构图 3 总平面
1. Axonometric drawing of the spatial structure of the complex
2. Plan of the spatial structure 3. Site plan

尽可能在设计过程中进行关联与互动，第一阶和第二阶的成果在总平面中结合成整体空间结构图。

Interrelationships and interactions are attempted wherever possible. The results of Phase I and II are presented in a structural drawing.

1 工作室“纸圈”模型 2 工作室空间结构模型 3 功能示意图 4 空间划分 5 空间限定 6 工作室平面

1. The "paper tube" model of the studio-workshop 2. The space structure model of the studio-workshop 3. Function diagram 4. Space division 5. Space definition 6. Plan

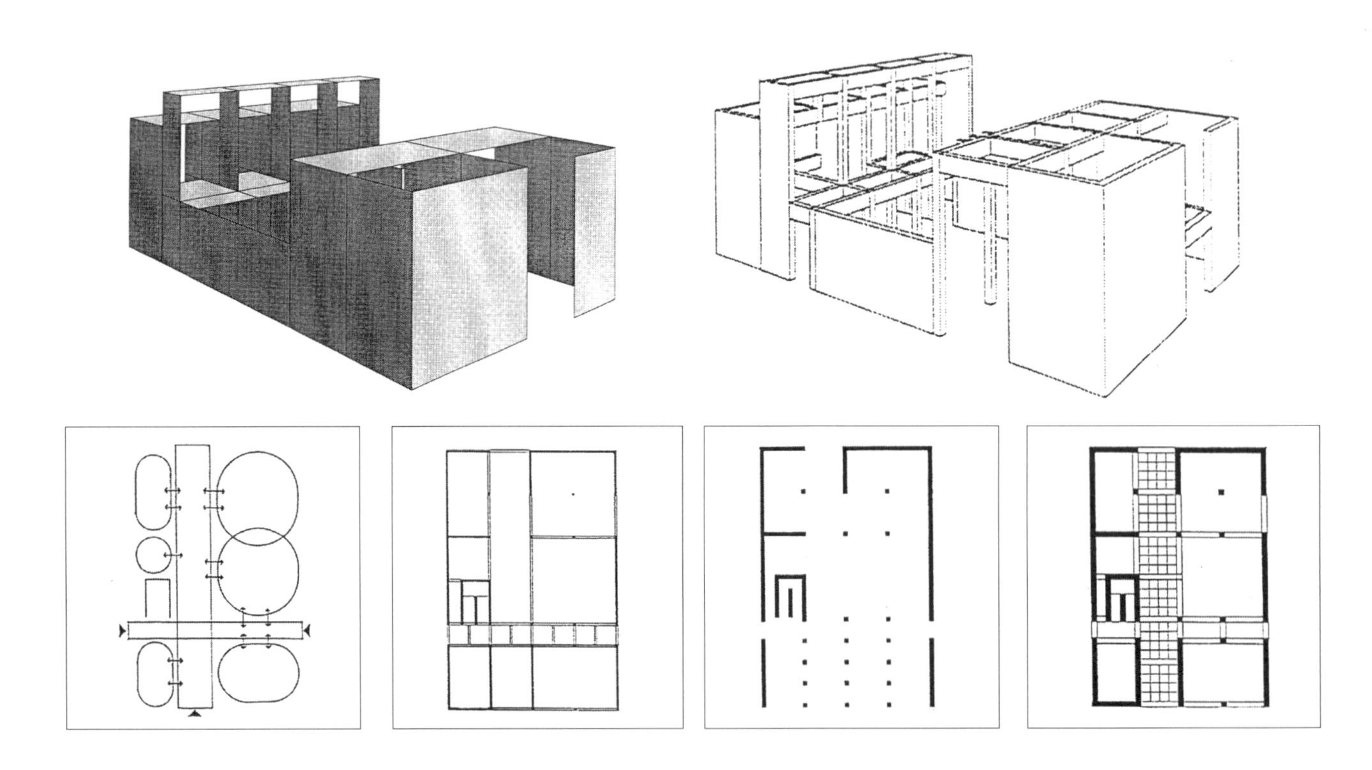

第三阶，工作室的使用需求多样化，不同净高及跨度的空间令设计问题更加复杂。

上图展示了单元的设计演变，从左侧空间元素的集合到右侧结构化的空间，但省略了过程中的不断试错。

In Phase III, the studio-workshops, with a diverse program requiring spaces of different height and span introduced a new set of questions and an increase of complexity. Here, it shows the transformation of the unit in steps, from the first aggregation on the left to structured space on the right. However, it does not show the trials and errors that had taken place in the process.

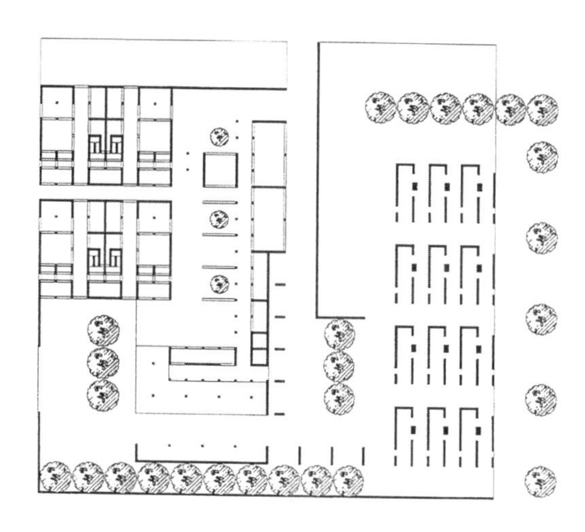

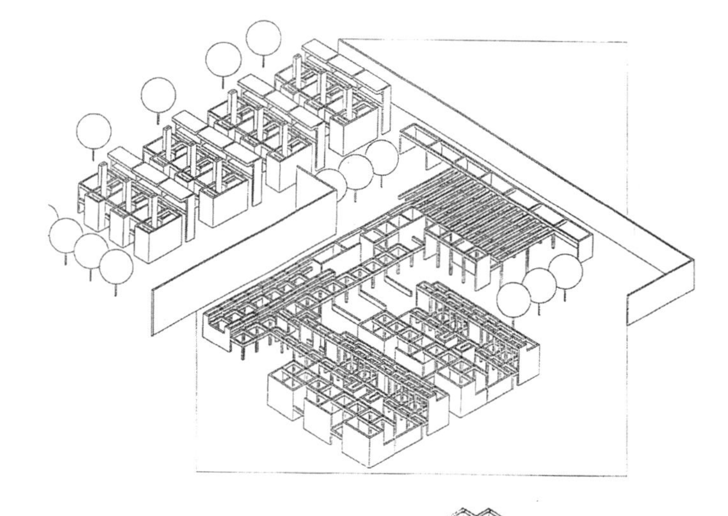

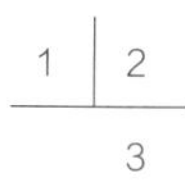

1 建筑群空间结构图 2 建筑群空间结构轴测图 3 工作室空间结构轴测图

1. Space structure plan of the complex 2. The axonometric drawing of the space structure of the complex 3. The axonometric drawing of space structure of the workshop

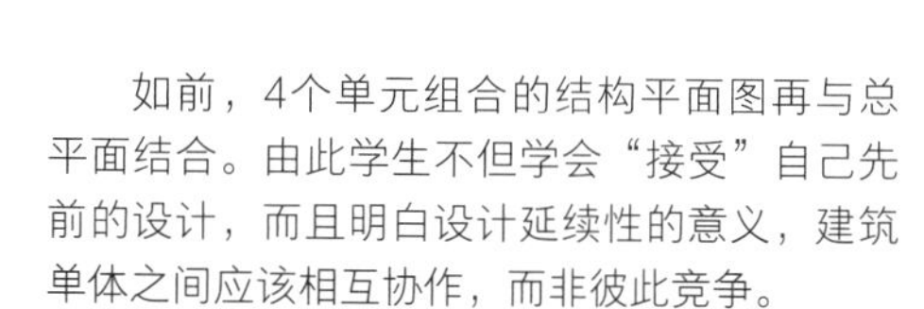

如前，4个单元组合的结构平面图再与总平面结合。由此学生不但学会“接受”自己先前的设计，而且明白设计延续性的意义，建筑单体之间应该相互协作，而非彼此竞争。

Again, the structural plan of the four units is related to the already existing environment (Phases I and II). Through it, the student not only learns to “live” with his/her (design) past but also understands the meaning of a design continuum in which individual objects cooperate instead of compete with each other.

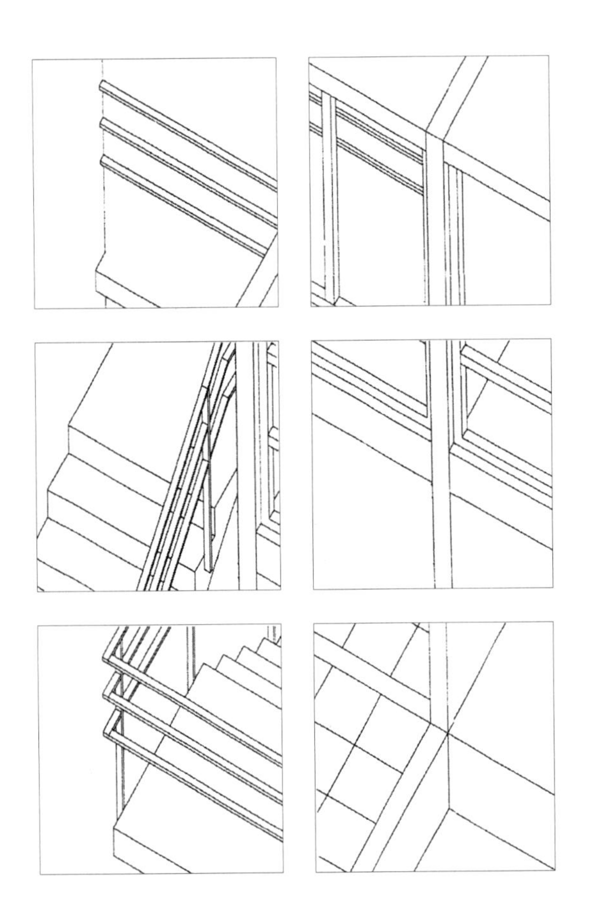

右侧是建筑群组一瞥，而左侧则放大了局部，呈现了建筑构件。我们不断强调3个尺度层级的交互关系，因此将构件作为建筑构造的组成部分，引入一个更小尺度的设计层级，或者说设计的范畴。

While the right side shows a view of the whole group of objects, the left emphasizes the notion of the component. Since the interaction of three levels of scale is seen as essential, the component, often seen as part of technology (construction) is introduced as yet another level of design, or design responsibility.

1 | 2 | 3 | 4

1 工作室细部轴测图 2 灰空间透视 3 单元立面剖面 4 工作室建筑立面

1. Axonometric drawings of details 2. Perspective of the in-between space 3. Elevation-section of workshop unit 4. Elevation of the workshop complex

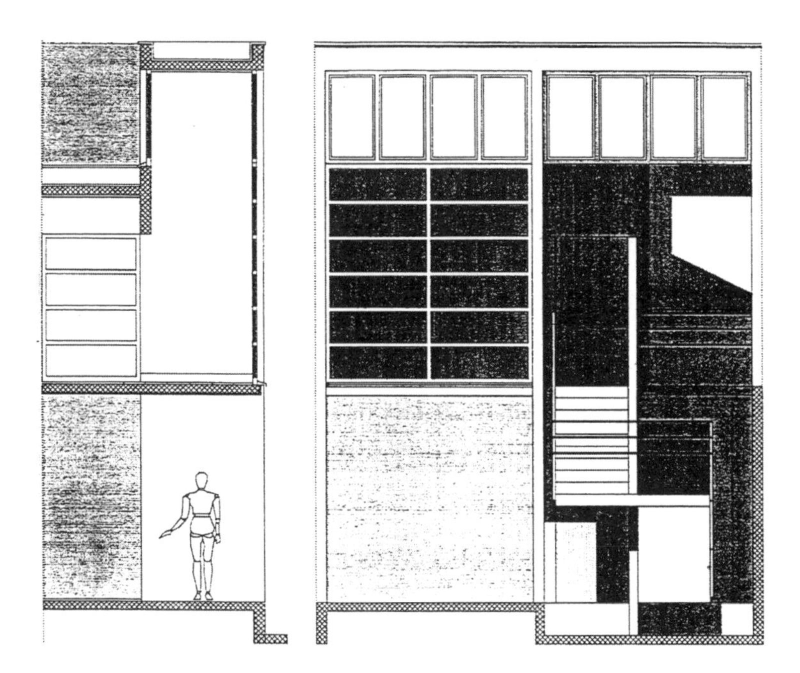

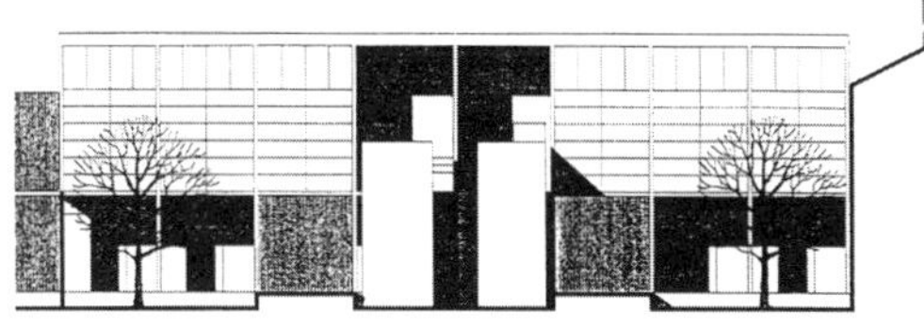

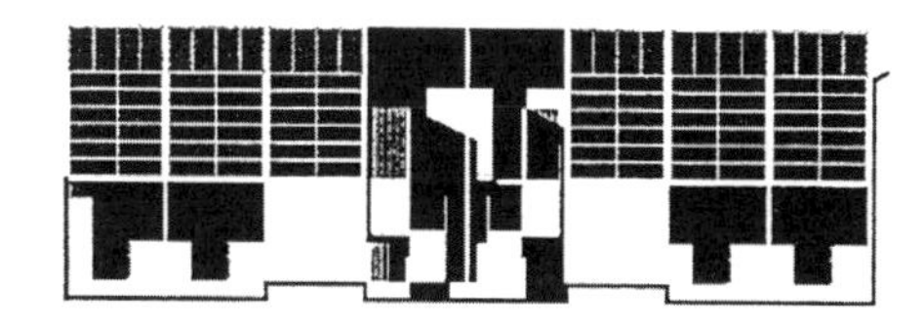

虽然单独绘制立面图容易使结果有图案化的倾向，但它仍然展现了设计的总体面貌，呈现一种氛围，并且确切地表明了形式关系及比例。

The drawing of an elevation in isolation is a questionable and often “graphitectural” affair. It can nevertheless show the overall spirit of the work, reflecting the atmosphere created, and certainly, inform about formal relations and proportions.

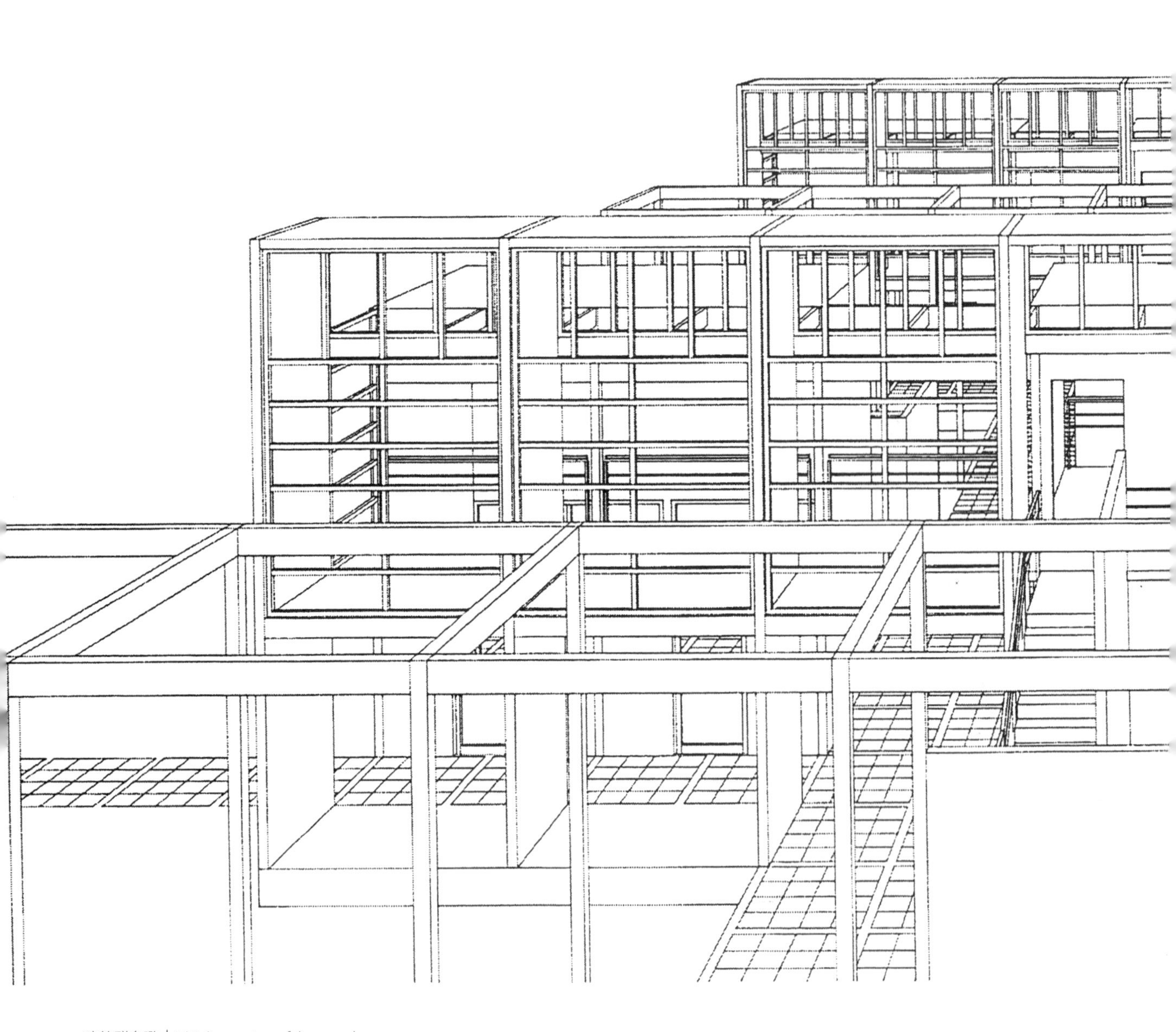

建筑群鸟瞰 | A bird's eye view of the complex

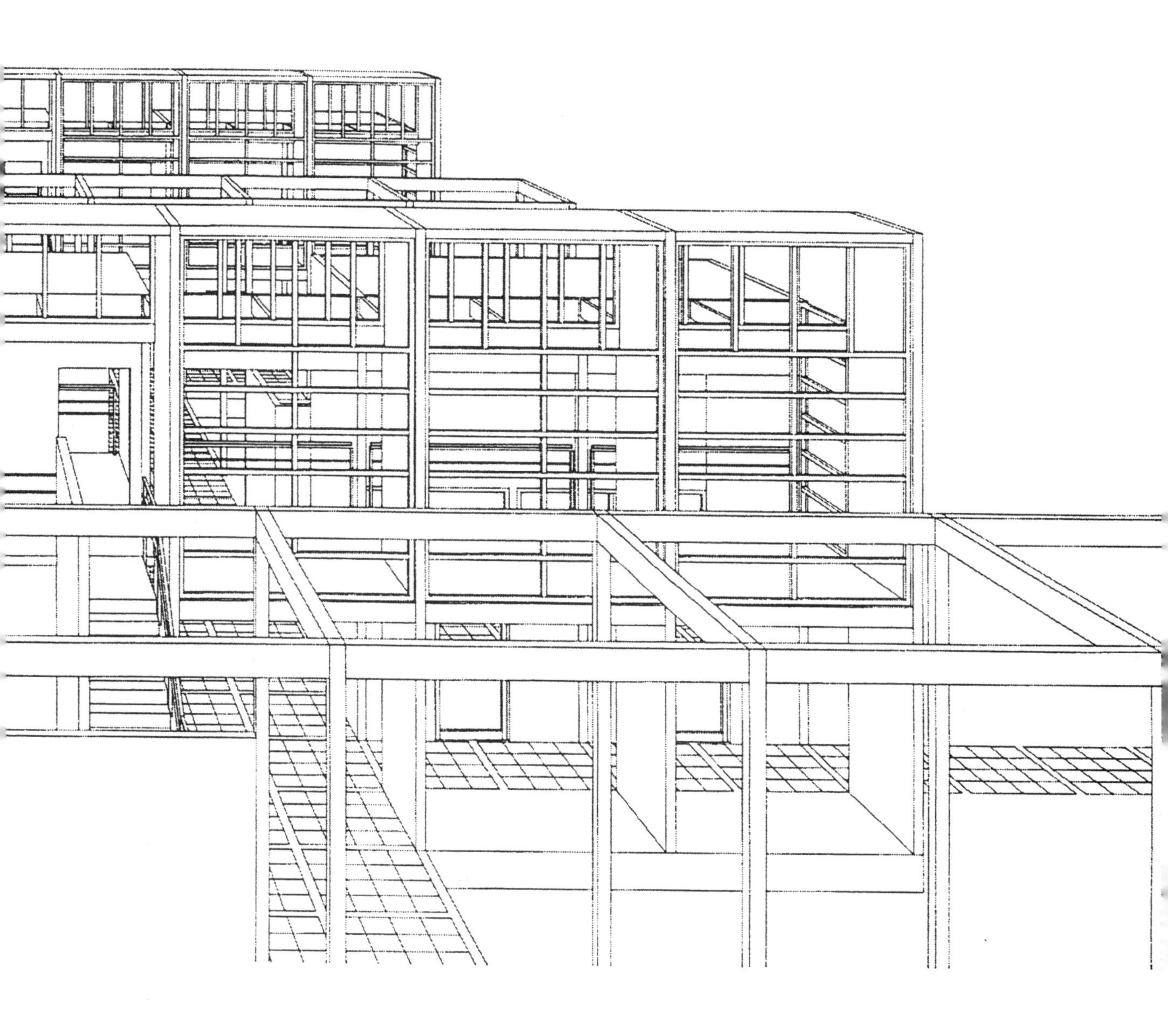

1998/99

旅馆，同里

Hotel Complex, Tongli

张彤 鲍莉
ZHANG Tong, BAO Li

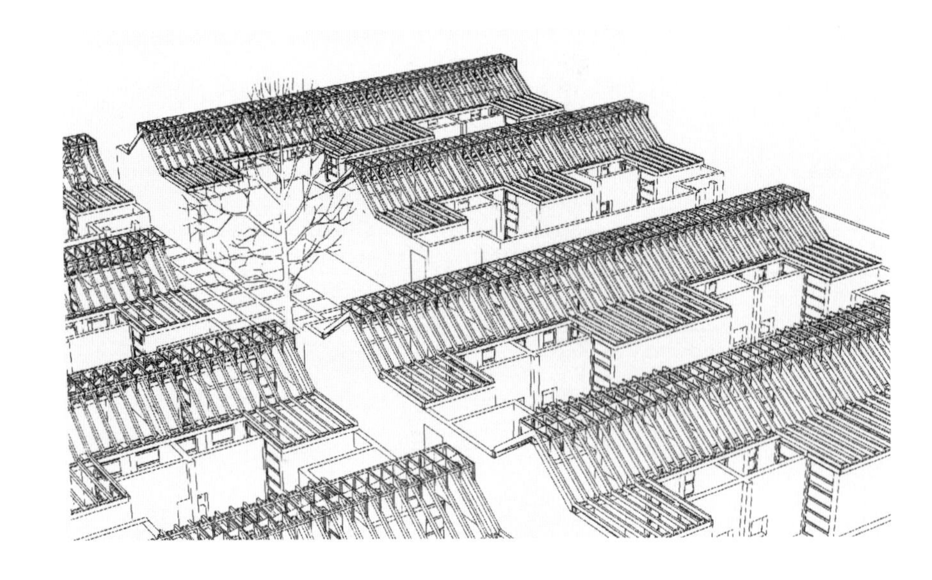

同里实验本质上是一个与设计方法学相关的实验。它是为建筑功能寻求形式及意义的尝试之一。支撑它的假设是参考环境的属性从本质上决定了建筑项目的概念生成。

此前的实验将撒比奥内塔作为参考环境。如果环境换在中国，又会如何？同里，江苏省南部的小镇被选作新的参照。正如撒比奥内塔，同里集中体现了这一地域的特点。它们作为两个不同地域的项目可以进行比较。

如前所述，同里实验（与所有其他实验，如撒比奥内塔）既然作为实验，就有成功或者失败的可能，但“失败或成功”本身应得到定义和解释，我们面临的是功能和意义的问题。

The Tongli experiment is in essence an experiment in design methodology. It is part of the experiment to give form and meaning to architectural function. Behind this attempt stands the hypothesis that the nature of the reference environment determines the concept of an architectural project substantially.

In past experiments, Sabbioneta had served as a reference environment. Yet the students of the Nanjing Exchange raised the question of the "Chinese". Because of it, Tongli, a small town in the south of Jiangsu province was selected. As in the case of Sabbioneta, Tongli represents the essence of a region. Again, as an experiment, the Tongli project can be compared with the Sabbioneta project.

As mentioned at the beginning, this Tongli experiment (like all other experiments, for instance Sabbioneta) is an experiment that can either succeed or fail. But "failure or success" should be defined and explained. We are dealing with questions of function and meaning.

1 设计场地区位图 2 用“纸圈”模型组织场地 3 住宅建筑组群的初步设想

1. Location map of the site 2. Site plan with the "paper tube" models 3. Preliminary ideas of the residential complex

1 | 2 | 3

场地选址在同里镇中心的南边，毗邻主要的运河，并与主要道路及镇的主入口相连。另一层考虑是原址上没有重要的建筑。

The site selected is a piece of land south of the centre of Tongli, next to one of the big canals and directly connected to a major road as well as the “entrance” to Tongli. Another favourable factor was the fact that no significant building occupied the land.

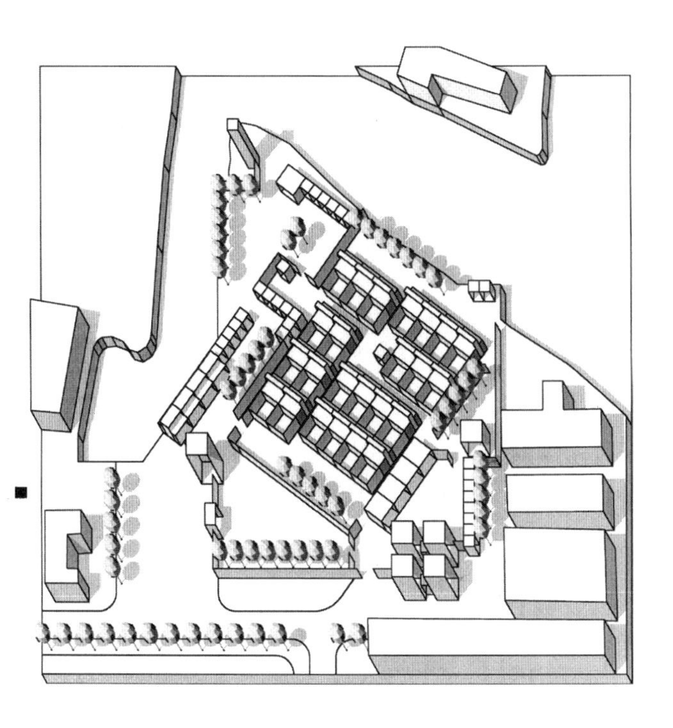

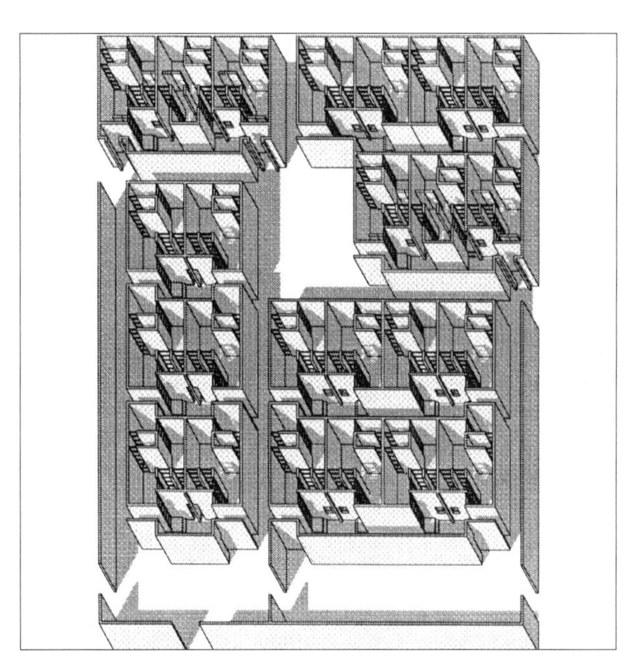

主题是一个能够容纳50至60人的旅店综合体（人数依据一辆旅游大巴的载客限额），以及容纳100人的餐厅。用简单的体块定义了若干不同使用功能的单元。首先用简易的卡纸模型组织场地，随后，将最初的总平面分解为不同功能的子项目，例如“住宅”，以便深入设计。

Since the overall theme of the project was a hotel-complex for 50-60 people (the bus-module) as well as a restaurant for 100 persons, a number of program units were defined in simple volumetric terms. The first attempt to organize them was formulated as a 3D-cardboard model. The decomposition of the site plan into functional units such as "housing" then enabled a more detailed investigation of each formal subset.

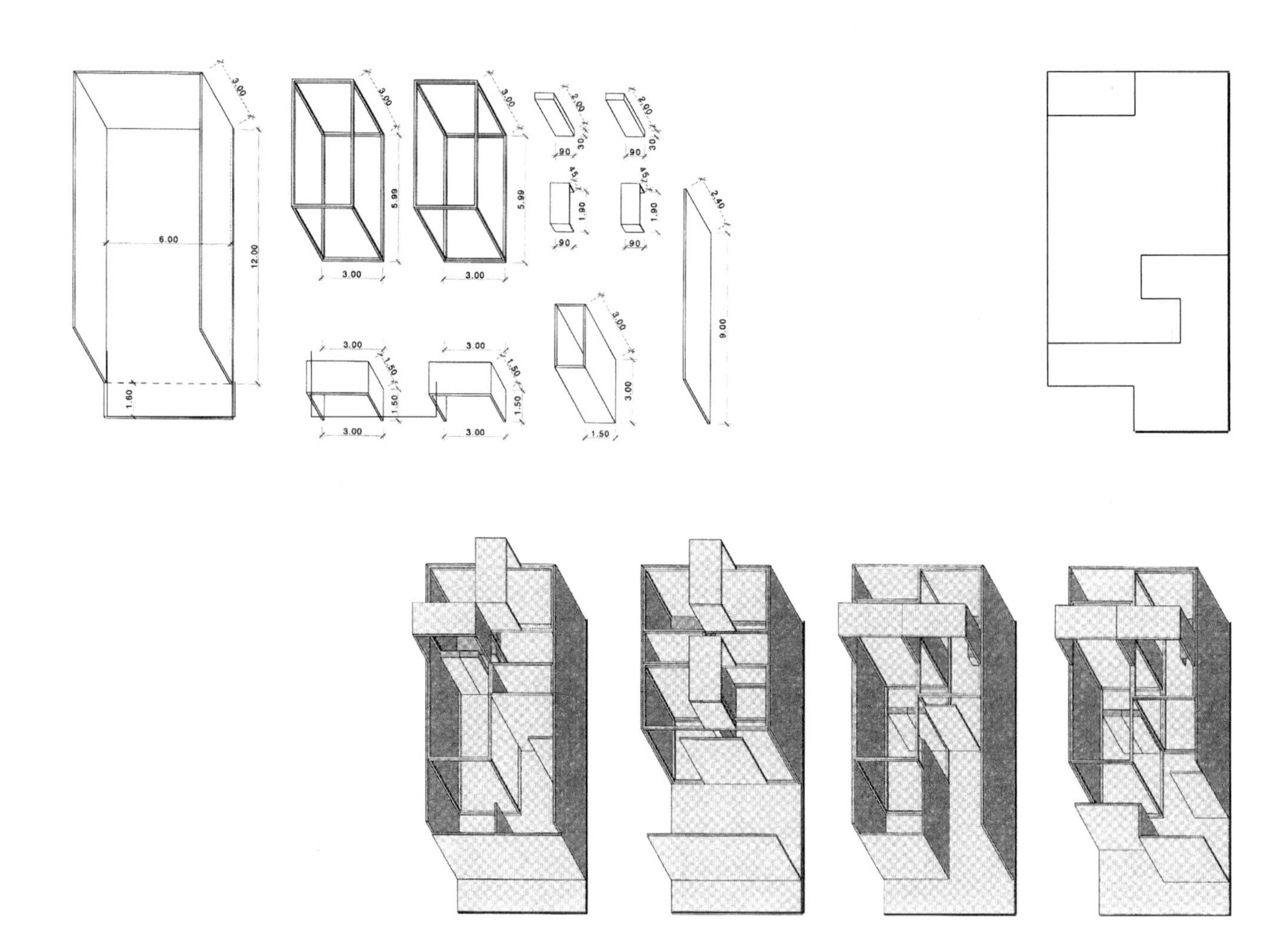

类似于总平面组织，居住单元设计的前期也提供了一些元素，诸如场地、墙体、天窗、盥洗室、附加墙体及一些家具单元。

本页展示了居住单元在“空间结构层级”上的4种变体。这种表达方式称为“结构模型”，与“体量模型”相对应。

Again, as in the development of the site, a number of elements have been given: the site, walls, skylights, a utility block, an additional wall and some furniture units. Shown below are 4 variations of a possible living unit on a “structural level”. The form of representation is called structural model, as compared to a volumetric model.

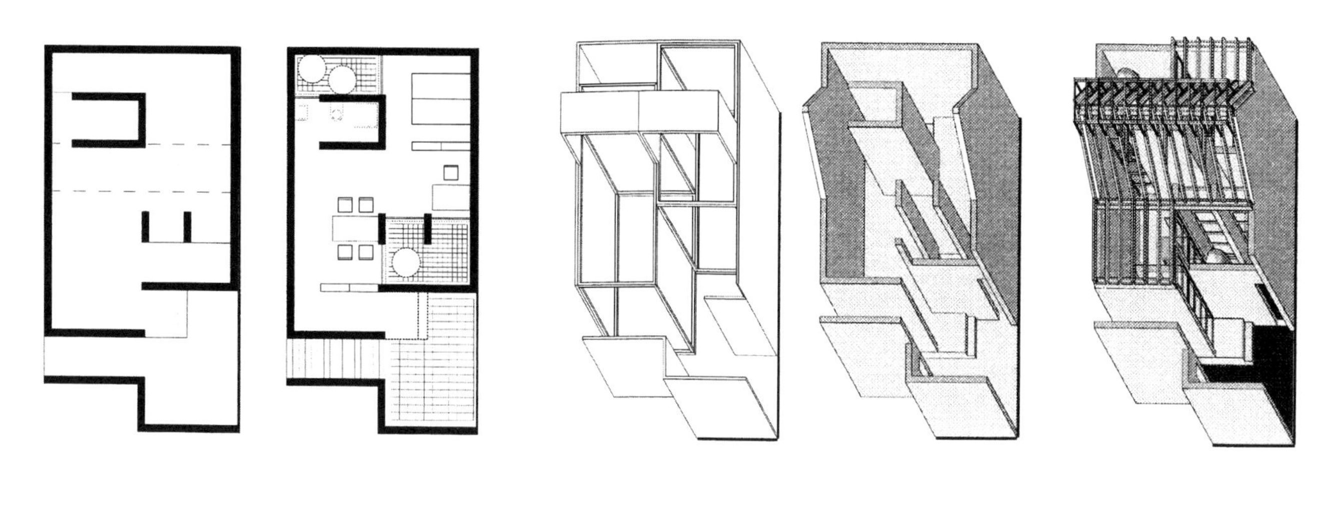

1	3	4
2	5	

1 给定的模块 2 模块的不同组合 3 平面的发展 4 模块组合的发展 5 室内透视图

1. Given modules 2. Compositions of the modules 3. Development of the plan 4. Development of the composition 5. Interior perspective

单元设计的推进。回应具体的场地，尝试从功能及意义层面回应“中国”问题，“屋顶”因而别具意义。

呈现了地面（地平或楼板）、垂直空间定义元素（墙体）、功能（家具）以及屋顶（作为水平的空间定义元素）。

Develop the unit. Since this exercise took place in Tongli, the “Chinese” question had been debated in terms of function as well as meaning. Because of this, the “roof” took on specific significance.

The ground (levels or floor), the vertical space-defining elements (the walls), the function (the furniture) and finally, the roof (as horizontal space-defining elements) are shown.

1 居住单元的组织 2 住宅建筑群局部鸟瞰 3 居住单元的镜像组合 4 形成内部庭院的 3 个单元组合 5 两对镜像单元的组合

1. Organization of the living units 2. A bird's eye view of part of the residential complex 3. A pair of mirrored living units 4. Three units grouping with courtyard 5. Grouping of two pairs of mirrored living units

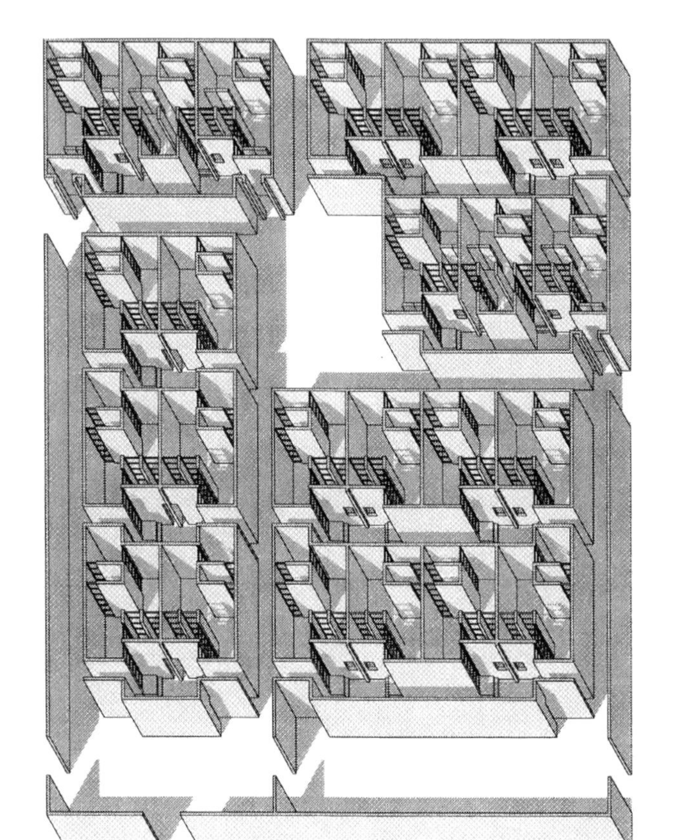

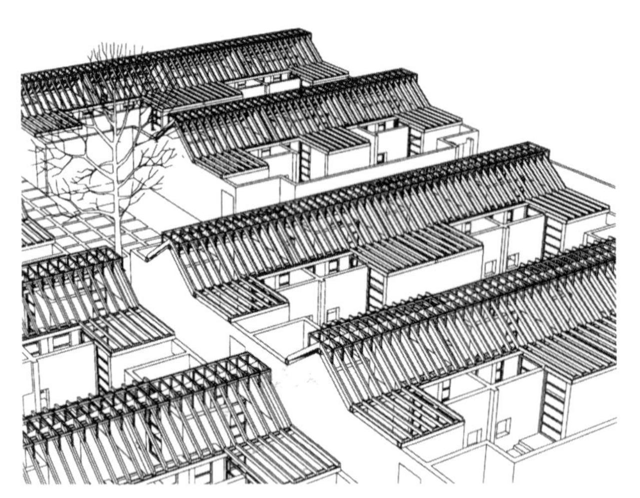

总平面显示居住单元以2至4个为组合。单元组合探讨了组织的原则。

对页是对多种组合方式的研究。单元与前院的不同组合方式带来了多样性。

The site plan shows that the living units are grouped in 2, 3 or 4 unit blocks. The grouping of units allows for the discussion of organizational principles.

The spread presents investigations that have been conducted and results that have been developed. The combination of the project space of the unit with a mediating space of the forecourt became rather important and added to the overall diversity of the solution.

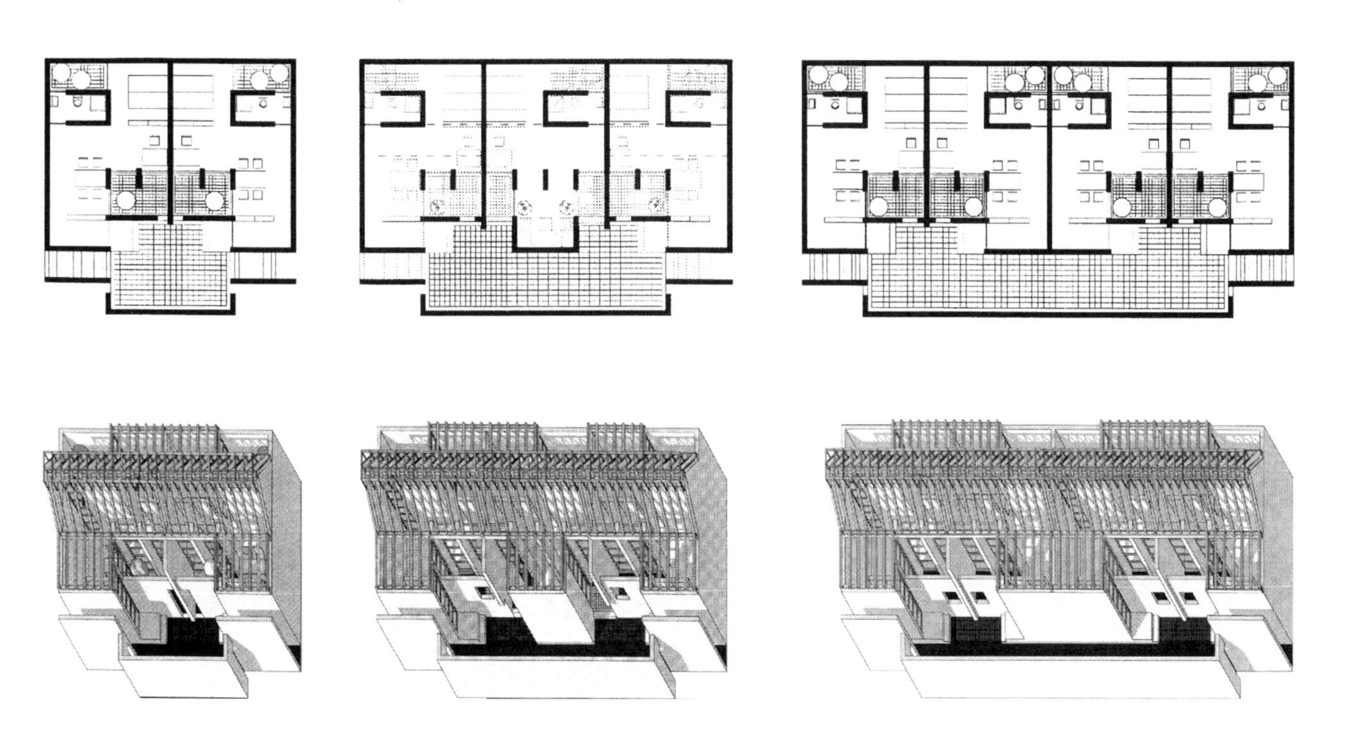

如图所示，4个单元的组合是由2个镜像单元的组合复制而来；虽然同样可以用4个相同的单元联排组合，但对称的方式显然更符合中国传统。组织的原则，屋顶及前院的关系直接受到作为参照环境的同里的影响。

It is quite clear that the unit of four is just the multiplication of the double unit using two symmetries. It would have been possible to design also a row of four. Yet it became quite clear that the use of the symmetry related more to the Chinese tradition. The organizational principle, the roof and the mediating forecourt thus related directly to Tongli as a reference environment.

1 | 2 | 3 | 4

1 回应场地的入口建筑组织 2 两种廊道布置呈现的空间转折关系 3 入口庭院透视 4 建筑入口透视

1. Organization of the BAUs for the "gate" building in site plan 2. Two manners of spatial transition by different corridor positions 3. Perspective of the entrance courtyard 4. Perspective of the entrance

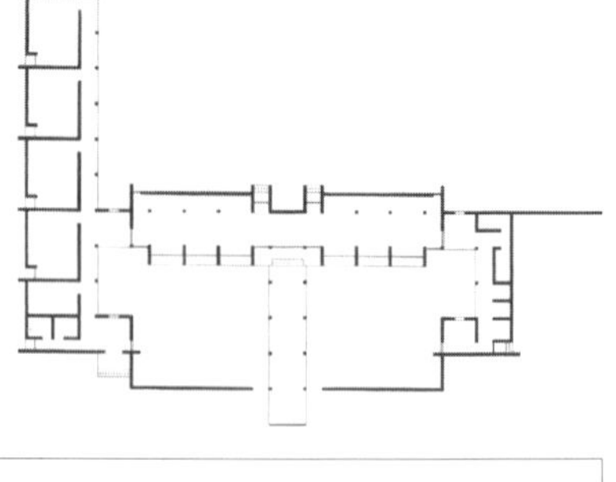

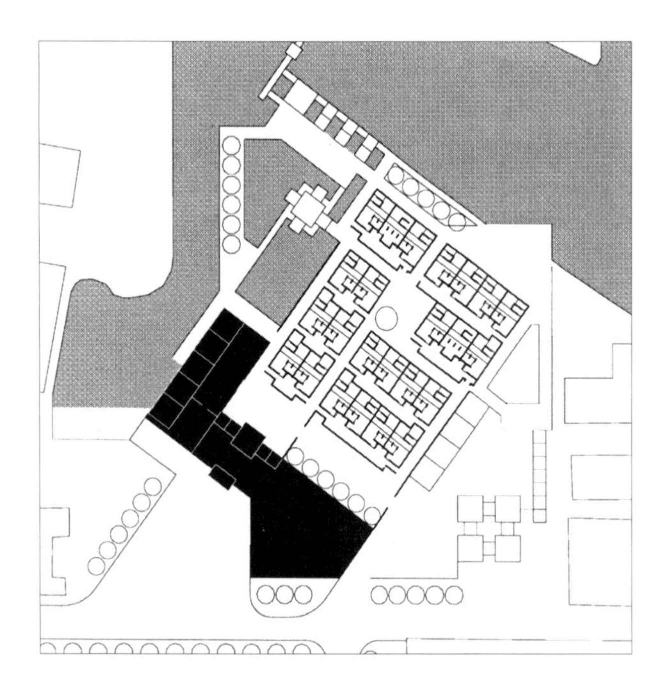

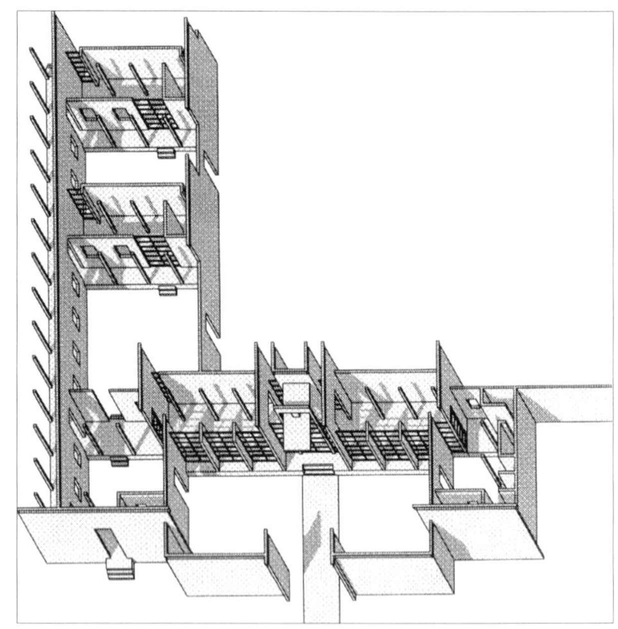

下一个子项目是包含游客接待处及行政中心的主入口。左图是子项目在总平面中的位置，右图展示了由空间定义元素所组织的功能与空间的两种方式。

Another subset of the overall plan involves the administration office, the reception (as a point of arrival) and the "gate". The left shows the subset in the overall plan; the right side shows two ways of organizing function and spaces through the arrangement of space-defining elements.

类似于居住单元的设计，考虑如何运用屋顶、连廊及墙面获得“中国特征”。图2下方的组织较优，因为在单元之间插入庭院使建筑群屋顶的起坡方向维持统一；将连廊移到临水的一面，与前院脱开，使后者更加完整独立，也凸显了连续墙体连接各个小庭院的角色。

Similar to the design of the living units, the roof, the courtyard, and the walls are introduced as important elements adding to the "Chinese nature". In this regard, the second arrangement in Figure 2 is better, because the directionality of the roofs is preserved with courtyards between units, and the forecourt is liberated with corridor unattached. The role of the walls as relating to the various courts is defined.

1 | 2 | 3

1 码头在场地中的位置 2 码头建筑轴测图 3 沿河面透视
1. Location of the harbour 2. Axonometric drawing
3. Perspective from the waterfront

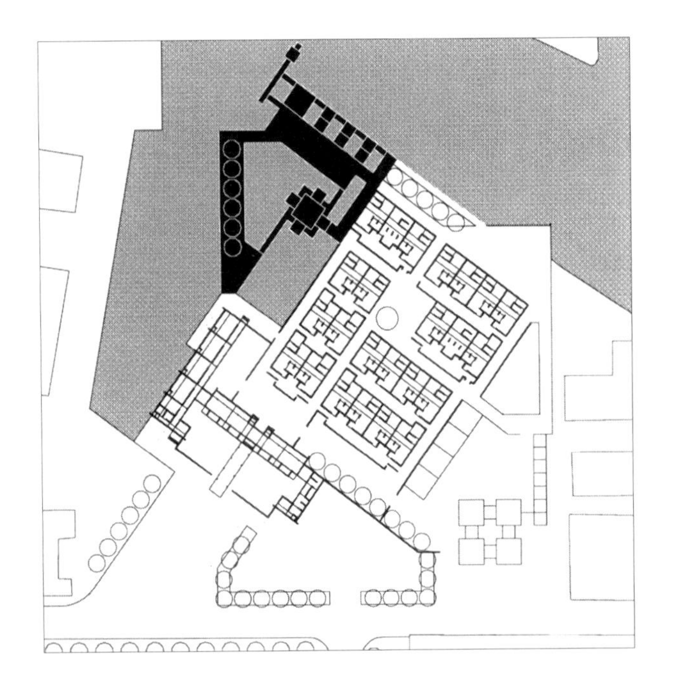

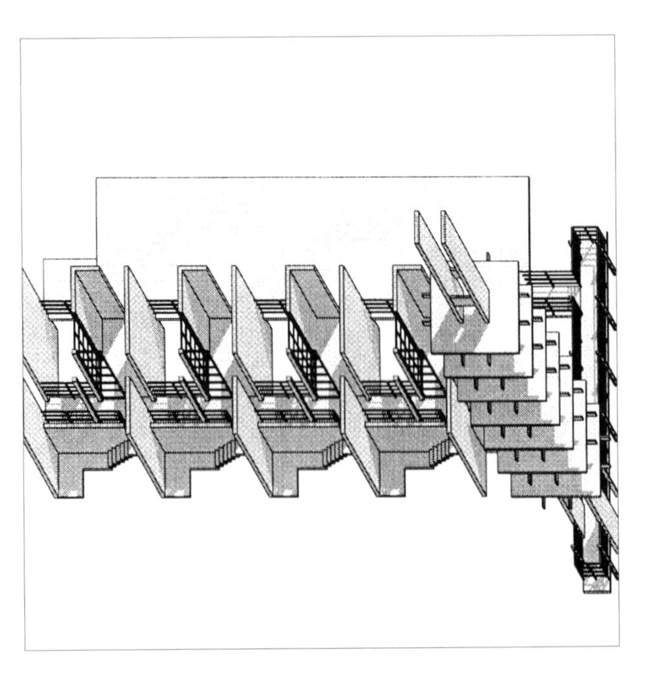

码头设计构成了另一个子项目。陆地与水体的交互是选址的特点，而码头强调了这一概念，它使旅馆直接与水路相连，可以方便地驾船进入运河。

The harbour constituted another subset in the overall plan. The interaction between land and water is one of the specific properties of the site selected. The "harbour" conceptualizes this relationship. Direct access to the water to boats and through them to the canals of Tongli and the connecting lake adds special qualities to the hotel under consideration.

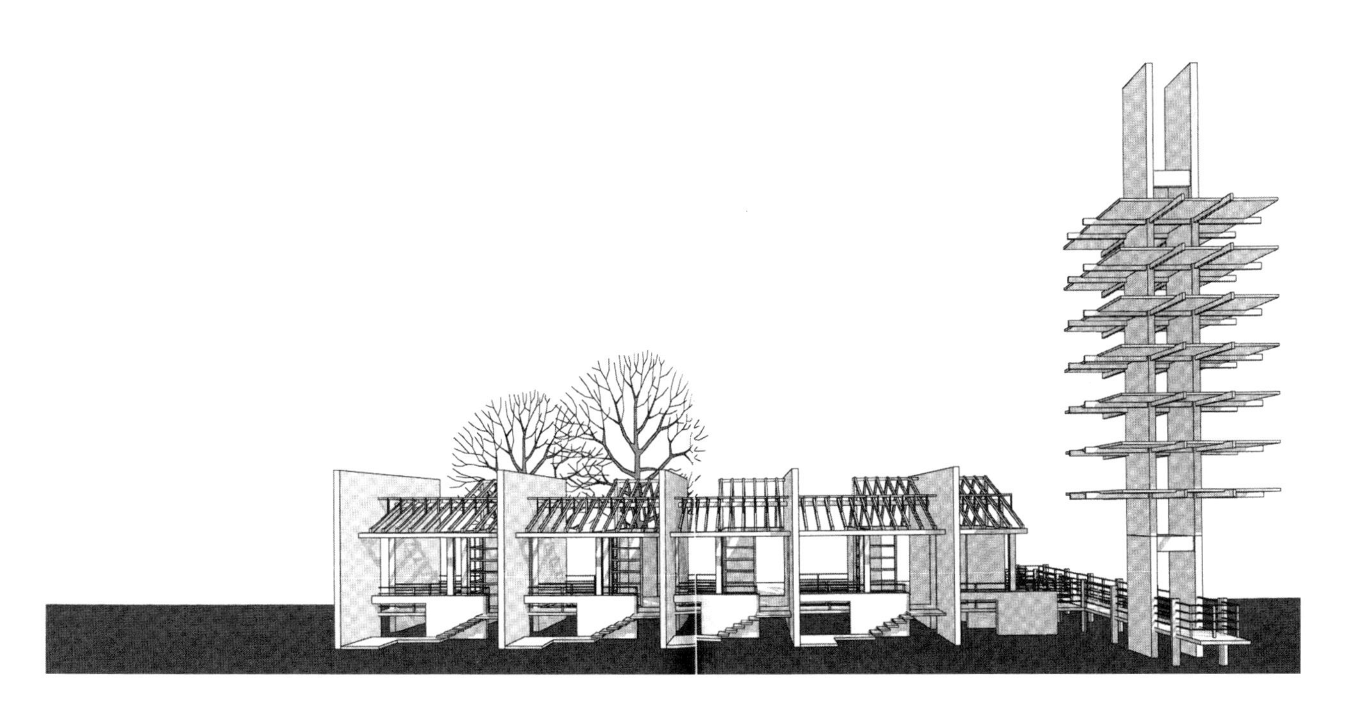

码头、河岸及塔相互关联，中国塔作为古典建筑形式的一种原型融入了旅馆综合体的设计。

The harbour, the waterfront and the tower as interrelated issues are presented. The image of the pagoda as an architectural prototype in classical Chinese architecture is adapted to the tower of the hotel complex.

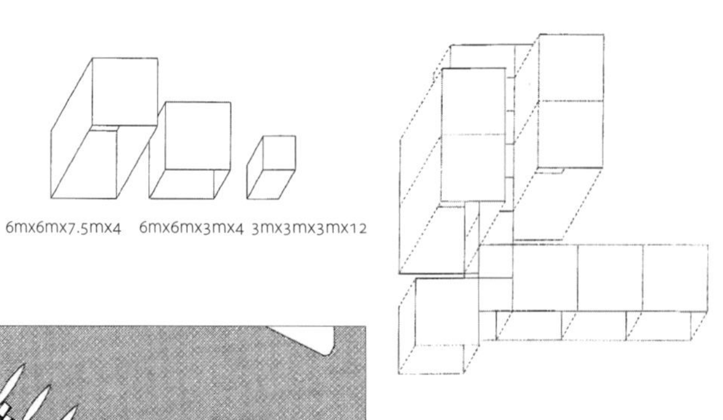

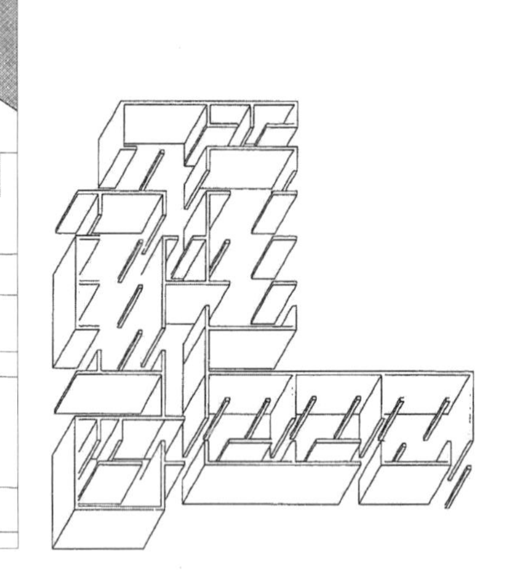

1 餐厅设计的 3 种基本建筑单元 2 依据使用需求的基本建筑单元组合 3 回应场地的基本建筑单元组合 4 基本建筑单元组合围合方式的具体化

1. 3 types of BAUs for the restaurant 2. Organization of BAUs according to utilisation 3. Organization of BAUs in the context of the site 4. Differentiation of space definition of the BAUs

1	2
3	4

餐馆设计前期给出了场地及一系列简单的要素。第一步必须阅读给定条件，通过体量的组织演化出“概念”，成为最初的构想。

As a point of departure, the “site” of the hotel and a very simple set of elements are given. The “restaurant” is the requirement.

As a first step, a reading of the given is necessary. The first hypothesis (the organization of volumes) evolves from the question of the “concept”.

1 空间组织 2 建筑体量 3 餐厅与居住单元的体量关系 4-6 结构系统的发展 7 结构系统与空间组织的结合

1. Spatial organization 2. building volume 3. The relation between the restaurant and living units 4-6. The development of the structural system 7. The combination of structural system and spatial organization

1	2	3	
4	5	6	7

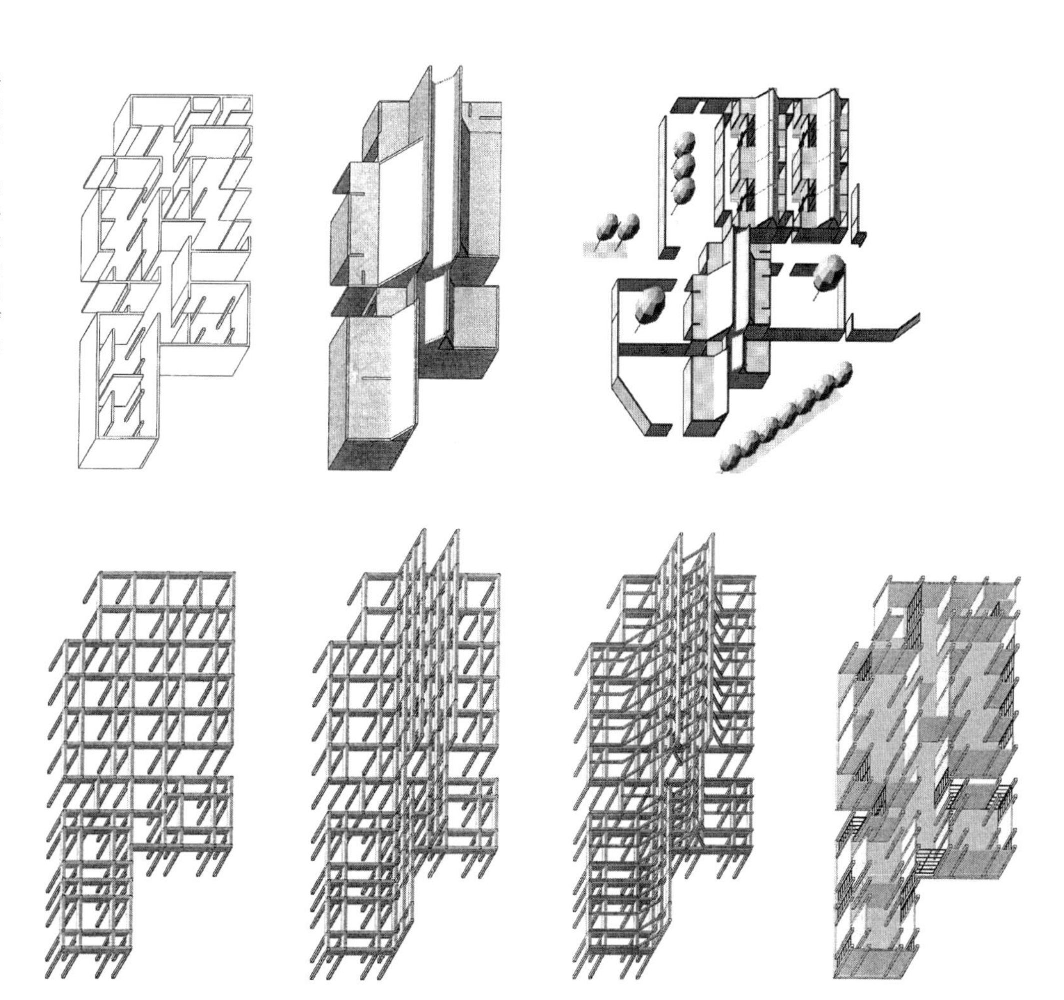

引入屋顶要素带来了多种参数，例如屋檐、屋脊的状态，既是问题也是设计的契机。同时，屋顶巧妙地连接了体量，并使建筑自然地融入到场所中。

基于空间/功能，建造/材料的相互作用进一步发展了结构系统。以概念化的方式模拟了建造过程，同时思考材料的概念化使用。

The introduction of the roof has given rise to various consequences, such as the condition of the edge and the ridge, posing problems as well as showing design potential. At the same time, the roof allows for an articulation of the volume and eases the integration of the building into the existing surroundings.

Based on the notion of space/function, construction/material as interactive concerns, the structural system has been developed. The process of construction has been investigated in conceptual terms. At the same time, the conceptual use of materials has been contemplated.

1 餐厅立面 2 餐厅与住宅体量关系轴测图 3 餐厅外部空间透视 4 餐厅内部空间透视 5 餐厅位置示意图 6 建筑群首层平面

1. Restaurant elevation 2. Axonometric drawing of the volumes of the restaurant and living units 3. Perspective of the exterior space 4. Perspectives of the interior space 5. Location of the restaurant in site plan 6. First floor plan of the whole complex

1		4	
2	3	5	6

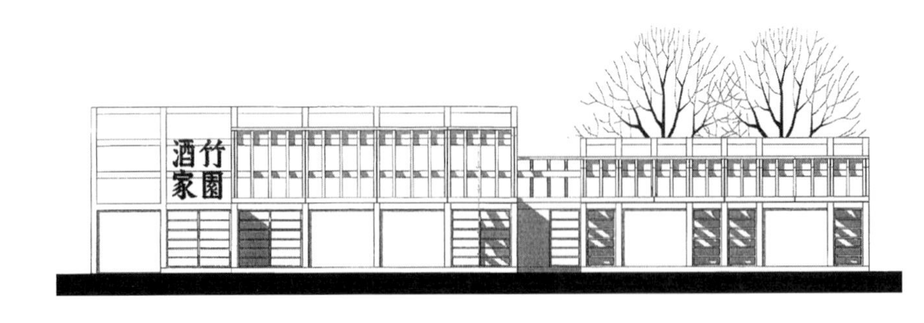

如前所述，在同里实验中建筑形式不仅是功能的，也是意义的载体。那么重要的是思考什么是中国的、地域的、适度的，因为这关系到结果的优劣。

希比尔·莫霍利–纳吉说传统建筑是灵感的源泉，但绝不应成为仿效的对象。我们希望这个系列的练习成为一个以中国建筑传统为灵感的当代建筑的探索。

As mentioned before, architectural form as the "carrier" of a function but also of meaning was the concern of the Tongli experiment. Consequently, questions of what constitutes the Chinese, the Regional and the Appropriate became important. At the end, the question of the "good" and the "bad" remains.

Sibyl Moholy-Nagy stated that traditional architecture can always serve as an inspiration but should never become the source for imitation. We hope this small sequence of exercises can be seen as an attempt to use the Chinese architectural tradition as an inspiration in the formulation of a building that aspires to be contemporary.

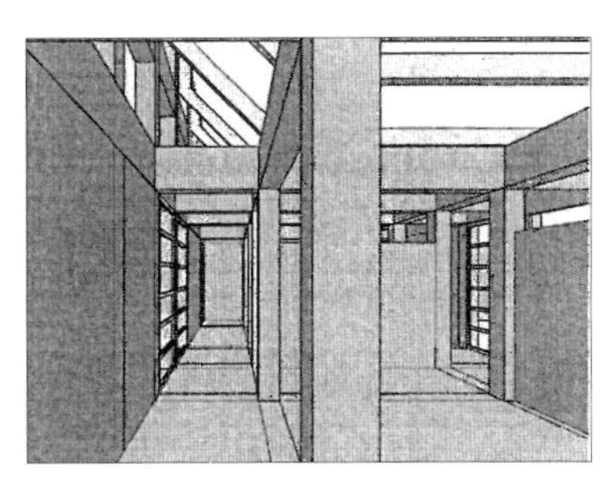

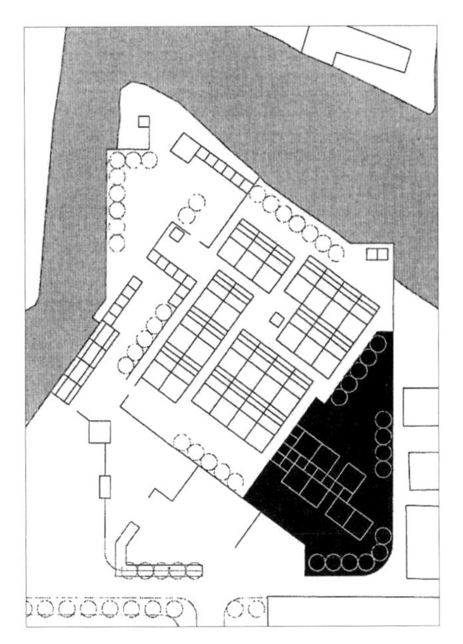

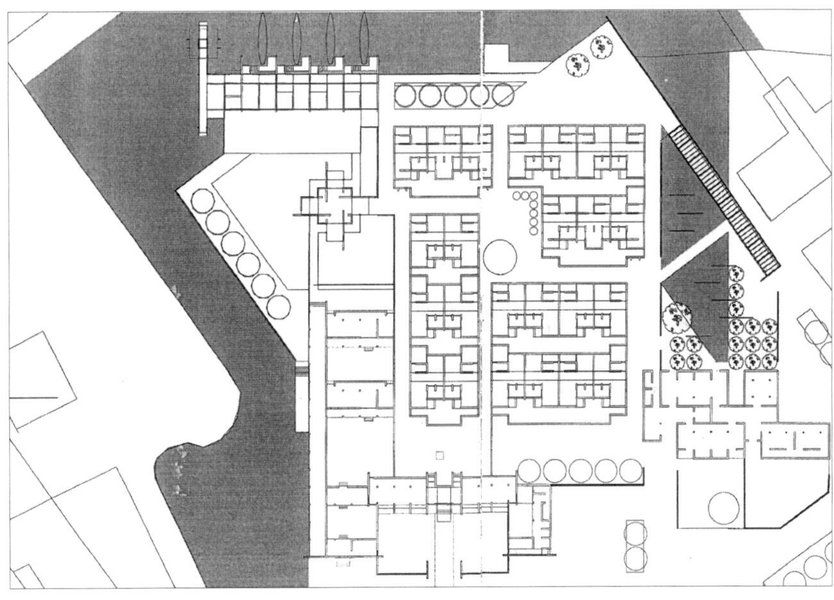

前面的设计已经提出了园林的设想。古典园林的要素经过诠释融入到场地中。

用墙、桥和步行道组织场地。因为步行道遵循了另一个网格走向，与既有的建筑网格产生了冲突及新的可能。首先想到的是为步道加上屋顶，方向的改变再次提出了问题。整体上看，设想稍显粗糙，主要的品质在于庭院的当代诠释。

In previous steps, a first hypothesis for the garden was developed. Essential elements of the classical garden have been interpreted and adapted to the site.
Various stages of development concern the "bridge", the walkway and the walls. Since the walkway follows a geometry of its own, a conflict or in turn, a potentiality was recognized. Adding a roof to the walkway seemed appropriate and possible at first glance. Again, the change of direction raised questions. Overall, the proposal is rather crude. Its main quality is the redefinition of the garden by contemporary means.

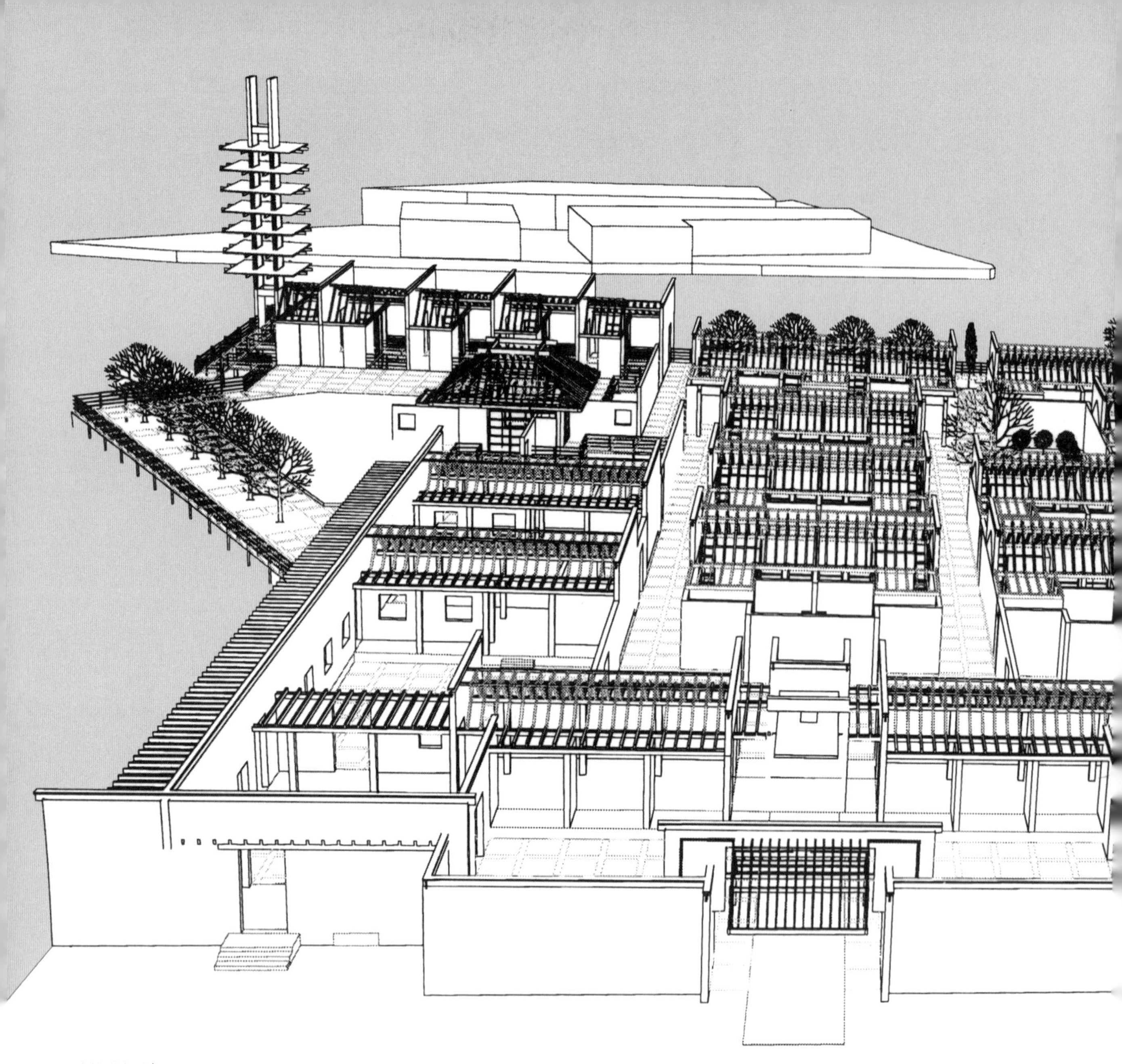

建筑群鸟瞰 | A bird's eye view of the complex

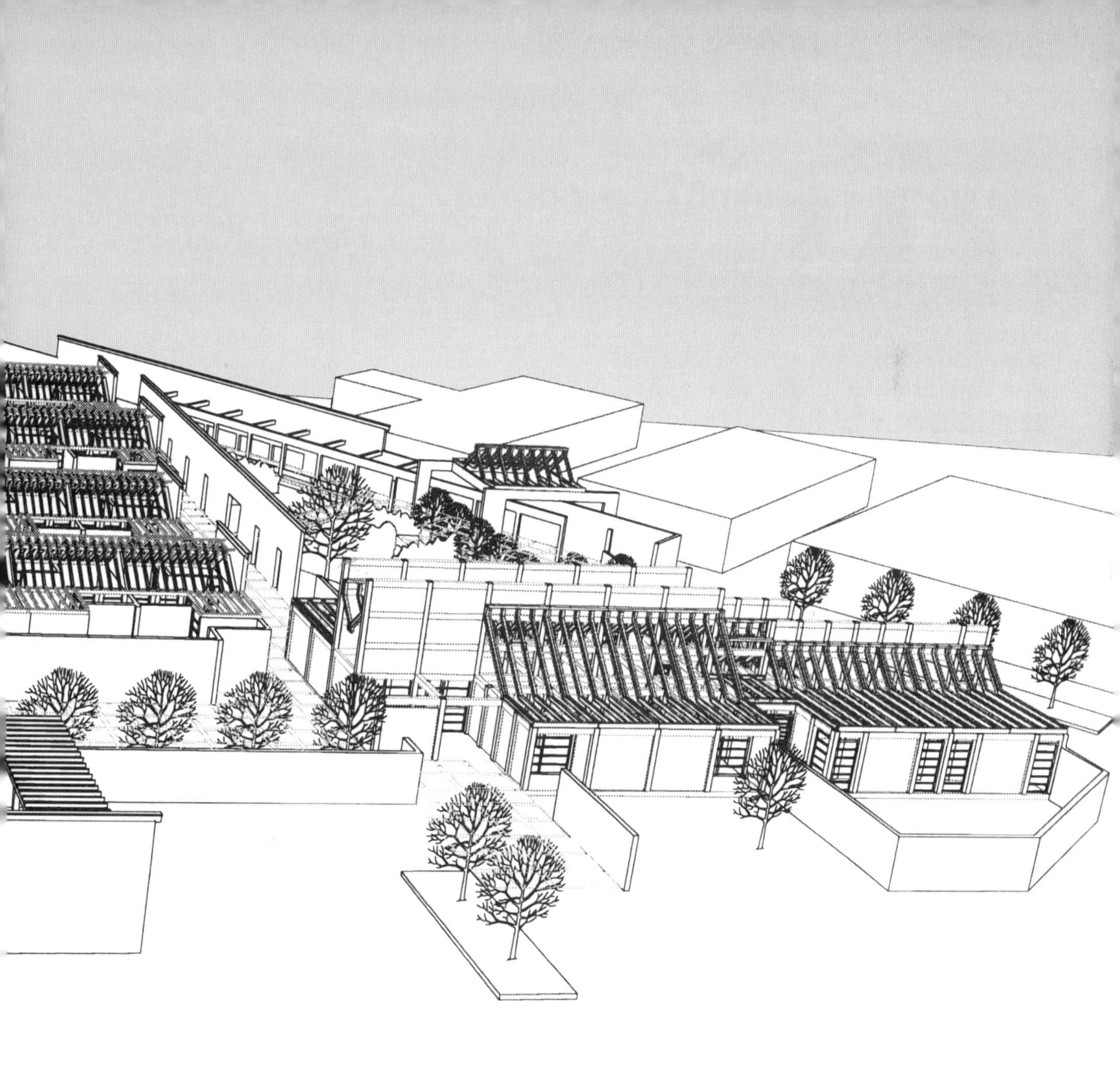

1999 年在 ETH 建筑系举办的《南京交流 1984-1999》展览，桌上特意放置的树枝寓意“立竿见影”。
The exhibition of "Nanjing Exchang 1984-1999" at D-Arch, 1999. The branch on the table stands for "Big stick, big shadow".

4 回响 | Resonance

1 南京交流 | Nanjing Exchange

2 对话 | Dialogues

单踊 | Shan Yong
顾大庆 | Gu Daqing
丁沃沃 | Ding Wowo
赵辰 | Zhao Chen
张雷 | Zhang Lei
龚恺 | Gong Kai
吉国华 | Ji Guohua
张彤 | Zhang Tong
鲍莉 | Bao Li

3 影响 | Impacts

东南大学 1990/91 一年级 | SEU Year 1, 1990/91
东南大学 1997/98 二年级 | SEU Year 2, 1997/98
东南大学 1999/2000 三年级 | SEU Year 3, 1999/2000
南京大学 2003/04，2004/05 硕一 | NJU MArch 1, 2003/04, 2004/05
香港中文大学 1996/97 一年级 | The CUHK Year 1, 1996/97
香港中文大学 2016/17 一年级 | The CUHK Year 1, 2016/17

4 解读 | Interpretation

1 南京交流 | Nanjing Exchange

瑞士的中立国角色使其成为最早承认并与中国建交的西方国家之一，瑞士民间组织则更活跃，很早就开始致力于对中国文化的了解及学术交流。中国自1978年开始向瑞士派出留学人员，但批次和人数都很少。直至20世纪80年代初，随着中国对外开放程度的加深，大规模的校际学术交流活动才开始启动。

尽管在1980年代初，南工的学生和老师已经通过一些交流项目开始了解瑞士的建筑设计和苏黎世联邦理工学院。但是南工与ETH建筑系之间正式建立联系，则可以追溯至1983年年底，时任建筑系副系主任的黄伟康教授作为南工代表团成员赴瑞考察，并在建筑系主任多夫·斯奈布利教授和海因斯·罗纳等教授陪同下参观洪格堡校区的建筑学院，对这个有着逾百年历史的学系有了最初印象。

1985年春，《苏黎世联邦理工学院和南京工学院交流协议书》签署。两个建筑学系之间的交流协议则于1985年年底由两位系主任——鲍家声和本尼迪克特·胡伯正式签订。以英语为工作语言。受当时国家政策对高校在读生的出国限制，原定双方各派送两名学生，改为南京方面派送两名年轻教师去ETH一个学年（10个月），而瑞方将派往南京的学生改为4名，仍为一个学期，以示对等。

1986年10月，南工的两位青年教师单踊和郑炘到达ETH，瑞士派出的4名学生也于次年3月抵宁。斯奈布利接待了单和郑。此前，交流协议并未明确由哪位教授担任项目导师。正当二人踌躇之际，克莱默派助手邀请他们加入自己的“基础设计”教学团队。克莱默流利的英语显然更便于交流，而且一年级设计课程很适合作为了解一个

Agreement

between

The Faculty of Architecture of the
Swiss Federal Institute of Technology, Zurich, Switzerland
and
School of Architecture of the
Southeast University, Nanjing, P.R. China

on the Exchange Program for Professional Degree Students
between the two schools

Preamble

It is the intention of both schools of architecture within their respective universities to establish a program of exchange.

This program is not intended to be a short range endeavour. It is built with the future of the next generation in mind. The intention is to promote an exchange of methodologies, of knowledge, of know-how, and of ideas through the exchange of persons that will endure and develop.

The next phase of the exchange program should again cover a three year period inclusive of the academic years of 2004/05, 2005/06 and 2006/07. As in the past at the end of this period its continued validity will be determined and the program be renewed or ended based upon the decisions of both partners three months before the agreement expires.

Agreement

1. This agreement covers a three year period including the academic years 2004-2007.
2. The program will allow the exchange of maximum two persons annually. Each school will select its own candidates.
3. Working knowledge of the English language is assumed.
4. The time and duration of the visits are agreed between both schools. The duration is one semester for two candidates or two semesters for one candidate.
5. Each school will appoint a counselor for the program. From 2004 until 2007 the persons in charge of the program will be Prof. Dr. DONG Wei at the SEU and Prof. Dr. Bruno Keller at the ETH Zurich.
6. It is assumed that fees as well as traveling expenses are taken care of by the participants in their home university.
7. All participants will benefit from insurance and medical care of the university in their host country.
8. The ETH Zurich will grant each Chinese guest a monthly scholarship of Swiss francs 1'650. — (including CHF 150. —for Health Insurance) in order to cover the living expenses. Lodging will be provided. In addition the costs for the two Seminar Weeks will be paid by the ETH.

   The SEU will likewise grant a monthly scholarship of RMB 450. --Yuan in order to cover the living expenses. Lodging will be provided. The costs of the official travel period at the end of the program will be paid by SEU.
9. Each candidate will hand in his application at least three months prior to the planned departure.
10. Open questions, program proposals and changes will be worked out between the schools of architecture of both universities.

Swiss Federal Institute of Technology

Prof. Dietmar Eberle
Dean, Faculty of Architecture
Zurich, 14 May 2004

Southeast University

Prof. Dr. Wang Jian Guo
Dean, School of Architecture
Nanjing, 07 September, 2004

2004年续签的“南京交流”《协议》| The renewal of the Nanjing Exchange Agreement in 2004

The neutral stance of Switzerland made it one of the first western countries to recognise and to establish diplomatic relations with the newly established People's Republic of China. Yet, the unofficial connection between the two countries preceded the diplomatic ones. Although China had started to send out scholars to study abroad since 1978, it was not until the early eighties that large-scale academic exchange activities resumed.

Students and teachers in Nanjing got to know Swiss architecture and ETH through several instances in the early 1980s. The first official contact between the two departments of architecture, NIT and ETH, was realised by the visit of Prof. Huang Weikang, the deputy head of the department of architecture at NIT, to ETH at the end of 1983. As a member of the NIT delegation, Huang also got a chance to have a first impression of this over-hundred-year-old department when the department head Prof. Dolf Schnebli and Prof. Heinz Ronner guided him around the Hönggerberg campus.

In the spring of 1985, the two universities signed the Agreement of Exchange. In the fall of the same year, Department heads Prof. Bao Jiasheng and Prof. Benedik Huber signed the agreement between the two departments. It enabled 4 students from D-Arch to study six months at NIT in the spring, and two teachers from the NIT side to spend one academic year (10 months) with an international student visa at D-Arch in Zurich, every year. Here, teachers were sent instead of students from the Chinese side because at that time, university students were not allowed long-term leave during their study period by the Chinese government. English was decided as the working language.

In September 1986, the first two teachers, Shan Yong and Zheng Xin, arrived at ETH. Four ETH D-Arch students arrived in Nanjing the next March. Prof. Schnebli received Shan and Zheng. However, the Agreement did not specify which professor to take responsibility for them. At that moment, Prof. Herbert Kramel asked his assistant to invite both of them to join the Basic Design course group. Shan immediately agreed upon the suggestion. He, as a member of the teaching reform team of the first-year program in Nanjing, realised that this was exactly what he would like to study there; besides,

克莱默与仲德崑（左一），鲍家声（右一）在东南大学，2011 | Herbert Kramel with Zhong Dekun (left) and Bao Jiasheng (Right) at Southeast University, 2011

教学体系的切入点，作为南京教改小组成员的单踊当即接收了邀请。

在此之前，克莱默虽从未到过中国，却早已与中国结下不解之缘。1980年，克莱默客座于哈佛大学研究生院，在当时盛行的后现代运动及对历史形式的热烈讨论中，其重视设计过程及结构构造、运用模块化构件的教学吸引了当时修读其课程的美籍华人肯尼斯·高，后者很快成为他在ETH的助教和博士生。1985年，克莱默作为哥伦比亚大学设计研究生院客座教授时，又认识了在此留学的南工建筑系毕业生温益进，后者亦追随克莱默到ETH成为其助教并修读博士学位，以安徽民居为对象研究文化与环境因素之于无名建筑的影响。而温益进，正是单踊的大学同学，克莱默正是让他向两位初到ETH的年轻教师发出邀请的。

克莱默的一年级基础课程对于单踊来说是一种全新的体验，他通过书信向南京教改小组的其他成员汇报他的见闻。顾大庆等青年教师发现克莱默的教学正是他们所寻求的、有别于南工传统教学方法及当时盛行的包豪斯构成练习、围绕空间与建造等建筑学核心问题展开的设计基础教学法。此后至2001年，先后来到ETH的大多数东南大学教师，如顾大庆、丁沃沃、丁宏伟、赵辰、张雷、龚恺、应兆金、冷嘉伟、冯金龙、吉国华、朱竞翔、陈薇、张彤、鲍莉、张宏、胡滨、李雱等，都相继学习工作于克莱默教席。

顾大庆作为“南京交流”的第二批教师于1987年秋来到ETH。从研究生时期开始，顾大庆便对设计方法及教学产生了浓厚的兴趣。延续单踊与郑炘的模式，顾大庆

王建国（左）与克莱默于苏黎世，2005 | Wang Jianguo (left) and Herbert Kramel in Zurich, 2005

Kramel spoke fluent English.

Before that, Kramel had never been to China but had already forged a bond with China. In 1980, Kramel was a guest professor at Harvard Graduate School. Amid the prevailing postmodern movement and the heated discussion of historical forms, Kramel's course stressed design process and structure, and the use of modular components. This attracted one of the graduate students, the Chinese-American Kenneth Martin Kao. The latter soon became Kramel's teaching assistant and PhD student at ETH. In 1985, as a visiting professor at the Graduate School of Design at Columbia University, Kramel met Wen Yijin, a graduate of NIT. The latter also followed Kramel's return to ETH as his teaching assistant and PhD student. Wen studied the impact of cultural and environmental factors on anonymous buildings in Anhui traditional housing. Wen had been a college classmate of Shan. It was no surprise, therefore, that Kramel asked him to send the invitation to the young teachers from Nanjing.

Kramel's first-year course was a totally new experience for Shan Yong. He wrote a long letter to his Nanjing colleagues about what he had witnessed at ETH. Before long, the NIT pedagogic group were convinced that Kramel's Basic Design course, which differed from the conventional teaching at NIT and the Bauhaus composition exercises due to its emphasis on space and construction, was the introductory teaching mode they had been looking for. Until 2001, most of the Nanjing exchange teachers had come and worked in Kramel's teaching chair, such as Gu Daqing, Ding Wowo, Ding Hongwei, Zhao Chen, Zhang Lei, Gong Kai, Ying Zhaojin, Leng Jiawei, Feng Jinlong, Ji Guohua, Zhu Jingxiang, Chen Wei, Zhang Tong, Bao Li, Zhang Hong, Hu Bin, Li Pang, etc.

Gu Daqing came to ETH in the fall of 1987 as the second batch of Nanjing Exchange teachers. Gu Daqing had developed a strong interest in design methods and pedagogy since his post-graduate studies. Following the same pattern as Shan and Zheng, that year, he did the studio exercises just a bit earlier than the regular students to test the feasibility of the program. Through doing these exercises, a close relationship was established between the two. Kramel dis-

“转型中的建筑教育”论坛，东南大学，2007 | Forum: Architectural Education in Transition, Southeast University, 2007

在第一年试做了一年级的全套学生作业，通常是比正常教学提前一步，以便发现问题。通过这一交流过程，他与克莱默之间建立了一种密切的学术联系。克莱默也发现顾大庆对教学法的兴趣，遂邀请他于第二年以助教身份继续参与“基础设计”课程的教学。于是，原本一年的交换生交流就演变成一加一的新模式。此一模式后来又沿用到其他南京来的年轻教师上。这对于年轻教师的训练来说是极其重要的，即第一年着重于建立一个关于建筑设计概念和方法的共同基础，第二年着重于教学实践的训练。第二年年末，顾大庆又明确了进一步修读博士的意向并得到克莱默的支持。克莱默意识到这可以是一种以“教育者的教育”为目的的、逻辑递进的三段式教师进修新模式：一，学习一种建筑语言；二，教授一种建筑语言；三，系统性地研究一个建筑问题。此一提议也获得鲍家声的积极响应。顾大庆于1994年获得ETH博士学位。在顾大庆之后，还有4位东南大学教师借此模式完成了博士研究。其他一些教师则在担任助教的同时修读“设计教育”的研究生课程，完成专题研究并获得“硕士后”学位。

克莱默对“南京交流”的贡献不囿于教席之内，他还致力于维系并推动两校官方层面的往来。1986年南工新任校长韦钰到访ETH时，克莱默积极作出安排并保持了良好的关系；1988年在温益进的协助下，ETH建筑系举办了齐康作品展。在20世纪80年代末中西方关系紧张时期，并不是所有曾经参与“南京交流”的南工（1988年南京工学院更名为东南大学）学系都能够保住这一纽带。两个建筑系之间的协议分别于1989年、1994年、1997年、2000年和2004年续签，持续地将东南大学的青年教师送

克莱默在东南大学，2011 | Herbert Kramel at Southeast University, 2011

covered Gu's interest in pedagogy and invited him to continue to participate in Basic Design as a teaching assistant for a second year. Therefore, the original one-year "exchange student" model transformed into a new "one-plus-one" model. Several young teachers from Nanjing have followed this new model. This is crucial for young teachers from Nanjing. In the first year, the purpose was to build up a common basis for architectural design through doing the same exercises as students and, in the second year, the focus was on gaining practical experience in teaching. Close to the end of the second year at ETH, Gu made clear his intention to pursue further studies for a PhD degree and this was fully supported by Kramel. Kramel then realised the third possibility of the exchange program. All these eventually formed a new concept of "the education of the educators". This is a logical progression of the three-phase development model: first, learning an architectural language, second, teaching an architectural language, and third, researching on an architectural issue. Bao Jiasheng also supported this as the basic pattern for the exchange. Gu Daqing received his degree in 1994. After Gu, four Southeast University faculty members completed their Ph.D. studies at ETH through this model. Some other SEU teachers completed thematic studies the same time as teaching assistants and received the post-graduate (Nach-Diplom) degrees on architectural education.

Kramel's contribution to the "Nanjing Exchange" was not confined to teaching, he was also committed to maintaining and promoting official exchanges between the two universities. When Prof. Wei Yu, the new president of NIT, visited ETH in 1986, Kramel made arrangements for her itinerary and maintained good relations with her throughout the years. In 1988, with the assistance of Wen Yijin, Kramel held an exhibition of Prof. Qi Kang's works at D-Arch. Not all the departments of NIT (which was renamed Southeast University in 1988) that had participated in the Nanjing Exchange were able to maintain this link during the tense period between China and the West in the late 1980s. The agreements between the two architecture departments were renewed in 1989, 1994, 1997, 2000 and 2004 to continuously send young faculty members from Southeast Uni-

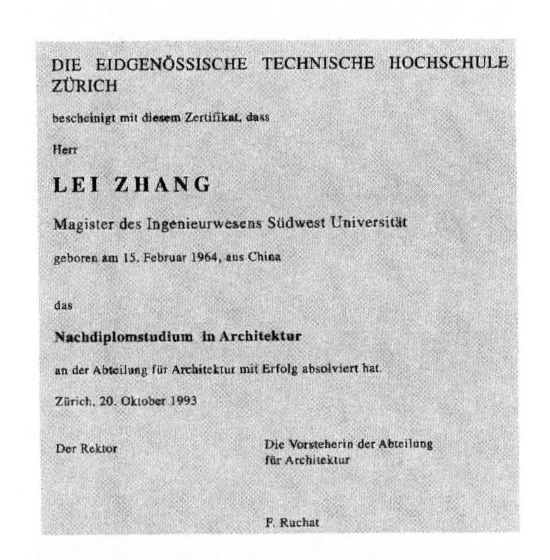

DIE EIDGENÖSSISCHE TECHNISCHE HOCHSCHULE ZÜRICH

bescheinigt mit diesem Zertifikat, dass

Herr

LEI ZHANG

Magister des Ingenieurwesens Südwest Universität

geboren am 15. Februar 1964, aus China

das

Nachdiplomstudium in Architektur

an der Abteilung für Architektur mit Erfolg absolviert hat.

Zürich, 20. Oktober 1993

Der Rektor

Die Vorsteherin der Abteilung für Architektur

F. Ruchat

张雷获颁的ETH建筑学硕士后文凭，1993 | Zhang Lei's certificate for his post-graduate studies in archtiecture at ETH, 1993

INTRODUCTORY EDUCATION IN ARCHITECTURAL DESIGN

THE DESIGN STUDIO: ITS FORMATION AND PEDAGOGY

A Dissertation Submitted to
THE SWISS FEDERAL INSTITUTE OF TECHNOLOGY ZURICH
For the Degree of Doctor of Technical Science

Presented by
DAQING GU
Master of Architecture, Southeast University, 1985
Born in June 30, 1957, Jiangsu, P.R. China

Diss. ETH No. 10306
Spring 1994

Accepted on the Recommendation of
Professor Herbert E. Karmel, Referee
Professor Dr. Kurt W. Forster, Co-Referee

顾大庆博士论文封面，1994 | The cover of Gu Daqing's dissertation, 1994

往ETH进修。为此，克莱默相继安排东南大学建筑系历任系主任——鲍家声、齐康、王国梁、仲德崑与王建国造访ETH并促成了学校层面的交流。克莱默本人也多次访问南京。他于1996年初第一次来到南京主持“基础设计”教学展的开幕式。次年他再次访问南京，被授予东南大学荣誉教授。2007年，他受邀在东南大学主办的建筑教育国际会议上做主旨发言。2011年，75岁的克莱默受邀出席了东南大学建筑学院举办的“南京交流”28周年纪念会。

“南京交流”的中方教师们归国后投入到本国的建筑设计教学改革中，他们把在克莱默教席的心得与教学方法先后运用在东南大学、香港中文大学和南京大学的不同阶段的设计教学中。昔日的师生关系也顺应地转换为教学和研究的合作者。克莱默对于年轻教师的教学给予不同形式的支持。比如，1997年他提议丁沃沃和张雷进行合作教学研究，两人在东南大学二年级的一个课程设计小组中试验新教案，运用计算机辅助设计软件建模及绘图，有关的教学成果以电邮传送给瑞士的克莱默，后者与当时在其教席的吉国华一起，根据对图纸的理解制作文本并将建议反馈回南京。当南京的课程设计结束时，一本相应的作品集也在苏黎世完成了，此也成为苏黎世和南京两方面教学交流的一个基础。他还开启了几个与中国的城镇规划和设计相关的研究计划，如“同里计划”和“张家港计划”等，皆在于尝试如何将他的教学法运用于中国特定的问题。

在近30年的时间中，克莱默帮助众多东南大学青年教师建立了对建筑设计与设计

1996年分别在ETH和东南大学建筑系举办的教学与研究的系列展览，1996 | The posts of the series of pedagogic and research exhibitions held at the architectural departments at ETH or SEU, 1996

versity to ETH for further education. To this end, Kramel facilitated the visits to ETH for successive department heads at SEU, e.g. Bao Jiasheng, Qi Kang, Wang Guoliang, Zhong Dekun and Wang Jianguo, to enhance university-level exchanges. Kramel himself also visited Nanjing several times. His first visit dated back to early 1996. He brought with him an exhibition on the Basic Design course to SEU and also proposed new joint research projects. He visited Nanjing again in the following year to receive the title of honorary professor of Southeast University. In 2007, he was invited to give a keynote speech at the international conference on architectural education hosted by SEU. And, in 2011, at the age of 75, he joined the celebration of the 28th anniversary of the Exchange Program.

After their return to China, the Chinese teachers of the Nanjing Exchange devoted themselves to the reform of architectural design teaching in their home country. They applied their experience and teaching methods in Kramel's Chair to the different stages of design teaching at Southeast University, the Chinese University of Hong Kong and Nanjing University. The previous teacher-student relationship had now turned into a new form of close collaboration in teaching and research. Kramel supported these teaching experiments in different forms. For instance, he proposed a joint project with Ding Wowo and Zhang Lei in the academic year 1997/98, putting a new program to test among a group of second-year students at SEU. Ding and Zhang adopted CAAD in their studio for producing 3-D models and drawings then sending results to Kramel via e-mail. The latter, together with Ji Guohua, who was in his Chair at the time, produced portfolios based on their understanding of the drawings and then gave feedback to Nanjing. When the Nanjing program was completed, a corresponding portfolio was also done in Zurich, which became a basis for further discussion of design pedagogy between Zurich and Nanjing. He initiated several joint research projects on China's urban and rural development and architectural design, such as Project Tongli and Project Zhang Jiagang. Through these projects, he had intended to combine his design pedagogy with China's local conditions.

For nearly three decades, Kramel has helped many young teachers

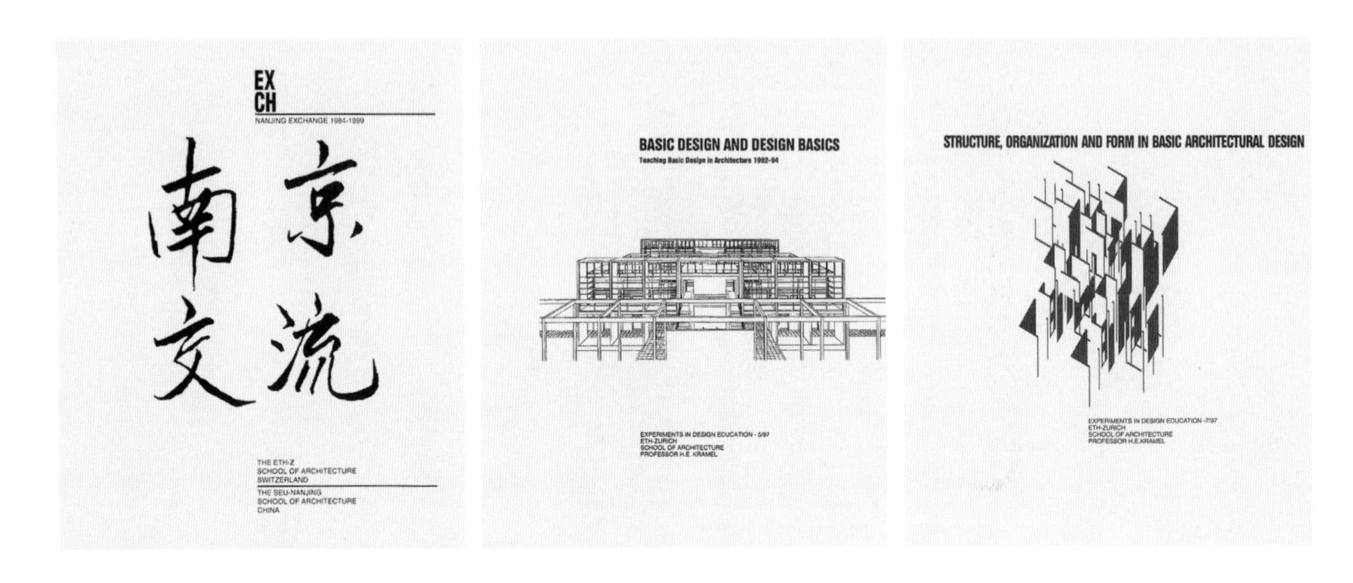

克莱默教席编辑的《南京交流》系列册子的部分封面 | Some of the book covers of the Nanjing Exchange series edited by Kramel's Chair

教学的系统化认识，维系了两个学系间教学与学术交流的纽带，支持了东南大学、香港中文大学和南京大学的教学法革新。“南京交流”的重要性及历史意义在于其对教育者的“教育”——不但传递设计的方法学，更传承设计教育的观念和教学法，这使“南京交流”的影响远远超出个体范畴，这是“南京交流”影响深远之关键。在30年前，远在瑞士的一位教授能够倾注如此大的心力，足见其远见。克莱默多年的坚守并未仅仅瓜熟蒂落，新一阶段的瑞中交流已经在新一代的师生中展开了。

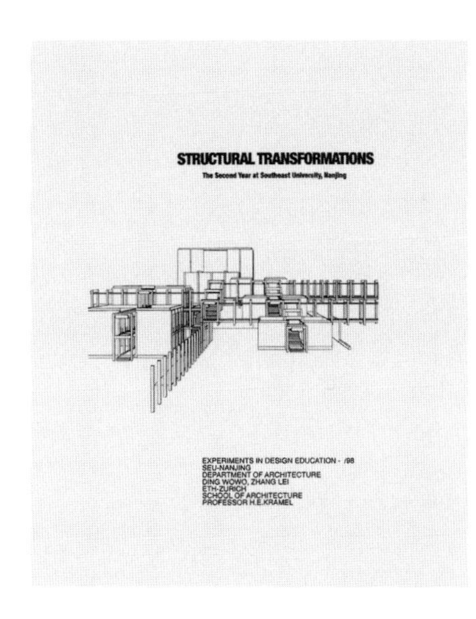

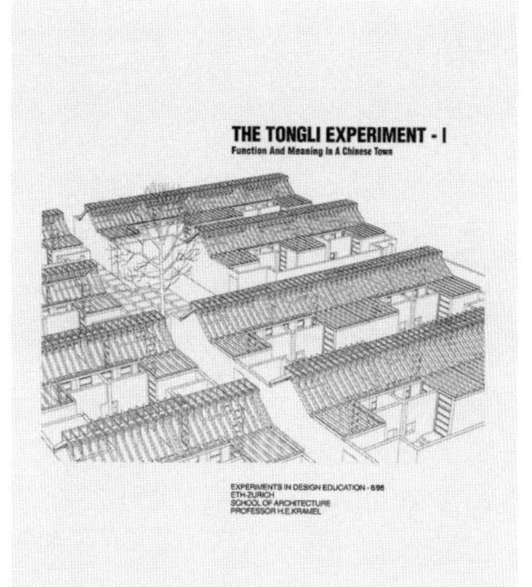

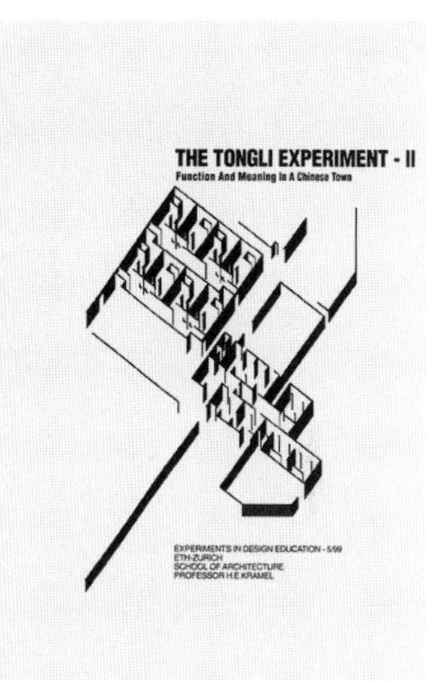

from Southeast University to establish a systematic understanding of architectural design and design teaching, maintained the link of teaching and academic exchanges between the two departments, and supported the innovation of pedagogy in Southeast University, the Chinese University of Hong Kong and Nanjing University. The importance and historical significance of the Nanjing Exchange lie in its education of the educator–not only transmitting the methodology of design, but also the concepts and methods of design education, which make the impact of the Nanjing Exchange go far beyond the individual. This is the key to the far-reaching impact of the Exchange. Thirty years ago, a professor in Switzerland showed his vision through unremitting devotion. Time has proved that Kramel's commitment is more than fruitful–a new phase of Swiss-Chinese exchanges has been launched among a new generation of teachers and students.

2 对话 | Dialogues

提问人 | 吴佳维 Interviewer | Wu Jiawei

单踊 Shan Yong

1986-87

问：您和郑炘老师当时第一批通过“南京交流”计划去ETH，对那里的教学整体印象如何？

那边的教学整体比较完整，比如一年级就叫设计初步与构造。马库斯·吕舍上构造课会当即现场画满个黑板的构造图。学生也很认真对待构造。瑞士人很有构造的意识。还有美术课，上课在大课室里面画素描之类，印象很深刻的是期末的时候，一个年级300多个学生，每人75cm宽一条，做抽象图形，字母、几何图形拼贴，然后左右跟别人的连起来，贴在一张巨大的白布上，构成一张巨幅，有好几天挂在庭院里面。太壮观了！美术训练除了绘画技巧之外对抽象及合作也是有训练的。他们有创意，但还是讲逻辑的。

问：当时您已经有一定的建筑设计经验，而克莱默教授课程的条件限制是比较严格的，那么，您所完成的课程作业和瑞士学生的会不会有很大差异？

我们当时完整地做了一个设计学校，比学生提前一步，相当于试做。模型跟绘图结合。都是手画的，图纸表达基本按照他的制图要求，不分线粗，然后我自己在总平面上加了阴影及内部线条反白。主题是火车站旁边的设计学校。一开始是一个门房，然后是一组教学楼，最后是宿舍。他们的办法一般是规则性的对位，在总图上做几个联排。我不想按照那样来，那时候，刚好贝聿铭做了香山饭店，我对折廊和围院很感兴趣。所以我做了一个风车状的平面，也有大的对位关系。我的方案拿出来之后，克莱默觉得很奇怪，“Chinese courtyard”，他一开始不以为然，后来觉得还蛮好玩的。有一次我周末之后回来，就发现他让助教把我的总平面复印之后通过裁剪的方式分成很多图层，分在很多张图上，将不同的元素分离出来。

问：瑞士学生作业之间的差别大吗？

学生作业的差异不算太大，等做到最后细部了才看出来一些不同。我对教授的教学法一开始是比较疑惑的，这种极其理性的方法，我们还不太习惯，尽管我个人在本科和研究生做设计的时候还是比较理性的。我会自觉地找一个逻辑，一个关系。我们传统教学经常凭感觉，说“这里这样舒服一点点”，原因呢，不知道，老师也不说。我发现有的老师原来做设计是没什么路子，没什么套路的，但是在ETH交流发现了这么严格的方法，回来之后就变得非常有条理，我们也觉得很吃惊。

问：克莱默教授的这套方法的优点是在于它比较理性吗？

应该是这样的，相对于中国的套路，中国式的思维来说。南京交流之前不画泡泡图，设计什么建筑就去参观什么建筑，对功能、材料、场地的关注可能是比较隐性的。东南大学还是比较注重技术的，体现在改图评图的时候，追求合理性。

问：在ETH第一次做建筑案例分析？

是的。那时克莱默教授好像就给了我一张平面图，然后我自己去找资料，找了很多照片及各种版本的图纸。（分析的方法有具体要求吗？）我听听他讲课也就知道了，玻璃啊，屋顶啊，比例和对位关系等。教授会建议我们周末去看建筑，送给我们火车票去提契诺的贝林佐那（Bellinzona）等地。我来来回回给国内的同事写了好几封信，第一封信我跟他们说“瑞士建筑看上去没有激动，但是很感动，很细腻，没有一看很不得了的，但很耐看，比例、材质、色彩关系，做法很细。瑞士在像做手表一样的做房子。中国人像在做房子一样的做手表。”

问：您去ETH之前，1986年在南工的教学改革参考了哪些已有的模式？

我们一开始是从鲍家声老师带回的MIT教案出发去探索，后来我去了ETH，顾大庆也去了ETH，我们才转向那边的教学法。那时不光是受了MIT的影响，还跟同济有关，那时的大潮流是做五大构成，平面构成、立体构成、色彩构成、肌理构成、空间构成，做得热火朝天，我们就觉得同济那个做得太过头，是工艺美术的东西，顾大庆硕士论文就开始关注了这个问题。王文卿、吴家骅在《建筑学报》上发表了文章，他们重新反思了传统，是其后30多年教改的开端。为了新教案，我们当时做了大量的试做，手稿应该有很多在顾老师那里，有些在我们自己手上。那时都是手写的，顾大庆为主，还有我、赵辰和丁沃沃。鲍老师在后面支持我们。

另外，ETH的影响最先是通过柏庭卫（Vito Bertin）带到南工的。他第一次是因为外国学生学习班过来南工，后来留下来做老师，因为说英语，就被分到刘先觉老师的历史组。刘老师让他教外建史，据说当时柏庭卫哭了，说他怎么可能教外建史呢？他是一个教设计的人。后来参加了历史组的一些设计，有一次设计一个园林，有个廊子，他带了个盒子来，一打开，发现整个是一个模型，有地形、廊子的。我还收藏了他1986年课程中制作的立方体模型。

顾大庆 Gu Daqing

1987-89, 1991-94

问：您较早就对原来的教学模式提出了质疑，这些质疑是产生于您自己作为学生时的困惑，还是1980年代初期对现代主义建筑了解渠道的增多（例如外籍专家的讲学、设计工作坊、外国学生的交流以及书籍展览的引入）？有哪些事件或者理论影响了您1986/87年的教案？

我想这不是我一个人发现了问题，那个时候大家都觉得我们的设计教学有问题，你讲的这些因素都有可能。就像现在这边（香港中文大学）的学生，也会在自己学的过程中发现问题，这是一个普遍的现象。具体就那个时候来说，一个是对原来教学的不满——师徒制，究竟学什么说不清楚，对自己学的东西不满足，困惑；另一个是通过一些参照系开始了解国外的设计教学情况，比如建筑杂志上的介绍文章，老师出国访问回来的讲座等。王文卿老师和吴家骅老师写过一篇关于《谈建筑设计基础教育》的文章，提倡理性教学，对我影响很大。设计教学要讲道理，但是究竟如何讲道理，也还是不知道。

为什么会对设计教学感兴趣？应该跟我的硕士论文研究设计方法学有很大关系。在研究的过程中我认识到设计方法学对于改进设计可能没有什么帮助，但对于教学可能有积极意义，因为它解释了设计的思维并提出了设计过程的模型。在教案的具体研究过程中鲍家声老师给了我们一本MIT的一年级基础教案，很有启发，此外当时系里流传一本介绍苏黎世联邦理工学院奥斯瓦德（Franz Oswald）教授二年级教案的书，里面的空间练习很有吸引力。我们还去无锡轻工学院和同济大学考察，得到的一个结论是不能搞构成那套东西。不过要摆脱构成的影响还很困难，1986/87学年的教案还是有不少构成的东西。

问：1986/87学年，您一边在南工施行一年级的新教案，一边通过单踊老师的书信接收到ETH那边的教学，会不会产生一种比较？对去瑞士的交流有些什么期待？

单老师的信起码是1987年初之后才寄来的，而1986/87学年的教案主要是在1986年上半年发展出来的，两件事情有时差，故没有直接影响。

我们当时还是根据我们自己的问题来设计教案的。比如我们在上学时的第一个作业是墨线和字体练习，用鸭嘴笔写字，发一张A3大小的蓝图，然后用铅笔在纸背涂黑再将稿子拓印到白纸上，或者用圆规的尖尖戳洞定位。这个过程没有任何设计训练的成份。我们希望改变这个状况，加入设计的训练。第一个线条练习作业是金元正老师做的模版，我们不直接给学生完整的图，而是一个结构图，学生用作图法把图画出来。第二个作业继续是线条练习，粗细线条，是从柏庭卫那里来的。第三个是立方体空间练习。第四个作业是外部空间设计，第一学期有4个作业。第二学期有测绘练习，有渲染，加了建筑分析的作业，然后是玄武湖公园的茶室设计。总的来说，有所改变，但是还是有构成和渲染的影响。那时瑞士影响还不特别明显，立方体练习和先例分析算是两个新内容。看到单老师对克莱默教授的一年级教学的介绍还是很震撼的，了解到基础教学还可以这样去教。

问：对赫斯利和得州骑警的关注是如何发生的?

写论文才有比较全面的认识，之前只是听说。赫斯利在ETH是个神一样的存在，他不仅开创了ETH建筑教育的一个新时代，新的基础课程，而且他的落幕方式也很奇特，去印度旅行就再也没有回去，据说是得病去世了。克莱默常常讲他和赫斯利的关系，但是很少讲透明性和柯林罗等。我写博士论文的时候有幸看了卡拉贡（Alexander Caragonne）关于得州骑警的书稿，才开始对整个的历史有所了解。我的教学没有什么练习直接来自赫斯利，他的空间作业，其实到那个年代也不是特别精彩了。觉得有几个挺有趣的，但是也没有拿来做。

问：第一次去ETH前后，对建筑学的认识上最大的转变是什么?

我感觉并没有认知上的巨大改变，倒是更加肯定了原先的一些认识。去瑞士之前看过单踊他们带回来的资料、奥斯瓦德的书，加上20世纪80年代引入的外文书籍，对现代主义建筑教育已经了解得比较多了，所以当时克莱默也觉得跟我谈论建筑时没有什么障碍。虽然一开始还是有因为文化上的差别造成的误会，后面互相了解了就好了。去了ETH后对建筑学的认识的最大收获是终于对基于现代主义建筑设计的入门教学有了一个直接的、亲身的体会。

问：您在克莱默教席的访学研习经历了几次角色的转换，可以描述一下?

其实我们去ETH的这些老师，第一年是做学生，主要的事情就是在真正实施之前试做一遍练习。这是一个很有效的学习方式，每个阶段都能学到一些设计教学的招数的，很受用。我跟克莱默聊得很多，因为其他助教绝大部分都是兼职的，所以没有课时办公室里人很少。他的讲座是德语的，我也就是去看看幻灯片，克莱默也不主张我们花时间去学德语。现在我觉得有点可惜。讲座围绕欧洲的现代建筑，尤其是与他的生活圈子密切相关的维也纳及瑞士的现代主义建筑。克莱默把设计课和构造课整合到一起了，每周两整天课。一般早上7、8点钟开始讲座，9、10点钟就讲完了，然后教授和助教回到办公室讨论接下来练习怎么做，之后助教跟学生讲练习要求，就到了午饭时间，真正开始做练习基本上是下午了。印象很深的是学期末的时候，克莱默用两个整天的时间，把全年级两百多个学生的图讲评一遍，当时感觉这个人的精力过人。评图结束后还要评分。他会拉住我一起给学生们打分，把所有作业又走一遍。要对每份作业的优缺点迅速作出判断，这是一个很好的训练。我对彼得·耶尼（Peter Jenny）教授带的一年级视觉设计课关注得比较多，他的办公室就在我们的楼下，上课也跟建筑设计用的同一个空间。交流第一年快结束的时候，才确定了我第二年做克莱默的助教。1989年我回南京之前又注册了ETH博士的学籍，这样就形成了交流的三阶段模式，即先是一年的进修，然后做助教和最后读博士。后来ETH又开设了“硕士后”（Nach Diploma）课程，克莱默也将“南京交流”纳入其中，第一个拿到此文凭的是张雷。克莱默对“南京交流”的贡献巨大，他认识到了“南京交流”这个事情的价值，一直坚持做下去，非常难得。

问：觉得克莱默教学中最有价值的部分是什么？

克莱默对我最大的影响是如何去教结构有序的教学法。是直接从设计入手的建筑基础教学，即便内容上我们现在做的东西已经和当时学的完全不同了。直接从建筑设计入手，而没有专门的作图或形式训练作先期的铺垫，这是很不一样的教学法。我们当时对抽象的空间练习感兴趣，但是克莱默的教学中并没有这类抽象的空间练习。形式和空间问题已经融合到建造、功能和场地的问题中。我们知道霍斯利的教学体系中有不少抽象的空间练习，克莱默在教学法上有不同的想法，他是反对霍斯利从抽象练习入手的方法，其中的一个理由是抽象练习和建筑设计不是一回事，抽象练习做得好，设计不一定做得好。不过我们在当时觉得霍斯利的空间练习是非常建筑化的，和国内那种完全抽象的构成练习完全不一样。所以要理解霍斯利和克莱默两人的教学法之间的不同，需要对特定的历史背景有所认识。

问：如何总结自己多年以来从东大到港中大的教学实践发展历程，认识上及经验上的？

回顾这些年的教学我觉得分为几个阶段。1988年发表在《建筑学报》上的文章《渲染、构成与设计》在当时是很有前瞻性的，建筑设计基础教学应该学设计，不是学绘图技法或者纯粹的形式训练。可是什么才是设计的教学模式？当时还不是很清楚。这是一个不断探索的过程。从整个时间跨度上来说，南京这一段应该算是移植。1986/87那个教案算是一个初步尝试，提出了问题，解答却不是很理想。1990年的那个教案明显吸收消化了克莱默的教案内容，此外把渲染表现的训练糅合到设计的主线当中去了。当时的环境还不能把渲染直接从教案中拿掉。不过设计还是广义的，有个小制作的练习，因此并不是直接从建筑设计开始。之后到了香港，集装箱建筑教案就比较纯粹了，也算是我和柏庭卫自己发展的一套教案，不过整体上明显延续了克莱默的教学法，即有一个场地将几个独立的设计联系在一起，所不同的是结合了香港本地建筑的特点。集装箱建筑教案一开始有6个阶段，包括开头和结尾的坐具和建造，后来改成4个阶段。中大当时一年级也有实体搭建的练习，当时是做树屋和地标等偏游戏性质的搭建。我们把它改成直接搭建木亭子，这和我们在克莱默那里获得的相关建造练习的经验有很大的关系。在实施集装箱建筑教案的过程中，我们也开始对克莱默教学方法的某些方面产生不同的看法。比如教案的会所设计中我们最初想延用克莱默的体块模型－空间结构模型－建筑模型的设计方法，但是这和集装箱的体块性质有矛盾，所以我们必须要有所改变。2001年建筑学院开始了新的教学架构，成立了建构工作室，我们的教学研究也进入一个新的阶段，我们设立的目标是让当代建筑变得“可教”，这个时期发展的“空间和建构”课程就与克莱默的教学有很大的不同，算是真正成熟起来了。而2009年以来我又重新回到基础课程的教学，分别对集装箱建筑教案和空间建构课程作进一步的整合和发展。比如建筑设计入门（一）就是将原先一整年的集装箱建筑的课程集成到一个学期的时间完成，同时还结合了建筑入门（一门观察记录的课程）和视觉训练课程，更加紧凑。而建筑设计入门（二）则继续发展了空间建构课程。总的来说，我的设计教学实践和研究大致上1986-1994是移植和学习期，1994-2001是整合和运用期，2001-2009是研究和发展期，2009至今算是集成和巩固期。

丁沃沃 Ding Wowo

1988-89，1994-96

问：20世纪80年代中期南京工学院建筑系师生对教学改革的呼声很高，在南京交流开始之前，其实已经有别的一些教学的案例，例如鲍家声教授从MIT带回的教案，对最初的教改起了什么作用？

鲍老师带回的MIT教案，也很好，我印象最深的是构造节点画得很细致，而我们的设计教学没有涉及构造的问题。去ETH后发现他们的设计教学对构造节点也非常重视，关注构造更关注材料。MIT的教学很有启发，但要落实到我们自己的教案并不容易。为了教改，我们采访了许多老教师，还翻阅南工早期的教案。我不觉得早期教学全然按照宾大，证据是刘敦桢先生的教案。那应该是解放初期的，文字像是日文译过来的，要点是强调结构和构造。我们当时感慨老一辈的教案有那么多工程知识，到我们学的时候，全部被虚化了，就剩造型，其实我们对渲染非常反感。由于教改砍掉了渲染，以至于很多老先生觉得我们的教学突然很陌生了。刘敦桢时期也有渲染，但是工程给排水构造那一块的讲课特别多。刘先生的教案很大程度奠定了老南工的基础，但后面慢慢没有了。王文卿老师当时拿来给我们看的很多教案偏绘画和形式构成，从平面开始，更多的谈修养。我们请教老先生的时候其实自己已经有很明确的方向了，不单是把大家的建议凑成一个整体。当时顾大庆说职业教育是要可教的，不反对品味（taste），但培养品位是精英教育，我们做的是职业教育，就是让每个学生都学得不太糟糕。

问：克莱默教授的教学方法最触动您的地方是什么？

空间。我觉得克莱默用九宫格模式把空间的问题说得非常清楚，全封闭、全开敞、半封闭的，很逻辑地把一件事情的规律做出来，那么这个教案就可以跟进了，其他的东西可以补充进来。在这之前，我们看了西方建筑大师的作品不知道怎么学，案例都是以历史的方法来讲，没有把现代建筑作为一种设计理论。包括现在，有些建筑评论往往都没有办法独立于建筑师具体的人去把设计说明白。空间图解是能够独立于具体设计的。此外就是克莱默教授关于形式构成三要素的论述，三要素图解是现代建筑的精髓，这是核心的部分，但是还有外延。它里面的每个元素都可以再诠释。

克莱默教我的是如何做一个老师，如何诠释形式问题，如何总结，如何教学生。

问：第二次去ETH参与克莱默教授的基础教学时，觉得跟5年前有什么变化？

就空间设计和重视构造而言没有什么变化，不同的是对城市问题越来越重视。克莱默一贯是以6mx6m的立方体为基础，因其可解决任何问题。从前能用方盒子做就用方的做，但后来他很喜欢的几个学生作业不是方盒子的，甚至屋顶也未必是平的，但一定不能乱来。后来他的计算机教学小组尤其想破掉固定方盒子作为基本单元的方法，试图用软件中的图库组件进行组合方案。我第二次去ETH时，假期的工作是将组件参数化，参数可调。模数是3m，1.2m，1.8m等。柏庭卫那时候在日本，我们用互联网交流设计。

做非洲项目时，克莱默介绍我看哈桑法迪（Hassan Fathy）的书。项目实施的时候，跟克莱默一起去了两次坦桑尼亚，一次我自己去的苏丹。克莱默向我介绍了荷兰后现代的结构主义，试图以结构主义的方法理解复杂建筑组合及其空间。后来我更认识到，在复杂的城市形态语境中结构主义更有意义。他让我研究结构主义，让我在研讨周带学生去荷兰，按照一本书前面的一张地图把所有结构主义的房子看一遍。凡艾克的建筑逻辑清晰而空间复杂——清晰的迷宫。克莱默觉得结构主义这套理论有坚实的哲学基础，是放之四海而皆准的。到后期，他试图用结构主义的方法演绎中国乡村形态的生成。然而我的博士研究证实了中国的农村村庄形态生成的机制并非如此，和几何结构无关，而是社会学的产物，倒是可以用后结构主义的理论去解释。

问：您两次从苏黎世回国后，在东南大学二年级进行设计课改革的思路是什么?

当时觉得一年级的教改蛮好，应该延续到二年级去。1989年开始进行二年级教改，开始只是一个小组。我更喜欢1989年的那一套教案。现在东大的周颖教授是我当时的学生，还有同济大学的陈勇教授和在北京的中国建筑设计院曹晓昕建筑师等。

我像在ETH时那样做教学文本，缩印学生作业，那个教案很原真地表达了当时的想法。二年级一共4个教案，第一个设计就不是方盒子，有地形限定的，用空间来说话，一个单一的空间——传达室或者小茶室。第二是单元空间的组合，小学校、幼儿园或者医院。组合类空间完全是受克莱默的影响，我理解了他的第二个练习。这种点到点的空间没有流线，学生进班级不需穿过别的班。这类空间讲韵律，开头如何，中间如何，结尾如何，克莱默天天讲的。第三就是综合性空间，做老年人公寓，需要一个门厅来处理流线。第四个设计学生要组织大中小的不同空间，集中人流。用幼儿园和老人院做题目的原因是，建筑师是为别人做建筑，要考虑到跟自己不一样的人。做幼儿园注意栏杆高度和样式；做老人院要考虑跟自己不一样体力的人，不是什么地方爬一爬就行的。交通类空间一般做火车站，架在上面的火车站，一个大跨度空间统筹很多个流线。我再去ETH之后他们就不教了，我回来之后又重新捡起来。到仲德崑做系主任的时候要求我再去执教二年级的设计教学，这时新的教案在二年级全面铺开。

赵辰 Zhao Chen

1990-92，1996-98

问：20世纪80年代，您从一名建筑学生的视角来看，瑞士建筑在众多涌入的建筑信息中是特殊的一支吗?当时如何谈论建筑?

当时分辨不清楚什么是瑞士，我们对先进国家的东西都比较盲目地崇拜，没有分辨能力。整个西方文化中细微的区别我们分辨不出来。1980年代的出版物，最早进入中国的瑞士建筑其实是提契诺。当时介绍美国建筑的出版物当然更厉害。当时以为瑞士的建筑都是博塔（Mario Botta）、加尔菲蒂（Aurelio Galfetti）、坎比（Mario Campi）等那样的。后来才搞清楚整个现代建筑跟瑞士的关系，瑞士的重要性，第二次世界大战以后如何接续包豪斯的薪火。从得州骑警转回瑞士，这才是现代建筑比较核心的东西，比那些表面的几个建筑师要重要很多。当年完全是很盲目的。

谈来谈去就是一些关键词。我记得很清楚，1984年，柏庭卫他们一帮瑞士和澳大利亚的建筑师和学生来南京，瑞士来的基本都是建筑师，是瑞中友好协会牵头的。有个意大利人，叫作安德烈斯（Andreas），是ETH毕业的，当时在那里做助教，他跟我说罗西、斯卡帕，但我完全不知道，我只知道博塔。他说："博塔不行，都是跟人家学的。"当时开了很多眼界。包括一些城市设计的东西。那个时候我们把美国的一切东西当作最先进的，真的是一切的一切。这样看瑞士对中国的影响是非常积极的，因为亚洲受美国影响太大，ETH通过"南京交流"好像是突破了一种以美国为主的垄断局面。

问：克莱默教授主持的木结构研究，是否为中国传统建筑研究提供了新的思路？20世纪80、90年代他在研讨周进行的建造实践课程，是否为您在南大的建构实验提供了借鉴?

克莱默主持了一个建造实践课程（Baukurs），以材料分，木、钢、混凝土、砖。其中木是最基本的。当时在瑞士有个推广、鼓励大家用木结构的运动（IP-Holz），跟可持续发展有关系，欧洲同期还有奥地利、巴伐利亚和北欧在做。克莱默是其中一个主要的教授，洛桑还有一个教授主要做结构方面的研究。那时克莱默刚好开始带木构的硕士后学位课程，带了好多学生，包括现在以木构为专长的建筑师卡米纳达（Gion A. Caminada）。我去瑞士之前已经对木构有相当的基础，但受制于研究方法的落后，且主要是中国传统木构方面的。我跟克莱默在这方面聊得很多,他对中国木构也感兴趣。在ETH的经历对我的帮助很大，帮我转变研究视角，改变研究方法。我觉得就应该从建构方面去理解，形成了后来对梁思成等人理论的批判。包括所谓"Tectonic"，从材料构造入手讨论问题，教学也可以从它入手，再复杂的建筑形式都可以顺着这个去理解，历史上的建筑也都可以顺着这个去解析。

问：以不同要素启动的设计方法,是否受到克莱默的启发？在高年级教学中强调某一个要素为出发点的意义何在？

我个人一直认为，我们接受ETH，克莱默的方法，是不容置疑的,但我也不认为我们是直接、简单地继承。我们根据中国的需要做自己的事情。我们的教改不是靠他们而做的，而是自己已经做了，到了ETH之后发现他们的教学，是我们需要的，才会融入到我们的教学中。

1999年我回来之后在东大做了三年级的教学改革，是在韩冬青老师的建议下开始的。因为中国学生的研究性比较差，高年级设计难度大了之后容易顾此失彼。我在教学中发现,前面几周时间浪费严重，因为他们什么都要想，但是什么都做不了。这跟低年级不一样，任务比较清楚，不容易浪费时间。我的想法是：对三要素的思考必须直接体现在设计教学过程中，而不是靠上某一门讲座课来学的，应该以某个要素先启动,或者是某种方法来实现；另外中国学生结构与构造方面特别弱，教学要有针对性。东大的设计教学是比较强的，但是形式训练偏多。学生到三年级画图能力都过关，但构造呢，虽然懂，但是融不到设计中，或者是被动融入，不是主动思考，不是设计原初的时候就考虑，这个ETH就很强。

让学生在完全不知道功能、环境等条件的前提下，直接进入到结构的思考。我们当时上来就要求做模型，以前头两个星期都是烂草图，我们则第一周出一批模型，第二周又一出批模型。学生就知道这个跨度可以这么来实现的。我记得那次克莱默教授和坎比教授一起来，我向他们介绍了教案，克莱默对教案很感兴趣，他说建构启动比较容易出效果。但坎比很不赞同，他认为建筑教学任何时候都是综合的，不能突出某个东西。我的理念是可以优先权不同，然后再综合，如果老是综合就总是专业的训练不够、研究性弱。学钢琴，有些练习曲就是专门练几个指头，就是这个道理。这个教案的模式有点像ETH老的课程系统里的选修课（Seminar），但他们靠的是教授的专长，我们这个靠的是教案的设置，逼学生一上来就思考结构构造，学生可借此理解跨度结构在建筑造型中的真正意义。高年级如果用整一个学年来做这个事情，是好的。

张雷 Zhang Lei

1991-93

问：您去瑞士之前是如何做设计的？是否受到当时某些建筑思潮的影响？

我当学生的时候，后现代谈论的比较多，还有解构主义，记得当时三年级课程设计做影剧院项目，从解构主义的作品中学习体量穿插的手法。知道瑞士提契诺学派是在读研究生的时候，通过瑞士现代建筑展览和《瑞士建筑70/80》这本书了解了瑞士建筑。

学生时代的课程设计是从渲染学起的，没有关于模数化的教学。读研究生期间正好顾大庆、丁沃沃等年轻教师从1986年开始教学改革，了解了理性的设计训练的方法，自己也非常认同。1988年4月份我研究生毕业后开始独立带88级的设计课，记得是图书馆和火车站，图书馆我自己先试做了一遍，还自己画了渲染示范图，并要求组里同学做大比例的模型。后来和丁沃沃一起带1989年二年级课程设计。

我们上大学的时候，其实和我们的老师是同时开始了解现代建筑的，“文革”期间他们也很多年没有接触到新的思想，甚至都没怎么接触专业。那时候刚刚对外开放，大家思想很活跃，年轻学生的反叛性很强，我们对传统的东西往往都持批评性态度。我们在学生时候接收了各种新的思想，但是也没有明确的方向。看到顾老师他们的一年级教改成果之后马上觉得它好，那些方盒子我觉得它很清爽，我自己比较喜欢简单的东西，再加上网格的控制就可以通过简单的操作来教学生。以前读书的时候看老师改图就觉得改完了感觉确实不错，但感觉这门学问是蛮深厚的，要靠自己慢慢领悟。

问：关于克莱默教授的“三要素”图解，您觉得 “场地—功能—建造”与设计的关系是否早已存在于东大原有的教学中？

三要素图解我最早是在设计基础教改里看到的，把设计要解决的问题说得很清楚。我现在讲研究生“基本建筑”课的时候还用。之前我们的老师在改图的过程中有提到这些要素，但设计教学主要还是按不同的建筑类型进行训练，没有明确将现代建筑作为理论基础，也就没有这么系统这么明晰。

问：在这种背景下，到瑞士之后会不会觉得克莱默的教学法没有新鲜感了？

去瑞士的时候确实对基础教学本身没有新鲜感，但还是觉得这套教学很好。当时是第一次出国，机会很难得，抱着虚心学习的心态，哪怕已经很熟悉，但还是跟着做一遍，看克莱默如何评价，因此在那里适应很快。

在课程之前，他会提一些新的要求要我去试做，他会反复看空间关系，要求通过模型和电脑模型反复推敲，对我来说确实比较程式化的。在瑞士的试做练习和之前在东大老师教的渲染和设计，都是一种经历。

问：作为教学助理的日常工作是什么？

我去的第二年带学生设计课。试做、排版、课程讨论，按部就班的教学。我觉得克莱默对教学的把控很厉害。他控制了输入条件和节奏。我跟学生之间的交流其实比较简单，不太需要我们教师个人有太多发挥。学生的作业差异性也不大，中国助教和外国助教带出来的也差不太多。

除了教学之外主要帮助克莱默做非洲的设计项目，是坦桑尼亚一座神学院的教堂、教室和学生宿舍等，我们中国助教经常在办公室加班，回去也没什么事做，不像瑞士助教到点下班。设计项目有点像是课程的延伸，讲求秩序理性、严格使用模数，这样在非洲才容易控制造价，把项目实施好。

问：在ETH掌握的CAD技术如何运用在教学上？

东大建筑系CAD实验室的第一任主任是卫兆骥，往后是赵辰、龚恺、我、吉国华等，除卫老师外都在ETH进修过。当时ETH使用电脑已经很普及，国内建筑系还比较少用。我1993年从瑞士回国之后在东大二年级带过一个使用电脑的教改组，是做茶室设计，效果还不错。但没有采用BAU（基本建筑单元），当时计算机组主要还是用CAD作为绘图和表现工具，研究空间还是以手工模型为主。

问： 南大的研究生教学偏重哪些方面？

研究生在理性的设计方法和材料构造两方面的知识和训练比较薄弱，除了这两点之外，我目前的设计课选择的基地往往是聚落环境，民居是模数化的、建构的和日常生活结合的，结合新的业态导入研究新和老的关系对于培养学生正确的价值观会有较大帮助。

我的课第一讲就是基本建筑。克莱默的三要素图解我仍然在用，后面主要通过具体的案例来讲解空间、材料和建造。我把这些内容和没有建筑师的建筑联系在一起，比如通过院落和聚落讲空间原型和空间组织，没有建筑师的建筑能够更好地把人和空间的关系说清楚，空间场所化了，建筑就生动了。

龚恺 Gong Kai

1992-93

问：1996/97学年一年级基础设计课程的总体思路和主题的编排顺序的依据是什么？

1996/97年教改主要是我搞的，我们之前用的是顾老师的教案。我有一个观点，教学只能改良，不能改革，我是个温和的改良主义者，我不希望教案十年没有变化，也不希望每年都重新来过。教改首先是发现问题，不是为改而改。所以这个教案跟原来的教案有很多传承，但是我加入了一个环节叫“认知”，每一个题目都从认知开始。我当时觉得，ETH的每个练习都是从抽象开始入手的，我希望反过来，从具象到抽象，而不是从抽象到具象。比较大的变化是这个教案中我们去掉了一些原来觉得还比较好的练习，例如立方体练习。方盒子的问题是，教师如何评判？洞开在这里跟在那里有什么区别？到最后无非又变成了传统的评判标准，如画得好不好，模型做得好不好的问题。3mx3mx3m是一个抽象的空间，里面承载什么样的活动？如何评判空间是太高或者太窄？没有功能就难以评判。中国学生的认知是从具象开始，太抽象的东西学生也不理解。所以新教案一开始是认识校园，做建筑测绘，案例分析也是一种认知。

问：教学改革过程中最大的制约因素是什么？如何克服？

最大的制约因素是反对的人太多，钟训正老师、齐康老师都反对。我取消了两个核心的练习，都得罪人的，我把顾老师他们的一些抽象练习拿掉了，另外把渲染拿掉了，在我这里开始东大彻底不做渲染了。我们那时候的渲染跟后面我们教的渲染也不一样了，所以你看到一件事情的发展趋势是恶化的话，就该壮士断臂了。

问：回顾这个教案，教学效果如何？

教学的收效还是可以的吧。教育面向大多数的学生，最好的学生不是教出来的，最差的学生也无能为力，学生群体的能力水平也是有大年小年的，不一定跟教案有关系。有时候跟大的环境有关。我认为，这里面很多东西是相似的，但你重新理一遍之后，就会上心。因此我管教学之后，要求下面的老师每年改动25%，因为按我自身的体会，一旦要改教案，自己就会上心；不然就会按原来的教，无所谓。但也不能一下改太多。我可能把一个开头，或者一个结尾改了。有改变，教师总会有劲头一点。

我一直认为东大不能是职业教育，否则就把自己摆低了，职业教育是技法教育。我们的教育是理想教育，做学校外做不了的事情。眼光要长远，不要做技术员的事情。哪怕从一年级开始也不该提自己是职业教育。

问：您觉得克莱默教授的基础课程是职业教育吗？

克莱默那套系统我是清楚的，我离开之后反而了解得多一点，包括构造讲师吕舍讲的技术性内容，克莱默其实没有讲什么技术性的问题。从我的角度来看我并不觉得克莱默是职业教育。职业教育是指强调技术方面的，解决当下问题的。

问：您做学生跟教学生的时候，教学模式有什么变化？

我学的时候基本都是老教师，“新三届”还没开始留校。教我们的那一批老师基本没有出过国。老师不怎么讲方法，还是偏重表现，表现技法好的学生成绩不会太差。不怎么重过程，看重结果。评图的时候，先看效果图，效果图不错了，再看看功能有没有问题。功能问题也不可能看太细，只看大的关系。

到我任教的时候，基本没有老教师了。老教师说的话学生比较愿意遵守，年轻老师的影响力就没那么大。顾老师当时说的话我们年轻教师都很同意，就是老的教师教学不一定要方法的，用什么题目都行，年轻教师就不一样了，需要依靠一套教案，使得施教的“人”变得没那么重要。顾老师当年做的事情是非常有意义的，我们跟当时其他学校不一样的地方在于我们有一套非常明确的教案，后来在别的学校传阅。另外，我明显感到教改之后，老师上课跟学生的沟通明显多了。

吉国华 Ji Guohua

1996-97，2000

问：您去ETH时，“南京交流”已经持续了近十年，在东大瑞士建筑文化的影响是否强于其他的国家？

瑞士建筑本身的影响并不明显，只是知道鲍家声老师支持从瑞士回来的老师做教学改革，主要在一年级，张雷老师回来后二年级开始改。我研究生快毕业时开始帮他们做一点计算机方面的教学，知道他们的教学受了瑞士的影响，东大的教学开始跟其他的学校不一样。

那时候大多数老师都有留学的经历，老一辈受瑞士的影响比较少。我1994年刚工作的时候，一、二年级基本是按教改的方式教，老教师跟着年轻教师的方法来。老教师比较关心建筑从简单到复杂，例如从不上楼到上楼。年轻老师注重的是一些专题的训练。这些是我做教师之后了解到的，做学生的时候并不知道。

问：您在真正接触克莱默教学的时候，是否感觉已经很熟悉了？

我在ETH第一年，接触到克莱默教授的那套教学的时候，并没有觉得对它很了解。我们原来的教学是不跟你说为什么的，只是去做，也没有明确的说如何去做；只是告诉你如何去改，改成这样或者那样。我毕业的时候对于什么是建筑，还没有一个清楚的认知。可以这么说，我对建筑更加全面的认知，是到瑞士之后。克莱默教授每次一谈建筑就会画一个大圈，里面若干个小圈是“reference”，或者“context”。一个箭头，随着时间的变化发展，建筑是属于这个参照里面的一部分。建筑设计的时代性与不变性就变得容易理解了。他经常提的“vital idea”应该是时代的价值观。

问：什么是“vital idea”？

这个词我一开始也不明白，他就问我中国现在热门的词。我说“change”，他说“change”这个词现在在中国是褒义的，但事实上它是个中性的词语，可以是褒义，也可以是贬义。因此，“change is good”就是中国这个时代的价值观。这些图示，包括三要素图解，我都是在去了瑞士之后才第一次看到。我一年级的设计课基本就是一些练习串起来的，没有配合设计课的讲课。老师上设计课只是带任务书去告诉学生要做些什么，不会说训练的目的。

从瑞士回来之后，我经常用那个三角形图解跟学生解释建筑。但这并不代表所有建筑，而是现代建筑，排除了对古典样式的讨论，把设计教学的范畴划清楚了。它剖析了现代建筑形式从何而来，从什么方面去讨论。这也是我们在南大研究生教学里面的“基本设计”的方法，那时候研究生来自全国的不同学校，基本都是传统的教学法出来的学生。我们的目的就是让他们重新理解建筑设计。

问：从瑞士回来之后，如何着手东大三年级的设计课改革？

1998年从瑞士回来带三年级，那时候二年级是丁老师和张老师带，我跟着冯金龙老师在三年级，要做的事情是把三年级衔接上二年级，题目还是按原来的。大的改革是，过去的教学以建筑类型为线索，我们想要做的是把它和问题类型结合在一起。某个题目偏重一个方向。例如“以建构启动的设计”，是一个剧场设计。把重点放在结构技术与建筑形式的互动上，把声学的东西去掉了，变成露天剧场来讨论“建构”。我们做的第一个是这个“建构启动”，1999年，韩冬青、赵辰、我和李飚等6个老师，每个人带半个班，大概15个学生。2000年下半年我离开了东大。之后鲍莉老师他们把文脉启动等接着做了下去。

问：有用克莱默的空间操作方法去做实际的设计吗？

好像没有。用在教学上了，我后来教研究生的“基本设计”课程，只有8周的时间，一开始做了一个比较类似的，和环境产生关联，由一些小房子构成的，通过每个房子的组合来发生各种各样的形式，因为本身的功能比较简单。后来又改掉了，做了一个没有文脉的，把房子做在楼顶上的加建，有单元的构成、有结构建造的问题，后来逐渐地想把问题集中，功能是跑不了的，主要是讨论建造的问题。总的思路还是受克莱默的影响，在教学上受他的影响还是比较大。

问：如果现在请您画一个关于建筑设计要素的图解，您会怎么画？

还是从那个三角形开始。我们之前其实一直都是受他的影响，现代建筑的一些操作思路方法，但我最近在反思，建筑现在其实又到了一个新的时期，是不是还要继续教这个所谓的“现代建筑”？现在出现的绿色建筑、数字建筑等，那一套东西可能已经无法应对了，完全通过这几条线来讲建筑的事情我觉得不够了。建筑形式从哪里来？如果我现在来画一个图解的话可能会很复杂，可能会在这三个关联中加入更多的关联，越来越多的因素，因素之间的关联。但作为教学来说，应该不会这么复杂。每个教学的设置有它一定的目标。作为自己实际的项目肯定是个复杂的状态。其实我们的教学也是复杂的，但是作为现代建筑的内核，它凝聚在这三点上。克莱默的教学其实在20世纪70年代时受到了后现代的影响，它在教学中也有“意义”的这个问题，有一种符号学的意义。他有些复古的，或者说古典的元素在里面，但这些没有出现在他的3个圆圈里面，圆圈里面没有“meaning”。

张彤 Zhang Tong

1998

问：您是1998年4月以访问学者身份到苏黎世并加入了克莱默教授的教席。此前两年，ETH建筑系代表团来宁，在《南京交流》协议中加入了“联合设计”(joint project)的合作方式，之后陆续开展了“安徽计划”(Project Anhui, 1996-98)，“江苏计划”(Project Jiangsu, 1996-98)等。您在苏黎世完成的《同里试验》第一部，是这些联合项目的延续吗？当时是如何确定以同里作为场地开展设计的？

我不太知道同里项目之前的来由，只是2017年冬克莱默教授来南京时谈起我去苏黎世交流的工作，他表达出对同里的兴趣。之后我专门去了两趟同里，针对性地准备了些资料。

问：从您发表的文章来看，您自1997年以来保持了对地域性建筑的兴趣，“同里试验”是否提供了一种新的应对地域性的设计方法？

在苏黎世交流的具体工作是设计运河边上的一座旅馆，大约是24个居住单元。我想对于克雷默教授，这是一次实验，他开始尝试把他的方法用于中国的环境；对我来说，这当然是一套新的方法，我逐渐认识到空间生成的结构性，以及在单元的组合和群体延伸与发展中显现出来的系统性。尽管如此，我仍然不确定这是一种应对地域性的方法，它的逻辑本质上并不因为在同里或是在撒比奥内塔有所不同。

问：您访问ETH时，“南京交流”已经开展了12年，东南大学一、二年级的设计教学也进行了一些相应的改变，1996年系里还举办了克莱默教授的“基础设计”教学展。那么，您在着手同里项目的时候，是否对“基础设计”所包含的一套设计方法已经很熟悉了？

我在2004年以前一直在建筑研究所工作，不直接参与建筑系的教学，会去看系里的展览和教学成果，感受到这是一种完全不同的方法体系，也对作业成果展现出的清晰的形式逻辑印象深刻。但因为没有置身其中，所以谈不上熟悉。事实上，在苏黎世12周的设计工作充满了困惑、摸索和冲突。

问：除瑞士外，您也曾经在多个国家进行访学或项目设计，如日本、法国、瑞典，那么ETH，或者说瑞士的建筑教育给您印象深刻的有哪些方面？

我去日本和法国完全是以建筑师身份去设计公司参加研修与合作设计，跟瑞典皇家理工大学的合作则是跨学科的交流，并没有机会去观察和比较他们的建筑教育，所以不太能够回答这个问题。

问：您认为“南京交流”对东南大学建筑系近30年来的影响发生在什么层面？

与ETH的交流是东南大学历史上对外交流持续时间最长、最为稳定和连续的一个过程。不包括已经离开东南大学的人员，目前在建筑学院教学的教师中有将近20位经历过这个交流，其他也有不少是早期参与交流老师的学生留校任职。尽管每个人参与的方式、获得的感受，以及之后传递出来的影响各不相同，但我想对于一个教学机构来说，没有比对师资整体组成产生的影响更深刻的了。它改变了东南大学建筑教育思想的本体，并且不限于此，这种改变渗透到整个中国建筑教育界。这个源自ETH交流的影响仍然在继续，随着时间的累积和时代的变迁，它的作用、成效和评价会在更大的比较范围内逐渐清晰起来。

鲍莉 Bao Li

1998-2001

问：您在东南大学取得硕士学位后，于1998年进入克莱默教授的教席进修，第二年成为他的助教，并先后获得了ETH的建筑学硕士学位及科学技术博士学位，可见这一段经历在您学术生涯的开端占有相当的分量。您会将它的影响总结为哪些方面？

在瑞士跟随克莱默教授的求学、从教及研究经历真正帮助我开启了学术之门。1998年进入教席时，除了第一周和赵辰老师有个交叠期，之后很长时间教席就只有我1位中国人，而之前最多时有5位同在，因此我与教授有充分的交流机会与时间，我们的交流和工作空间常常从工作室延伸到咖啡馆、食堂和餐馆。克莱默教授给了我亦师亦父般的教导与关照，作为学术和精神导师他对我的职业与人生的影响是全方位的、持久的。我深感幸运的是，在学术生涯之初找到一个契合并启迪自我的立足点。

这段经历于我，最重要的是学会理性的思维及工作方法，“理性之方法乃捷径”（rationis modus brevis），“Logic”也是教授常挂在嘴边的词。无论是教学、研究或是工程实践，都讲求严密的逻辑、章法与进程。教学中的体现即是阶段性、层次化地拆解重点和问题，形成结构化的设计教学体系和练习化的设计课题设置，从而实现“建筑设计是可教的”。

这段经历对我学术生涯的具体影响，主要体现在三个方面：

首先，是对西方建筑和现代建筑的理解。虽然当时国内已有不少书籍期刊的引介翻译，但总体对于现代建筑的认知还是零星和片面的。教授给我的第一课，就是从建筑语言的学习开始，只有掌握语汇修辞，才可以在一个频道上交流。这个于我其实就是对西方建筑，尤其是现代建筑的补课。补课的方式有传授、阅读、讨论和游历，各个环节相互关联支撑。游历的补课是从奥地利开始，从阅读萨尔兹堡古城和维也纳环形大道开始。既是因为教授自身的奥地利背景，更是因为奥地利，尤其是维也纳在现代化转型中的特殊地位和贡献。

其次，是对整体与部分、历史与当下、城市与建筑关系的认知。教授推崇20世纪中叶兴起的结构主义思想，熟稔意大利城市与建筑文化，也与瑞士意大利语区即提契诺学派建筑师交往甚密，因此他的设计基础训练讲求对建成环境的认知，其中意大利理想模式小镇撒比奥内塔更是成为其之前设计基础教案的基地。而教案中的每一个阶段性练习都是部分与整体、当下与历史、建筑与城市的对话，从认识整体中局部入手，最终回到整体层面形成完整的设计成果。这大大拓展了我对建筑的时间性与在地性的认知，也是真正建立起城市-建筑（Stadtbau）的概念，理解建筑与其环境的相互依存和促进的关系。

第三，培养国际视野，关注本土实践。虽然出国是学习西方先进思想和经验，但其目的是学成归来，将学识与见识用于应对中国快速发展中的本土现实问题。在瑞士时，我的设计与研究较多关注江南城镇与乡村建设、关注当代居住与绿色设计等，这些主题大多延续至今。我进入教席的1998年，教授已经转到高年级授课，课题是发展中国家的建筑设计，但他对东南的青年教师依然是从设计基础入手来学习现代建筑语言和设计教学方法。大约是1996年教授来访中国时去了趟同里，非常喜欢，于是1998年的设计练习——“同里实验”就由同里替代撒比奥内塔作为基地环境，尝试在中国现实语境下开展设计研究与“意义”的探讨，之后瑞士交换生也按这个思路做了设计和论文。1999年我做助教和读硕士后的时候，还做了一个张家港的乡村建设项目，今日来看依然是具前瞻性和启发性的。这些也间接地促成了我后来博士论文最终选题的确定。选题期间有一个选项是比较中意这两个具代表性的小镇，为此我专门去撒比奥内塔做了实地调研。但因为当时教师人手紧张，系里希望我尽早回去，就放弃了这个比较研究的方向，转向解析同里的江南传统水乡城镇系统研究。

问：您在瑞士期间参与了克莱默教席在非洲的设计项目，可以介绍一下团队协同设计的模式吗？

在瑞期间，先后参与了教席在坦桑尼亚的两个实际项目，都是受意大利天主教会的委托，在当地做的公益性项目，1所女子培训中心，1座教堂。前者我参与了设计的全过程，和同事一起完成了建筑设计及文本制作，之后到现场与荷兰的施工图深化设计公司及来自中国的工程建设公司交底。期间，我们住在教席之前设计的旅舍中，参观了教席已建成的学校、教堂等项目，对教授的设计思想和空间追求有更深的体会。

实际项目的设计思想、流程与设计练习如出一辙，从场地环境、功能策划与空间类型、营建技术三个方面入手，设计分为体量模型、结构模型与建筑模型三个阶段。与练习不同的是，第一个阶段设计就结合了项目预算与建设时序计划；结构模型阶段除了结构-空间体系的深化外，还加入了环境适应性考量，并落实到了气候边界的形态及材料构造设计；建筑模型阶段深化了重点空间和家具设计，以及建筑构件系统、墙身屋顶构造等，并整合了水电等设备系统的设计。整体推进的设计流程有清晰的框图，充分体现了教授一贯的设计研究性。有趣的是，我按照中国的作图习惯，标上了轴号和三道尺寸，使得交底工作，尤其是与中国的工程建设公司，变得十分顺畅。

问：您在去瑞士之前已经是东南大学建筑系的教学秘书，后来又担任二年级设计教学的组长，到克莱默教席之后，您觉得当时东南大学和ETH的教学有什么不同？

去瑞士前，我教了1年的一年级和2年的二年级，一年级是改革过的整体基础教案，二年级是平行于教改实验班的传统教案。我是在从瑞士回来后的2001年，开始担任二年级设计教学的组长。

到教席之后，首先是从学生的角度来理解设计基础教学法的，第二年的硕士后项目也是教学方法论方向，同时作为助教，直接参与了高年级设计课程前期的教案准备、课题试做、过程讨论和评图，试做最终发展成一份完整的设计文本。相较于第一年以建筑语言与设计方法为主的学习，第二年从教与学两方面的体验让我对教学的差异有了更多认知。彼时东南大学的设计教学已经历了十余年的教改，二年级丁沃沃、张雷两位老师的局部教改也在热火朝天地推进中。但我之前对教改的理解还是停留在成果表层，执教时对教案的理解也很局限。来了瑞士之后，才真正明白设计教学可以不是传统的依凭个人经验和灵感的师徒制传授，而是基于共同的语言与清晰的逻辑、围绕建筑的基本问题而不是基于建筑类型展开的理性化教学，设计可教，设计教学亦有章可循。这个认知不只让我对设计教师的角色重新定位，而且也极大地影响了自己对设计教学与研究的理解。

问：回国后，对二年级设计教学的组织受到克莱默的哪些启发？有跟他进行后续的讨论吗？

2001年回国后受龚恺老师安排，我开始接手二年级设计教学的组织工作。在龚老师的支持下，与朱雷、陈秋光等老师合作，我们延续了丁沃沃、张雷老师教改的理性化思路，重新构建以空间为核心、以类型为载体的设计教案，“空间、环境、建构”三线并进，以由简入繁的4类空间类型为线索组织系统化的教学内容，并在教学实践中着重探索与之相对应的结构化的教学法、多元媒介的设计辅助手段和有序的教学组织，使教与学的过程具有更好的秩序性、可操作性和弹性，力求帮助学生建立整体的建筑观和有效的设计思维方法。当时教学组多为年轻老师，尚缺乏教学与实践经验和成熟的设计思想，因此一个现代理性、操作性强的教案是培养年轻老师也是保证人才培养质量的有效手段。

2003年12月东南大学召开了“建筑教育国际论坛：全球化背景下的地区主义”，克莱默教授应大会之邀做了主题发言，我做现场翻译，同时我也在会上发表教学实践的初步成果《教与学的有序组织——建筑设计的教学方法初探》；之后于2005年在全国专指委的年会上发表了《结构有序的教学法研究与实践——关于二年级建筑设计教与学》；2007年出版《空间的操作——东南大学建筑学院建筑系二年级设计教学研究》一书。同年，我开始负责三年级设计教学。

克莱默教授教学思想的启发不仅仅体现在我个人对设计教学的理解以及之后对二、三年级设计教学教案的架构与执行上，还间接地促成了东南的建筑设计基础理论课程的重构。2002年，在韩冬青老师的主导下，原先以“建筑设计的基本原则和方法为主要内容”的《公共建筑设计原理》课程，被扩展成4门贯穿二、三年级的《建筑设计基础理论课程群：空间论、建构论、环境论、方法论》。不难看出教授的三元图解，建筑基本问题及其关联是课程群架构的基本参照。2004年我接替孙茹雁老师承担了第三部分主讲，课程更名为《建筑环境与设计》，确定了课程内容重点围绕场地设计与场所营造（site planning, place making）展开，也是依循同一个逻辑。

那些年，无论是设计还是理论课的框架都先后有与教授当面交流和汇报过，他对理论与设计课程的并行互补也表示赞同，认为学生通过理论课程进行的“补课”也十分有助于设计课的理性教学。

问：您认为克莱默的“基础设计”教学法当今有何现实意义？

任何人为的事物都有其产生的背景和局限，教授的教学法建立在现代建筑理论与实践的基础之上，并承袭了得州骑警的现代建筑教育思想。而现代之后50年来建筑又经历了更为多元的发展，其教学法的基本“空间-环境-建构”的三元图解已显得单薄而局限，“体块-结构-建筑”的三阶段模型法也容易陷入操作的程式化窠臼。但我认为其对当下的设计教学依然具有一定的启发性。其一，以当代建筑理论与实践为背景，关注建筑的基本问题，抽象出核心关键词和空间原型，并发展出相应的设计方法，使设计与教学成为有章有序的逻辑过程；其二，强调建筑是多维度、多要素的整合，在任何一个练习阶段，始终在整体框架下探讨建筑空间与环境场所、结构构造的依存关联；最后，教学中借助模型为媒介推进设计，在设计进程中强调不同比例下媒介工具（如模型、图纸等）的尺度、材料等运用，要与设计问题与设计深度相匹配。

3 影响 | Impacts

《关于基础设计》封面，1989 | The cover of *On Basic Design*, 1989

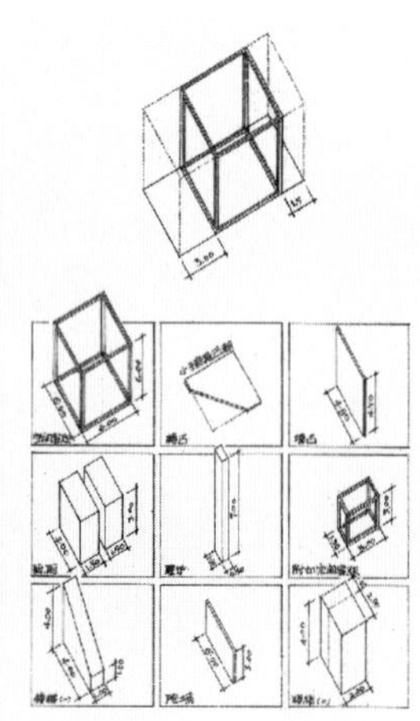

《建筑设计基础教程》, 1991 | *Basic Tutorial for Architectural Design*, 1991

顾大庆于1989年秋离开ETH，离别之际，克莱默将多年的教学体悟以短诗形式写成4页的《关于基础设计》寄语他。回到东南大学，顾大庆基于过去两年的经历着手新一轮的教案设计，首次将克莱默的“基础设计”教案引入中国。由于当时新学年已经开始，因此在原1986/87年教案的基础上嵌入了菜场建筑测绘及以“基础设计”为蓝本的社区中心设计（包括煤气站，菜场摊位及小商店等）。其突破之处在于对建筑空间形式问题的讨论先于功能、场所及经济因素，而其后又能与诸因素关联起来。

1990/91学年的每一个设计练习开始时都以任务书的形式明确了设计内容、给定的元素、工作过程、评判标准及时间表。这套教案所含讲座及学生习作随后编成《建筑设计基础教程》。形式上，1989-1991年的教案集合了传统的渲染练习、平面图形研究及克莱默式的设计问题。1991年秋顾大庆再度赴瑞，但教案仍在东大使用。

1997年，龚恺从美国访学归来，开始负责一年级设计教学，此前他也曾在克莱默教席访学一年。龚恺调整了练习的顺序，使学生从较为熟悉的环境和物品进入设计的范畴，最终进入建筑空间的研习。因此1997年教案的每个设计题目之前都设置了观察、测绘或案例分析类的作业。原教案中设计问题所严格设定的条件也变得宽松。在第二学期的结尾加设了轻质地标设计与搭建，以呼应学年一开始时的文具架制作。

一些“准”建筑设计练习则被取消，如“立方体练习”。因为龚恺在教学中发现，由于功能条件的缺失，教师越来越难评判这些抽象作业的优劣；同时，经过好几届学生的探索，新的形式可能性几乎已被发掘殆尽，学生因而越来越重视设计表现，而非

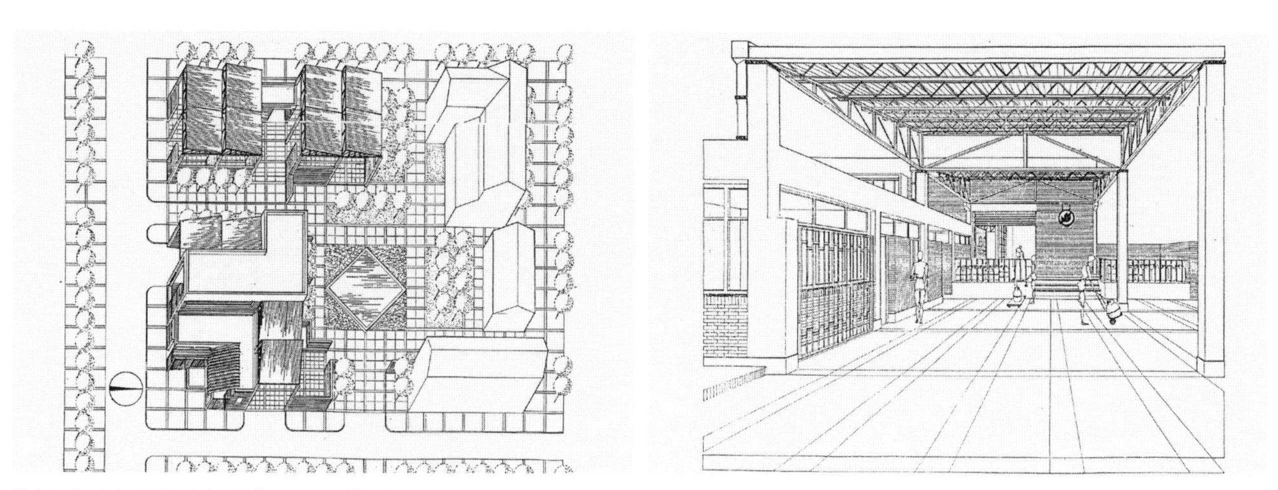

学生作业《建筑设计基础教程》, 1989/90学年 | Students' work, Basic Tutorial for Architectural Design, 1989/90

Just before Gu's first return to Nanjing in the autumn of 1989, Kramel wrote him a four-page long poetic note on architectural education titled "On Basic Design". Upon his arrival, Gu commenced a new program based on his experience in Kramel's teaching group. The 1989/90 program was the first attempt to transplant Kramel's Basic Design program. The exercises inserted into the old program included the building survey of the market on Shanxi Road and the design project of a community centre based on the setting of the Basic Design program with a gas station, market booths, and a store. It was a breakthrough from previous forms of teaching in the sense that it was able to discuss architectural formal operations prior to factors of function, place and economy, yet was later able to connect with these factors.

In the 1990/91 program, an assignment sheet describing the content, the given elements, the working process, the criteria, and the schedule was distributed to students at the beginning of each exercise. The lectures and students' work from these two years were edited into a booklet titled "Basic Tutorial for Architectural Design". It was a combination of the traditional rendering exercises, graphic study, and design projects akin to Kramel's Basic Design program. The program was continued by other teachers when Gu left Nanjing for ETH again in the autumn of 1991.

In 1997, Gong Kai became responsible for the first-year design studio when he came back from his academic visit in the USA. He had also been Kramel's teaching assistant. While retaining most of the content of the previous program, Gong changed the sequence of the exercises in order to start with more familiar surroundings and end with architectural space. In the 1997/98 program, each design project was preceded by tasks of observation, measurement or precedent study. The rigid setting of the kit-of-parts in the projects became flexible. While the making of a stationary at the beginning of the program was kept, the building of a light-weight landmark was added to the end of the second semester.

Some of the quasi-architecture exercises were removed, for instance, the cubic exercise. Gong observed that it became increasingly difficult for teachers to critique the output of the exercise due

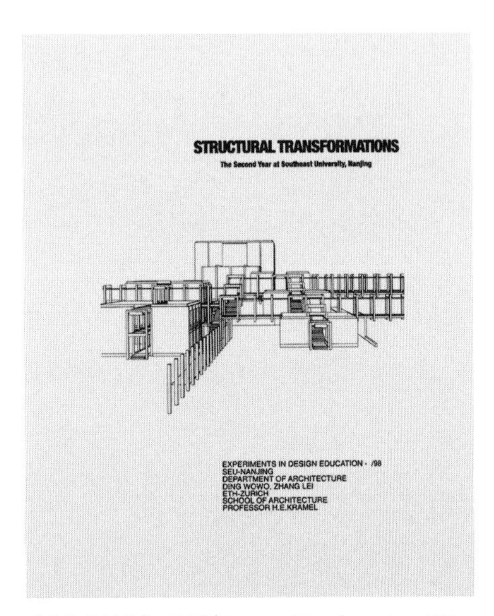

《结构性转化》，1998 | *Structural Transformations*, 1998

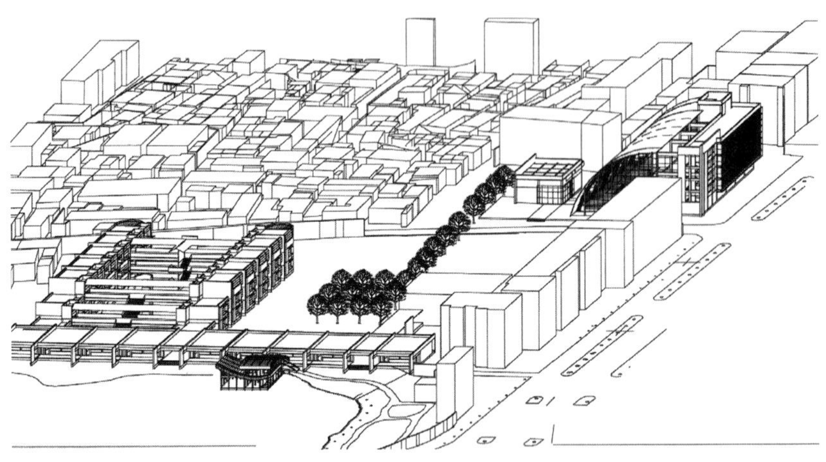

学生作业，东南大学二年级，1997/98学年 | Student's work, 2nd year studio at SEU, 1997/98

过程本身。此外，最终撤销了愈发与新教案格格不入的渲染练习。

丁沃沃于1989年夏末从苏黎世回宁，此时新的一年级教案已经施行了几年，取得一些效果。如何在后续教学中巩固形式的理性思维方式？她从最基本的设置开始：给出具体的场地，以模型作为设计工具、强调结构和构造等。

最大的转变是题目设置从按建筑类型变为按问题类型。围绕克莱默的建筑要素图解中的三个方面，新的教案由4个复杂度递增的设计项目组成，训练相似的设计方法：1. 空间限定（基于基本几何体量的单一空间）；2. 空间序列（重复式空间组织）；3.多种功能空间的组织；4.复杂流线的综合空间组织（交通建筑）。场地的复杂程度也渐次提高。构造在各个项目中体现，在各阶中要求绘制建筑围合、楼梯、铺地或结构设计的详细图纸。

然而这一教案仅在丁沃沃所带的学生小组中短暂试行，直到1997年她再次从ETH返宁后才和张雷一同继续打造。参加1997/98新一轮试验的6名学生同时借助了计算机辅助设计。主要的设计工具仍是实体模型，每个子项目要求制作1:20的细部模型。当时学生在计算机使用CAD R14软件进行三维建模，根据不同阶段生成体量模型、结构模型和建筑模型。图纸用电邮发送到苏黎世，克莱默和当时在其教席的吉国华依据自己对图纸的理解进行编辑并提出反馈建议，再传回南京。教学试验完成时，苏黎世方面也完成了作品集《结构性转化》。

经后续调整，1998/99年教案的设置回归克莱默式的条件限定，类似于其居住单

设计条件设定（左）及学生作业，东南大学二年级，1998/99学年 | Given conditions (left) and student's work, 2nd year studio at SEU, 1998/99

to the absence of function. As time went by, most of the resulting possibilities had been explored, thus students focused more on the presentation of the design than the process. The rendering exercises, which was no longer compatible to the goal of the new teaching, was eventually eliminated.

When Ding Wowo returned from Zurich in the late summer of 1989, she attempted to change the second-year design studio, starting with a few basic settings, such as by provideing a specific site, using physical models as design tools, emphasizing the structure and construction aspects, in order to reinforce the idea of logical development of form established in the first-year program.

One of the main alternations was to switch from the building-type mode to the problem-type mode. The program built on the three basic issues listed by Kramel. It included four successive projects of increasing spatial complexity that train similar design methods: 1. space definition (the development of a single volume space out of basic geometry); 2. spatial sequence (the organization of repetitive spatial units); 3. space organization of a complex program; 4. synthesis space with complex circulation (a communication building). The level of complexity of the site condition increased in the projects. In terms of construction, detail drawings were shown in different stages of the program to exemplify the construction of the building envelope, stairs, exterior pavement and structure design.

The program was first implemented among a small group of students and only for a short time. It was not until 1997, when Ding returned from her second stay at ETH, that the experiment was picked up again and modified in cooperation with Zhang Lei. The experiment was made with a small group of students in the academic year 1997/98, working with computers.

The design process was also based on physical models. A 1:20 detail model was required for each project. 3D modelling was used in different stages to generate volumetric modesl, structural models and architectural models. The drawings were done on PCs with CAAD R14 and then transferred to Zurich via e-mails. Ji Guohua and Kramel organized these drawings into documentation according to their own understanding and gave feedback to Nanjing. The documen-

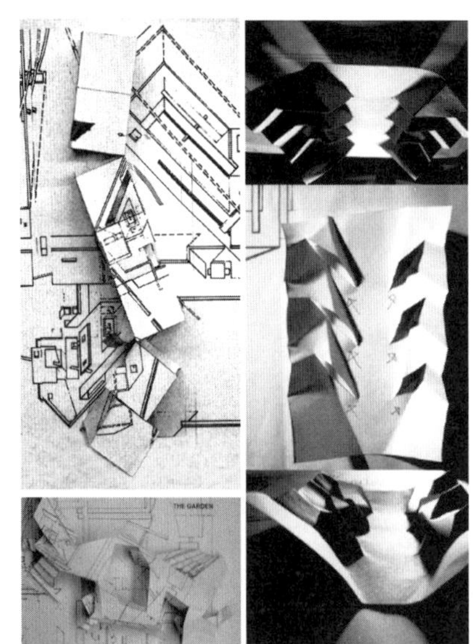
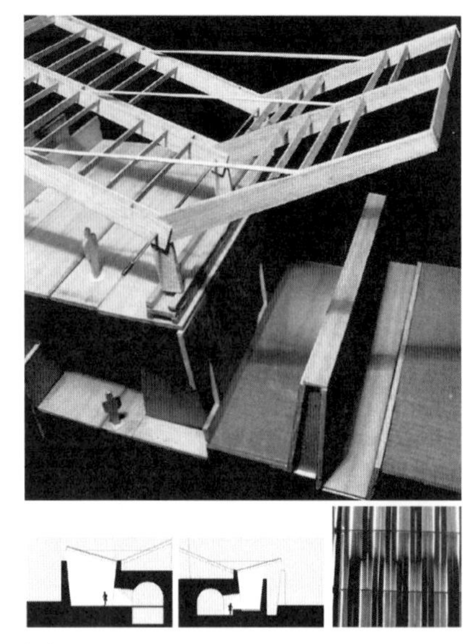

南京大学建筑研究所2003/04学年研究生设计课程学生作业：（左起）《基本设计》《概念设计》《建构研究》| Graduate students' work: (from left) Design Basics Studio, Tectonic Studio, and Conceptual Design Studio, GSA, 2003/04

元设计：框架（18.0mx7.5mx7.5m），基座（18.0mx7.5mx1.5m），1至2个更小的体量（作为功能块），以及若干水平向和垂直向的板片。元素的摆放需基于模数1.5m。

从学生作业来看，一旦被模数化的结构框架所规限，操作几何体块或装饰性构件的倾向便大大收敛。学生的注意力转向了建筑的功能块与外围护结构之间的关系、板片的放置与其生成的空间、引桥联系茶室与水岸的方式。立面上丰富的元素也非出自视觉上的意趣，而源于功能或建造。

至此，东南大学一、二年级的设计教学都实现了转型，高年级的教学亟需改变。鉴于一年级引入了建筑设计的基本概念和方法，二年级着重于多种空间的组合，那么三年级应面向有所侧重的综合设计问题。韩冬青、赵辰、吉国华、李飚等教改小组成员决定以4个专题设计为切入点。每一个设计题目针对克莱默的三对建筑要素（功能/空间，场地/场所，材料/建造）之一，而其余两对要素则作为修正因子。

南京大学建筑研究所GSA的设立为它的创建者们提供了一块难得的白板。他们对于建筑及建筑教育认识的统一程度实为国内其他院校所罕见，基于建筑作为一门学科及一种职业的想法建构了研究生教育的框架。新系统的设计与他们在ETH的经历息息相关。与一般院校中研究生从一开始就明确了导师并很快参与工作室的活动不同，南大研究生入学后首先经历若干个高强度的专题设计，以此重构不同本科背景学生的知识体系，提供研修的共同基础。“基本设计”围绕三对建筑基本要素展开；“建构研

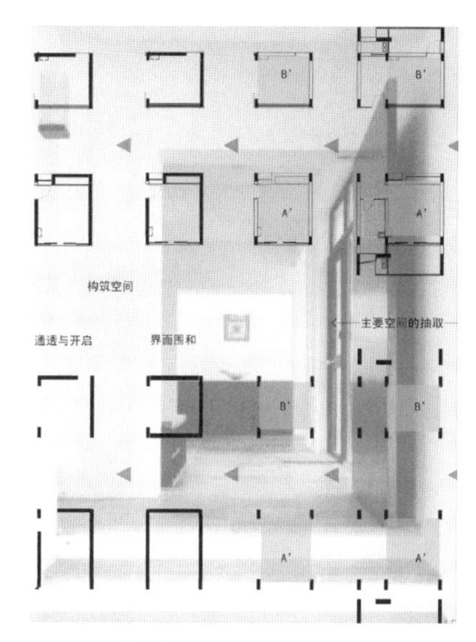

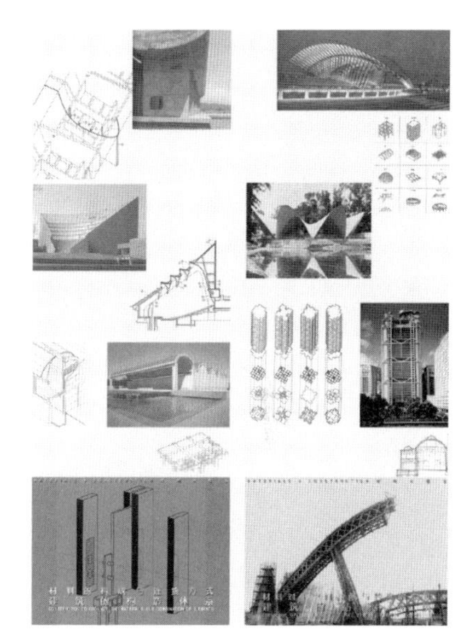

南京大学建筑研究所研究生讲座课程：（左起）“建筑设计方法论”，“现代建筑设计基础”，“材料与建造” | Lectures for the graduate program at GSA: (from left) Design Methodology, Preliminaries in Architectural Design, Materials and Construction

tation was finally edited into Structural Transformations. Kramel represented an educational experiment, interacting between SEU and ETH. After a few attempts, the final version of the assignment setting in the 1998/99 program returned to Kramel's tradition of strong restrictions, resembling that of a two-story living unit. A frame (18.0mx7.5mx7.5m), a foundation (18.0mx7.5mx1.5m), one or two small volumes and a number of horizontal/vertical planes were given as input. The position of the elements was based on 1.5m modular.

Once restricted by the modular structural frame, the intention to manipulate geometric volumes or decorative components was diminished. The concern was shifted to the relation between envelop and core, the position of the planes and the zones generated by them, and the way the entrance bridge connected the shore to the building. Elements that enriched the facade were added not out of visual interest but for functional or construction reasons.

Until 1998, the freshman and sophomore year program at the department of architecture at SEU had been reformed. The senior year, however, had remained unchanged. Given that the freshman year introduced the basic concepts of architectural design and design process and the sophomore year dealt with space configurations with increasing complexity, the third year was to tackle synthesis design problems with different priorities.

The teaching team was led by Zhao Chen, Han Dongqing, Ji Guohua, etc. They decided to arrange the program thematically into four phases. Each of the first three phases took one of the three determinants (function/space, site/place, material/construction) as the dominant factor, appointing the other two as correction factors.

The establishment of Graduate School of Architecture at Nanjing University provided a tabula rasa for the group of reformers from SEU. The high consistency shown in the concept of architectural education shared by the founding group of GSA and the recruits they selected was rarely seen in other architectural departments in China. They constructed a new framework for graduate education based on their understanding of architecture as a discipline and as

香港集装箱建筑 | Container architecture in Hong Kong

究”系列课程以多个专题指向材料与结构、建构与视觉、构思与建造之间的相成及冲突；“概念设计”则反映了克莱默之后ETH的建筑教学影响。系列讲座课程提供了相应的知识背景。

顾大庆从1994年开始于香港中文大学任教。两年后，柏庭卫结束了在日本三年的教学再次与之成为同事。此前他们曾在克莱默教席长期合作，建立了深厚的友谊和默契。2003年朱竞翔也来到中大。在此，他们的教学与研究工作相得益彰。

顾大庆和柏庭卫希望建构香港环境中的教学，因为单纯依靠遥不可及的西方经典建筑难以引起本地学生共鸣。他们发现集装箱建筑是理想之选——它首先是现代工业化制造的产物，而后经不同使用个体之手又转变为乡土建筑，从而勾连了原型与场所。集装箱单体是一个空间单元，它同时又可视为空间定义元素。克莱默的基本建筑单元在这个港口城市找到了完美的对应物。

集装箱建筑应用于教学的想法产生于1995年，经过两年的酝酿后正式实施至2000年。与“基础设计”教案的结构相似，“集装箱建筑”包含不同尺度上的设计问题，从单元内部空间到建筑群组织，在建筑入门教学中涉及了地域文脉、无名建筑及日常观察等问题。装配部件的思路在2012学年开始的“建筑设计入门1”中延续。

在研究和教学中，顾大庆和柏庭卫对建筑设计形成了新的理解，认为建筑形式本身是设计的核心要素之一，与场地、使用及技术一样，有其自身的知识体系和逻辑。

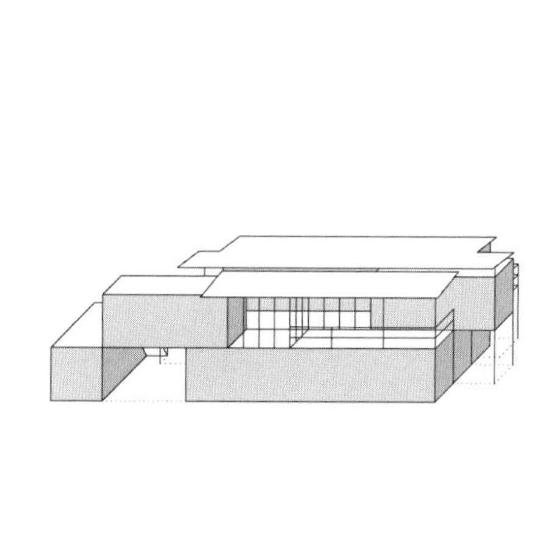

集装箱组合的基本方式 | The manipulation of containerss

a profession. The construction of the new system is closely related to these founders' experience at ETH. Different from the conventional graduate education, it built on correlated design studios and workshops that fostered intensive thematic studies. By doing so, the school reconstructed the knowledge system of students with different under-graduate educational backgrounds and established a common ground.

The Design Basics studio revolved around three pairs of form-deciding elements. The Tectonic studio addressed the inter-dependency and conflicts between material and structure, between tectonic and visual expressions, and between conceiving and constructing from various aspects. The Conceptual Design studio reflected the influence of the teaching at D-Arch after Kramel. A series of lectures provided the appropriate knowledge background.

Gu started teaching at Chinese University of Hong Kong in 1994. Two years later, after three years of teaching in Japan, Bertin became Gu's colleague again. They had built up a life-long friendship and rapport through years of cooperation in Kramel's Chair. Zhu Jiangxiang came in 2003. What they have established at CUHK is a beneficial equilibrium between teaching and research.

Gu and Bertin intended to develop a structured program with reference to Hong Kong because the classic precedents in western culture were too remote for the local students to perceive. They saw the container architecture as an ideal means– it was a product of the modern mass-fabrication, yet was turned in the hands of individuals into vernacular architecture, thus relating a prototype to specific conditions. The container bears the duality of a spatial unit and a space-defining element. Here, Kramel's Basic Architectural Unit found its counterpart in this port city.

The idea of container architecture was introduced in the program of 1994. After two years of experiment, the program was implemented till 2000. With a similar structure to Kramel's Basic Design program, the Container Program involved various issues at different scales spanning from the space inside a unit to the organization of a cluster, integrating the regional context, anonymous architecture,

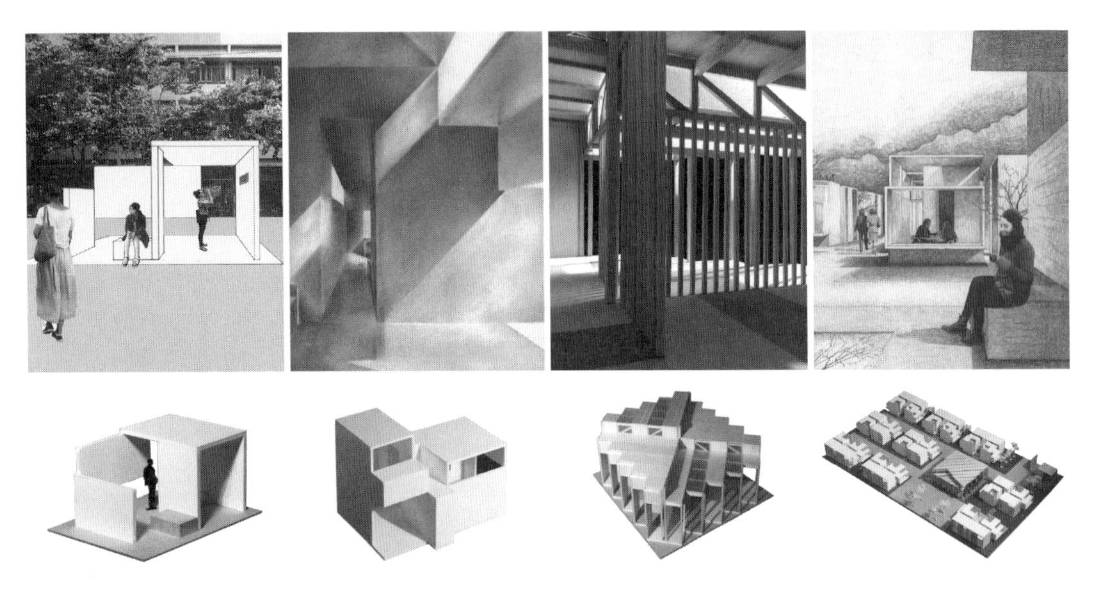

香港中文大学2016/17学年本科课程设计1，学生作业 | Students' work, U1 Studio at the CUHK, 2016/17

“集装箱建筑”之后，受马库斯·吕舍在ETH构造教学的启发，顾大庆和柏庭卫开始“建构工作室”的实验。设计操作的启动不再依赖部件化的方法，同时又秉承了从模型制作中找形的理念。设计变为在把玩基本建筑形式构件（杆件、板片或体块）的过程中发现形式规律，进而在使用需求、场地、结构和材料等条件约束下不断演化的过程。此入门教案对形式语言连贯度及空间效果的强调远重于对使用功能的满足，促进学生更主动地运用想象力及研习先例。

“建构”教案结构经历了几个阶段的演变。起初由于学院当时的垂直工作室制度，教案是一些抽象练习和设计项目的结合。后来顾大庆和柏庭卫尝试将整个学期编排成一系列主题相关的练习。在2008/09学年，垂直工作室实行的最后一年，教案回到克莱默式的结构——用进阶的设计练习完成一个设计项目，融入场地及使用的问题。这种“练习化”的设计课程模式继续在2009年之后顾大庆主持的“建筑设计入门2”及2012年开始的“建筑设计入门1”中发展。近年来“建构”教程通过持续的教学与学术交流影响着国内院校。

上述这些教学尝试的共同之处在于均以理性及批判性的思维探索建筑设计的核心问题：空间与形式。在这种意义上，它们挑战了建立在美学熏陶及经验积累基础上的传统建筑教学，实现了建筑教育的现代转型。以下展示其中的7个教案。

香港中文大学2016/17学年本科课程设计2，学生作业 | Students' work, U2 Studio at the CUHK, 2016/17

and daily observations into the introductory design education. The kit-of-parts method involved in the Container Program has been applied again in the Unit 1 Studio since 2012/13.

In their research and teaching, Gu and Bertin have formed their own understanding of architectural design, holding that architectural form itself is one of the core elements of design. And, like site, use and technology, it has its own knowledge system and logic.

After the Container Program, inspired by Markus Lüscher's construction program at ETH D-Arch, Gu and Bertin started Tectonic Studio. The initiation of the design operation was liberated from the kit-of-parts method, yet the essential idea of pursuing formal concepts through model-making was retained. Design became the process of discovering formal orders in the operation using basic architectural forms (sticks, slabs or blocks) and in turn, the evolution amid restrictive conditions of program, site, structure and material. In this introductory program, the emphasis on functional use gave way to the consistency of formal language and spatial effects. With the new program, students were more motivated to use their imagination and draw support from precedent study.

The tectonic program went through several phases of development. In the beginning, it was a combination of abstract exercises and design projects. In the next phase, Gu and Bertin tried to organize the whole semester with thematic exercises. In 2008/09, the last year of thematic studio, they returned to Kramel's method by making up a design project with a series of exercises and naturally embedding in them issues of site and use. This pattern of "exercised" project was further developed in the Unit 2 Design Studio after 2009 and the Unit 1 Studio after 2012. The tectonic program was brought to mainland China through a close bond between Gu and the School of Architecture of SEU.

These pedagogic endeavors based on logical development and critical thinking challenged conventional beliefs in time-consuming aesthetic cultivation and long-term experience. In this sense, they have contributed to the modern transformation of China's architectural education. Seven of these programs have been chosen to be presented here.

东南大学 1990/91 一年级 | SEU Year 1, 1990/91

建筑设计基础 | Basic Design Course

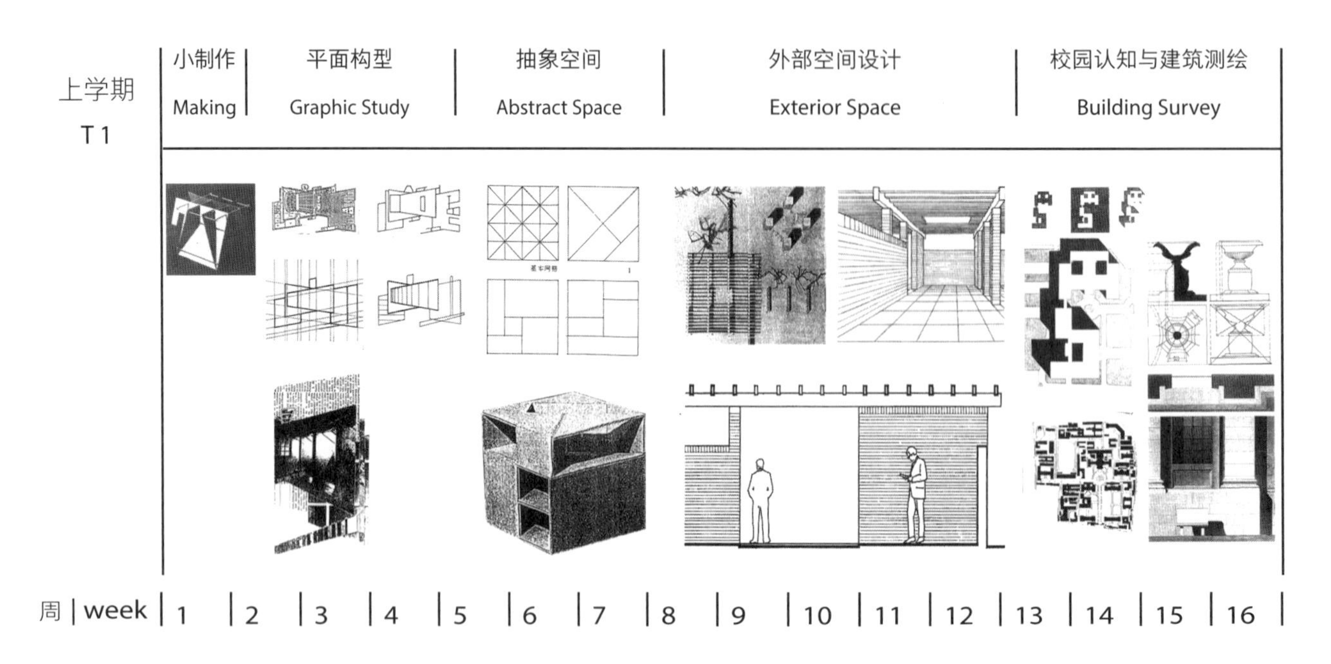

顾大庆将建筑入门教育的目标总结为4个方面：培养设计意识、训练解决问题的能力、发展赋形能力、训练表现技能。对应的，教案围绕3条线索——建筑设计原理，入门设计练习以及基本表现技法。第一阶强调设计与生活的关系，呼吁从日常观察中学习设计。第二阶培养形式敏感度。第三阶是对1986/87年教案中“立方体练习”的改进。为建立评判优劣的法则，新教案要求实现三对关系：垂直与水平分隔，实体与虚空，围合与开敞。第四阶类似于克莱默教案中的第一阶“入口设计”。校园观察要求学生在不同尺度上运用所学概念（体量、空间、空间限定、人体尺度、材料、建造等）。第二学期关于设计方法。初次尝试了克莱默用一个场地统领多个子项目的方式。案例研习注重挖掘设计品质的来源——对组织方式的研究。

Gu summarized the mission of preliminary architectural education in four aspects: the cultivation of design awareness, the training of problem-solving abilities, the development of form-giving abilities, and the training of representation skills. Consequently, the program had three clues: the principles of architectural design, the preliminary design exercises, and the basic representation skills. The first phase was an introduction to design and its relation to daily life. It aimed to arouse students' awareness of design in their daily experience. The second phase helped students develop their sensitivity to form. The third phase was a revision of the Cubic exercise in the 1986/87 program. In order to establish the criteria for selection, the new exercise required three pairs of relationships: vertical and horizontal divisions, solid and void, enclosed and open. Phase Four resembled Phase I in Kramel's program. In the next phase, students were asked to observe the campus using the concepts they've learned (mass, space, space definition, human scale, material, construction, etc.) at different levels. The second semester moved on to design methods. Kramel's holistic program mode (e.g. to organize all the buildings on a single site) was used. Precedent study aimed to explore the source of quality, which led to the study of organization.

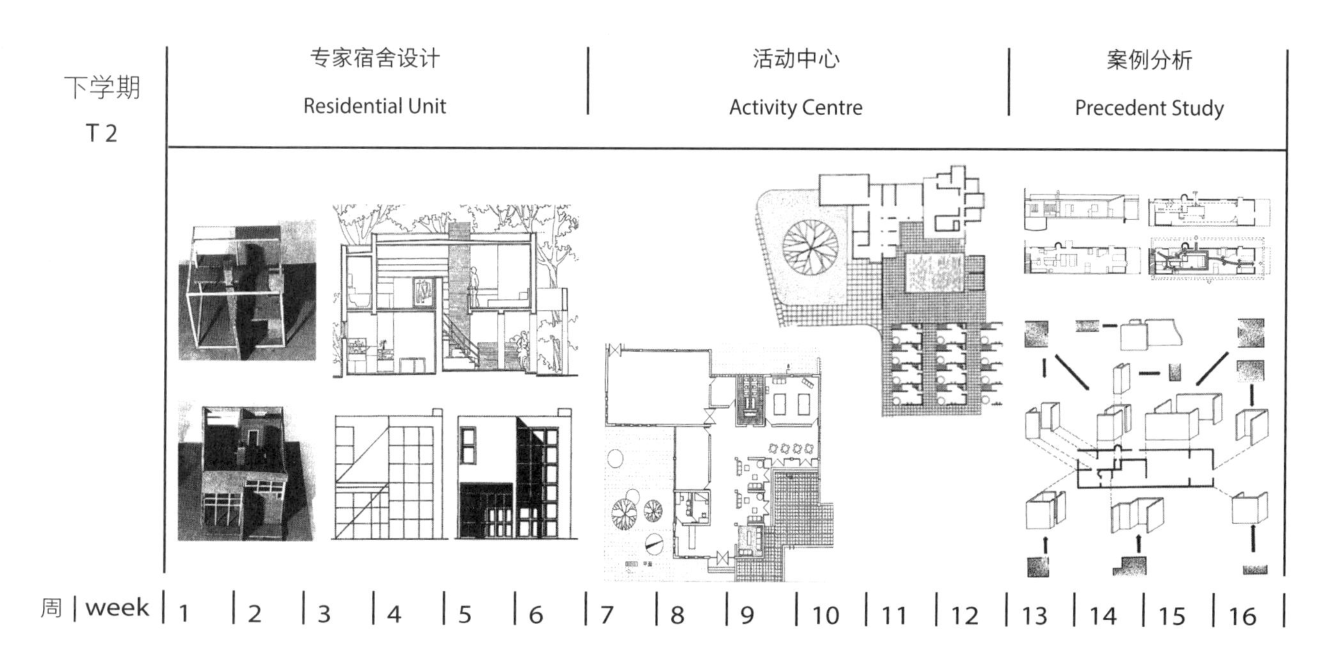

东南大学 1997/98 二年级 | SEU Year 2, 1997/98

课程设计 | Design Studio

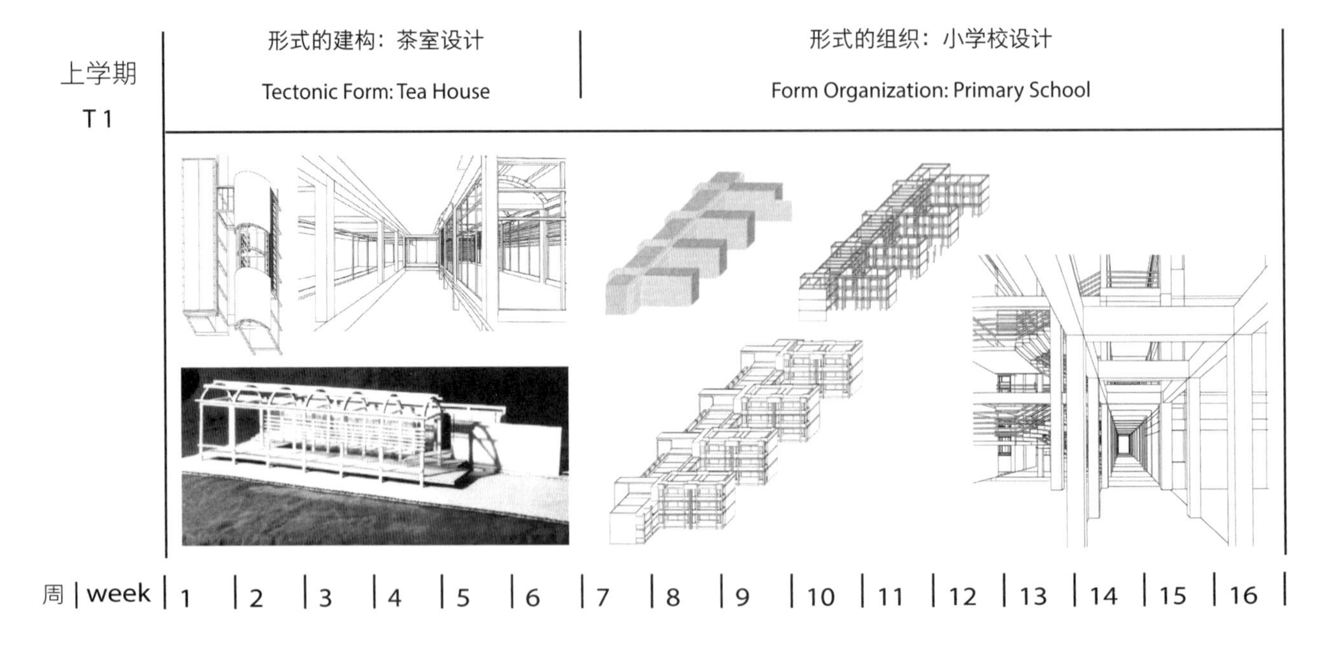

这一个教案中4个题目的共性来自于模块的使用及结构框架和作为填充物的建筑围护之间关系的连贯性，换言之即相同的形式语言。学生实际上是在尝试预设形式系统的多种变体。

题目一关注点在于建造——木建筑节点的建构表达。题目二是重复性的空间单元组合。设计需满足多种活动相应的视线及光线要求，并以合理的流线联系。场地设在一个平坦的城市街区。虽然普通砖结构适应的结构跨度及灵活性不大，但因为这个项目的尺度及空间特性，选用砖结构呈现特定的材料感。题目三设计的主导因素是场地：城市边缘的陡坡。老年人中心有类似题目二的空间重复性，但流线更复杂。这个题目不再提供功能体块作为前提。题目四的图书馆设计选址于两条主要城市道路的交汇处。学习尝试不同的基本建筑单元组合方式来回应场地。之后将这些体量图解转化为具有不同围合度的空间单元的体量模型，空间最后通过建筑构件来实现。

In this program, the commonality between the four phases was created by the use of modular and the consistent relationship between structural frame and in-filled envelope. In other words, the common ground was created by an identical formal language. Students were working on variations of a given formal system.

In Phase I, the issue of construction became dominant. The joints of the wooden construction gained tectonic expression. Phase II worked with repetitive spatial units. The design must satisfy the sight and lighting requirements of multiple activities and connect them with the proper circulation. The site was a flat area inside a city block. Despite its limitation in span and construction flexibility, masonry was appointed as the construction method for its inherited suitability for repetitive structure and material expression. The site condition, a steep slope, became the decisive factor of Phase III. The theme was a home for the aged with a repetitive mode similar to that of Phase II, but more compact and diverse in function. For this phase, no preset volumes were given. The site of Phase IV was chosen in flat areas at the street corner abutting a major city intersection. BAUs were used to explore possible volumetric operations relating to the site. These volumetric diagrams were then transformed into space cells of different characteristics, depending on the program. The space was, in the later steps, realized by space-defining components.

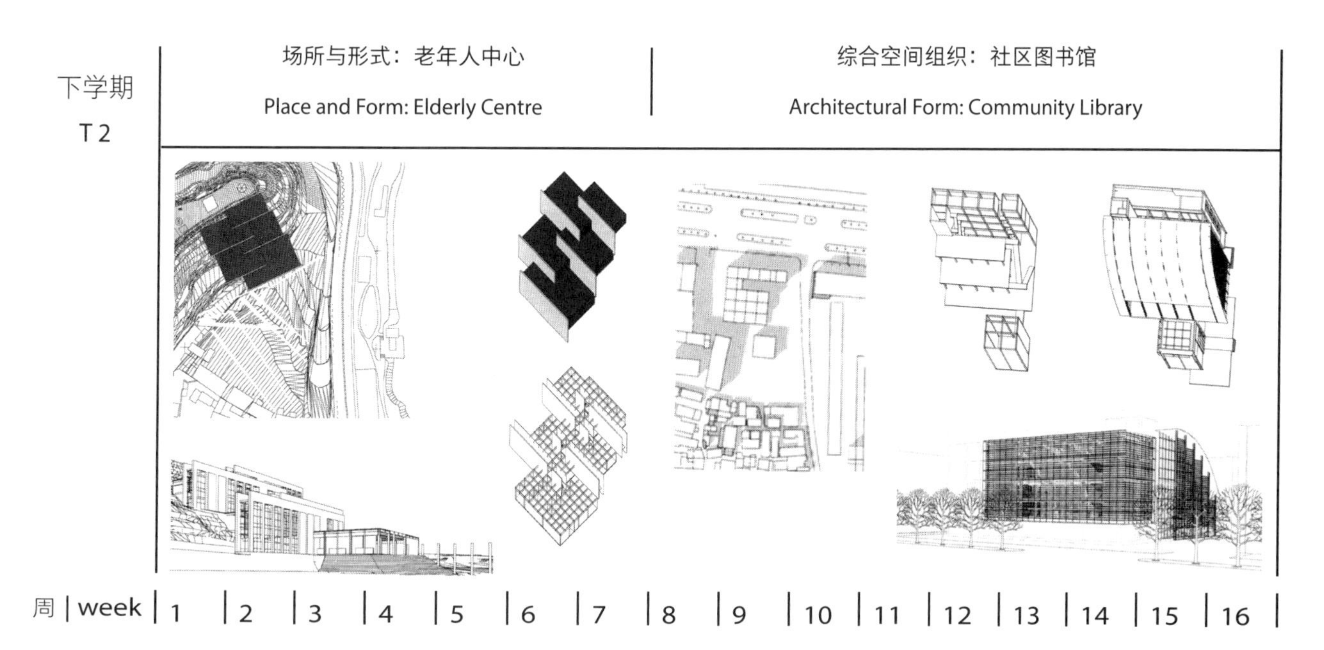

东南大学 1999/2000 三年级 | SEU Year 3, 1999/2000

课程设计 | Design Studio

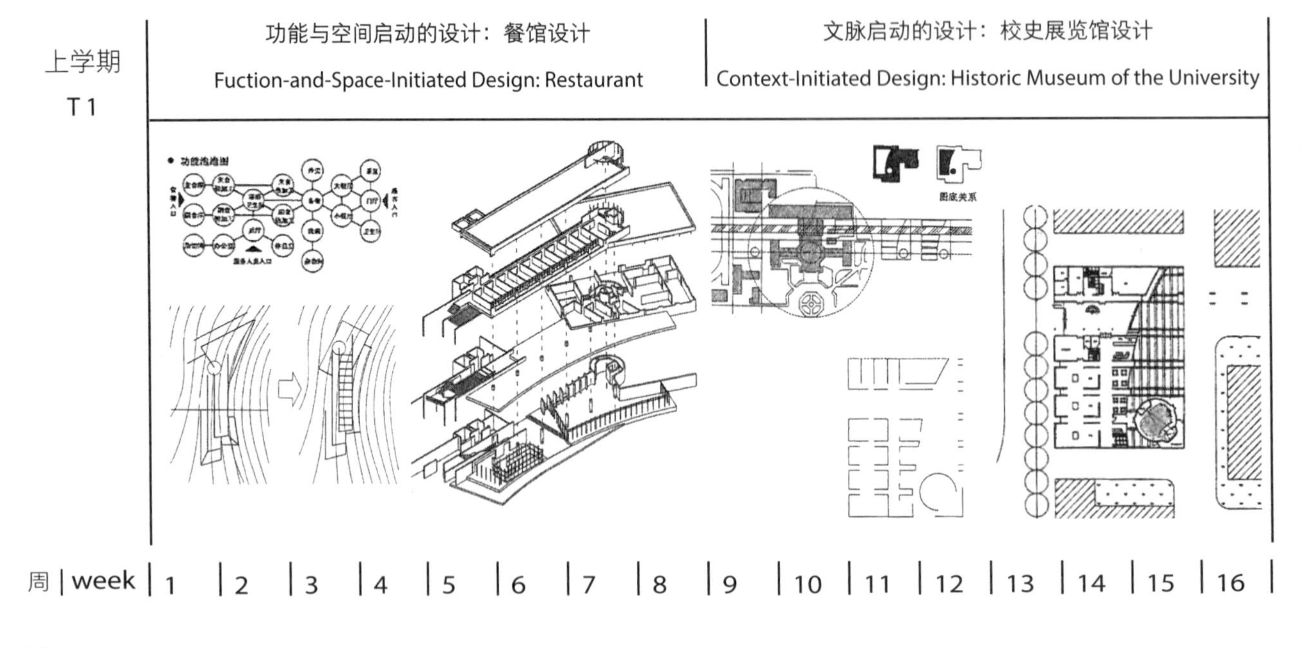

题目一以功能和流线较为复杂的餐厅为题。设计过程始于功能泡泡图和正交组织的方块图解。空间秩序来自于功能单元的组织。然后将这个正交系统叠加到坡地上，在尽量保持原有功能和流线关系的前提下适应地形。题目二则在文脉结构中寻找空间组织逻辑。学生在校园规划、建筑间距及绿化的规律中发现空间秩序的线索。作为校史展览馆，还必须符合流线、视线和光线的需求。题目三最能体现这个教案的特点，探索了技术性因素在赋形过程中的主动角色。为了凸显建构的表达，题目将往年的剧院设计改为露天影剧院，以减少声学热学方面的限制。设计前期分析大跨度建筑案例，再加入具体任务书和场地条件。设计过程中关于大跨度建筑、声学、计算机辅助设计的讲座不仅提供技术性的支持，更涉及了建筑史中的结构意义。实体模型是主要的设计工具。题目四则往往与当年的全国大学生建筑设计竞赛题目相结合，完成一个综合设计。

For Phase I, the theme of a restaurant was selected for its complexity in program and circulation. It started with a bubble diagram of function and a block diagram of orthogonality. The order of the spatial configuration came from the organization of functional cells. Following that, the scheme was positioned onto a slope, and then distorted to fit into the landscape. Phase II found the logic of space organization in the structure of context. Students searched for the clue about the spatial order in the campus planning, the spacing of buildings, and plantations. As a university gallery, it must satisfy the requirements on circulation, sight and lighting. Phase III was the signature of the program. It explored the positive role of technical factors in the form-giving process. To make the tectonic expression more prominent, the theme changed from an opera house to an open-space theatre, reducing the restrictions from acoustic and thermal aspects. The process involved first a case study of large-span constructions. Then, a detailed program of the theatre and the situation was introduced. Lectures about large span structures, architectural acoustics, and computer-aided design not only supported students in technical aspects, but also discussed the meaning of structure in architectural history. This phase used physical models as the dominant working tool. Phase IV is often combined with the "National College Students Architectural Design Competition" to complete a comprehensive design.

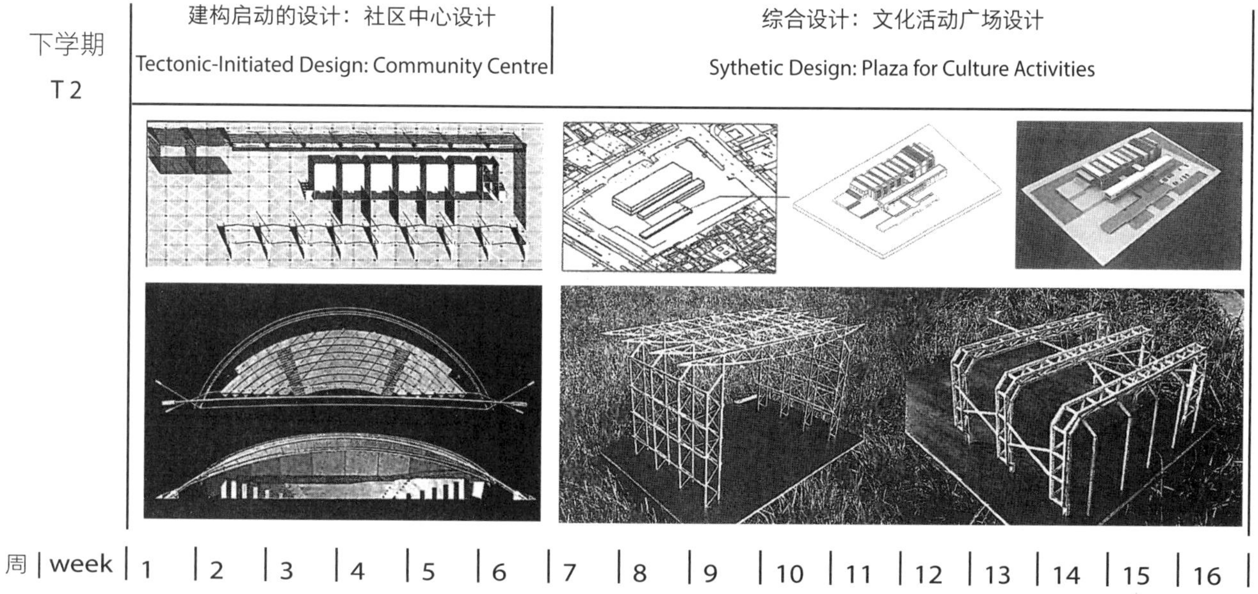

南京大学 2003/04，2004/05 硕一 | NJU MArch 1, 2003/04, 2004/05

"基本设计" | Design Basics

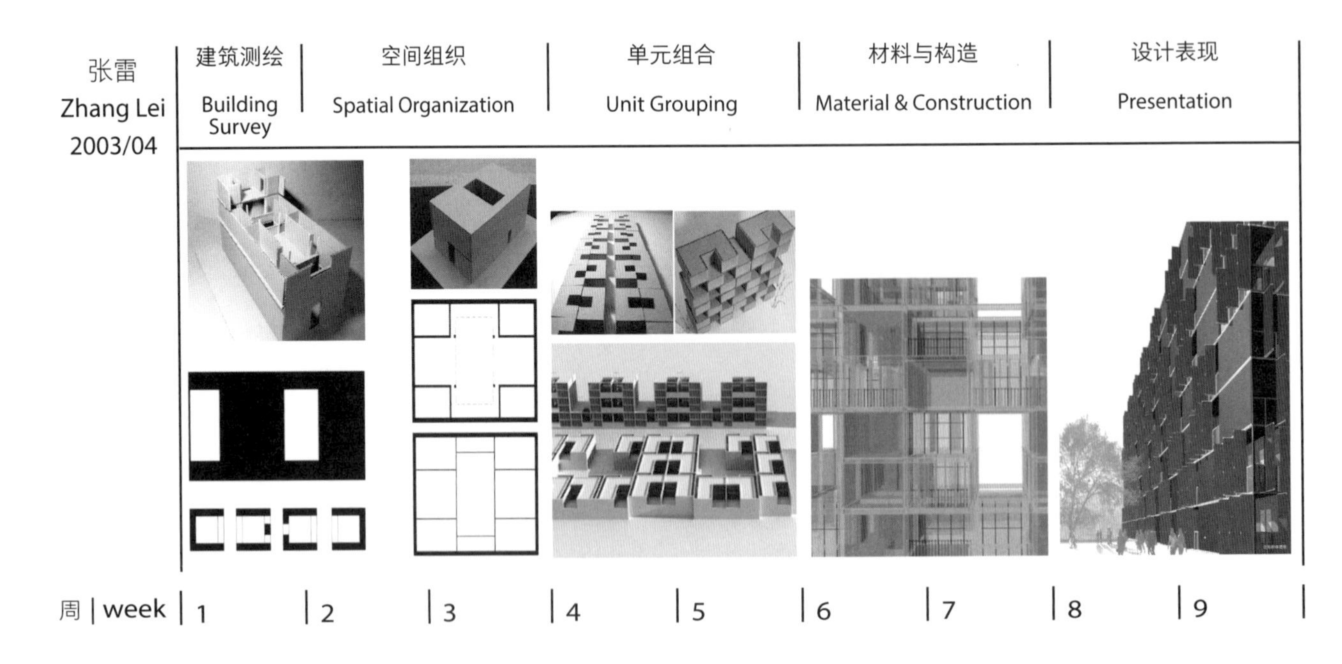

张雷的“基本设计”在教案的不同阶段分别讨论不同的设计要素。如2003年“城市院宅”以传统民居为参照，学生依次从单元内的空间组织、单元组合形成新一级空间，到材料的交接，用不同形式的模型推敲。场地组织则以3人为组，每人一种单元类型去构成建筑群，此时单元可以在水平或垂直向上复制叠加。

吉国华的“基本设计”以20世纪90年代克莱默教案中的居住单元设计为蓝本，逻辑地串联一系列设计问题，以联排式建筑为题。给出的空间框架和功能模块成为设计的起点。学生首先在没有外加条件下组织模块，然后才转向场地调研并制定任务书。此后按照这些条件调整先前的空间组织，再完善功能分区、选择结构体系、材料和建造方式。教案通过功能和场地的变化来调整。如2004/05年把场地设在某建筑的屋顶，由此限定了结构的间距和建造方式；2006年题目是狭长地块上的东西向3层住宅小区。

Zhang Lei's program referred to the core design issues at different phases of the design process. The project of urban houses with courtyards took its reference from traditional residential buildings. Students sequentially worked on the spatial organization inside a unit, the combination of units to form in-between space, and the joints between different materials, with physical models at different scales. For the unit organization on-site, students worked in groups of three, each contributing one type of unit. The assemblage of units could develop vertically.

In Ji Guohua's program, Kramel's living unit project was used as a tool to bind together a series of design issues and foster a logical development. For the design of a group of repetitive buildings, a set of frame and volumes was prepared. After the first manipulation with the given modules, students turned to site surveys, wrote briefs for their own projects, then revised the model accordingly. It was followed by further articulations of functional zones, the selection of structural system, material and construction. Ji introduced variation to the program by changing functional requirements or site conditions each year. For instance, the site for the 2005 program was changed to the roof of an existing building, which posed limitations to the structural alignment; the 2006 program was the design of a residential cluster with three-floor area types on an elongated plot along the north-south orientation.

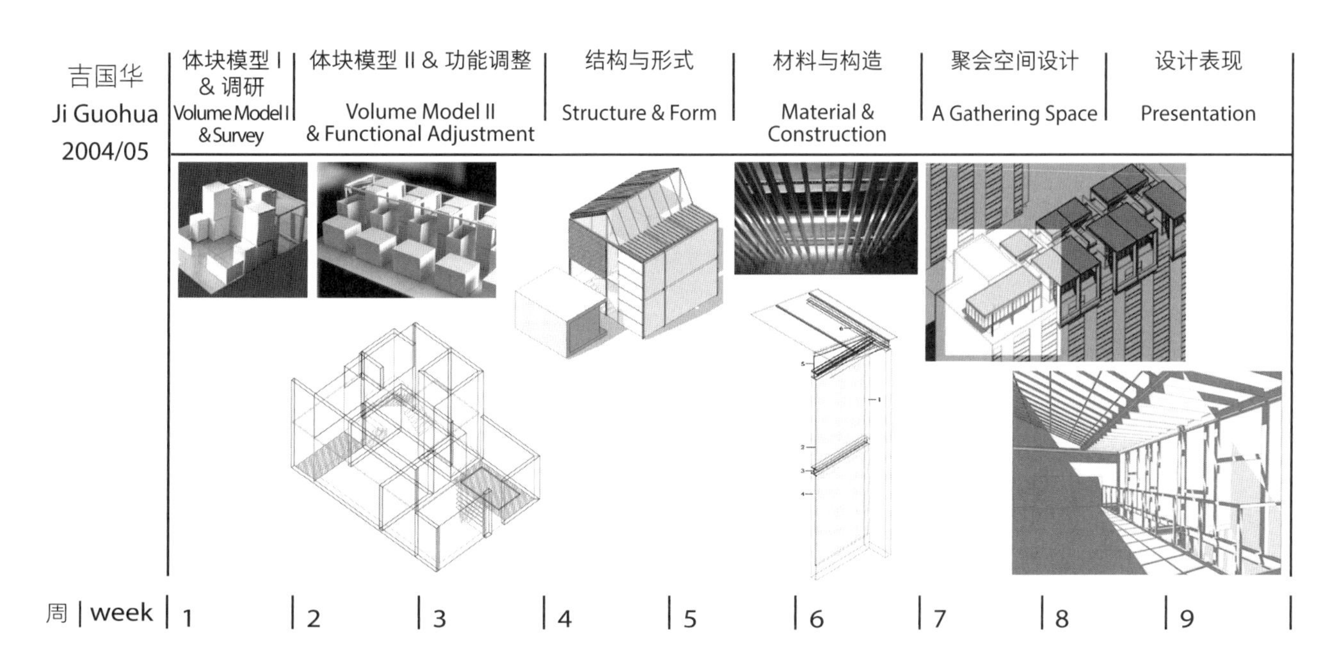

香港中文大学 1996/97 一年级 | The CUHK Year 1, 1996/97

集装箱建筑 | Container Project

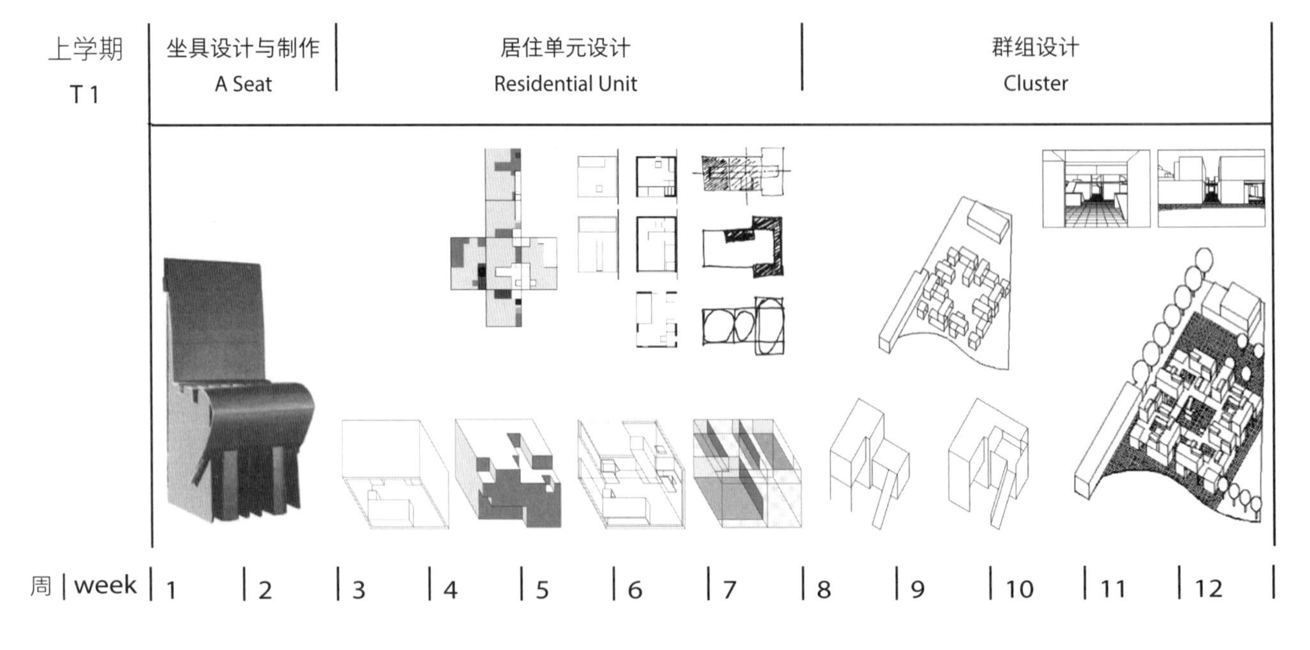

教案包括5个阶段。类似于早年在东南大学的教案，以制作物件开始，不同的是，包含了人体尺度，是集装箱建筑设计系列的准备阶段。

“集装箱”系列从居住单元切入，这些单元在第三阶用作场地组织的模块。观察与反馈贯穿了设计过程。实体－虚空的反转模型把由表皮和家具所定义的空间实体化。概念模型帮助提炼和验证空间想法。每个步骤以模型带动，最后用图纸记录。第四阶运用了从体块模型到空间模型的方法。为强调集装箱建筑特定的材料与构造方法，第二学期首先进行了建筑测绘及调研。集装箱建筑的立面设计具有一定的进深，考量与其后室内空间的关系。1:20的局部剖面和立面阴影图刻画了构件的层次及波纹板的特殊肌理。最后一个阶段呼应学年开始时的坐具制作——真实尺寸的木构亭子，也可视为大型的室外家具。从非抽象的操作开始，以对建筑空间的实际搭建结束，在第一年的学习周期中再次明确了设计与制作的连贯性。

The program comprised five phases. Similar to Gu's early experiment at SEU, the program started with the making of an object, however, this time, it involved human scale, serving as a preparation phase for the container design series.

The container architecture started with a living unit, which became the basic unit for the site arrangement in the third phase. Observation was woven into the design process. A solid-void model was built to solidify the space defined by envelop and furniture. Conceptual models helped to recognise and verify the consistency of spatial ideas. Architectural drawings were required at the end stage as the precise record. Phase IV applied Kramel's method of evolution from block model to space model. In the second semester, in order to address material and construction issues, students first conducted building surveys of container buildings, associating with real-life and site conditions. The exercise of facade design concerned protruding or recessing spaces with a certain depth and their connection with the interior space behind. Partial section and shadowed facade drawing in 1:20 portray the layering of components and the unique texture formed by corrugated panels. The last phase echoes the first phase with the design and construction of a wooden pavilion— a large scale outdoor furniture. To begin with less abstract operations and end with the actual construction of architectural space accomplishes a loop in the first-year learning that addresses the consistency of design and making.

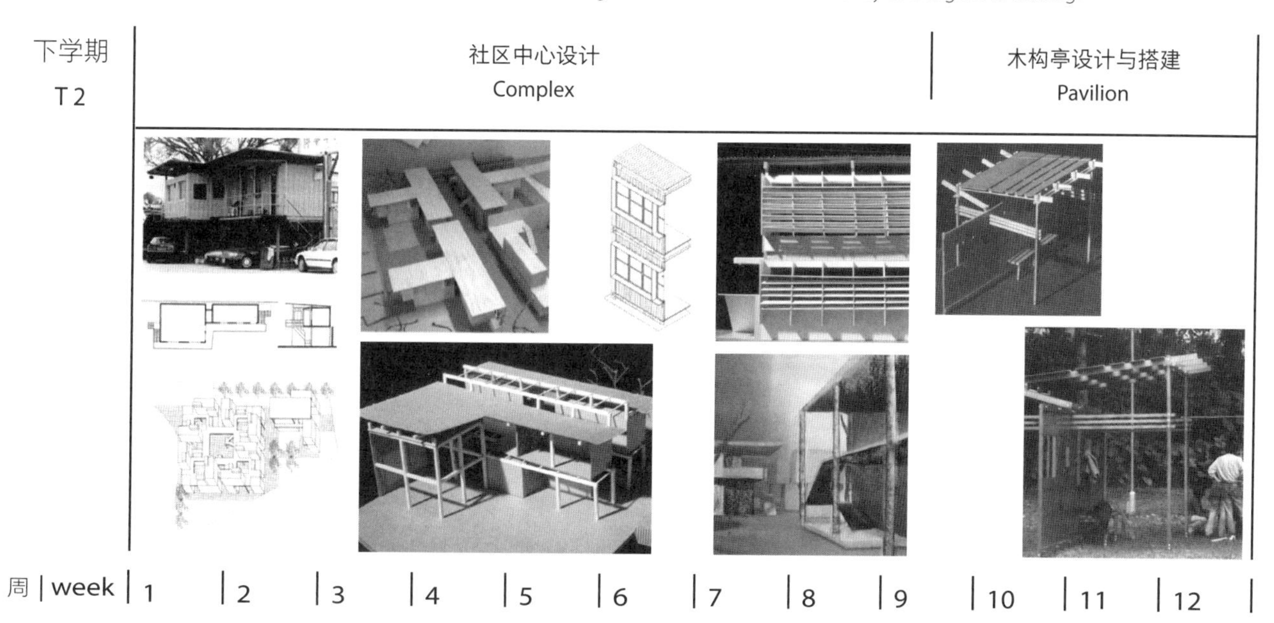

香港中文大学 2016/17 一年级 | The CUHK Year 1, 2016/17

建筑设计入门1+2 | U1 & U2 Studio

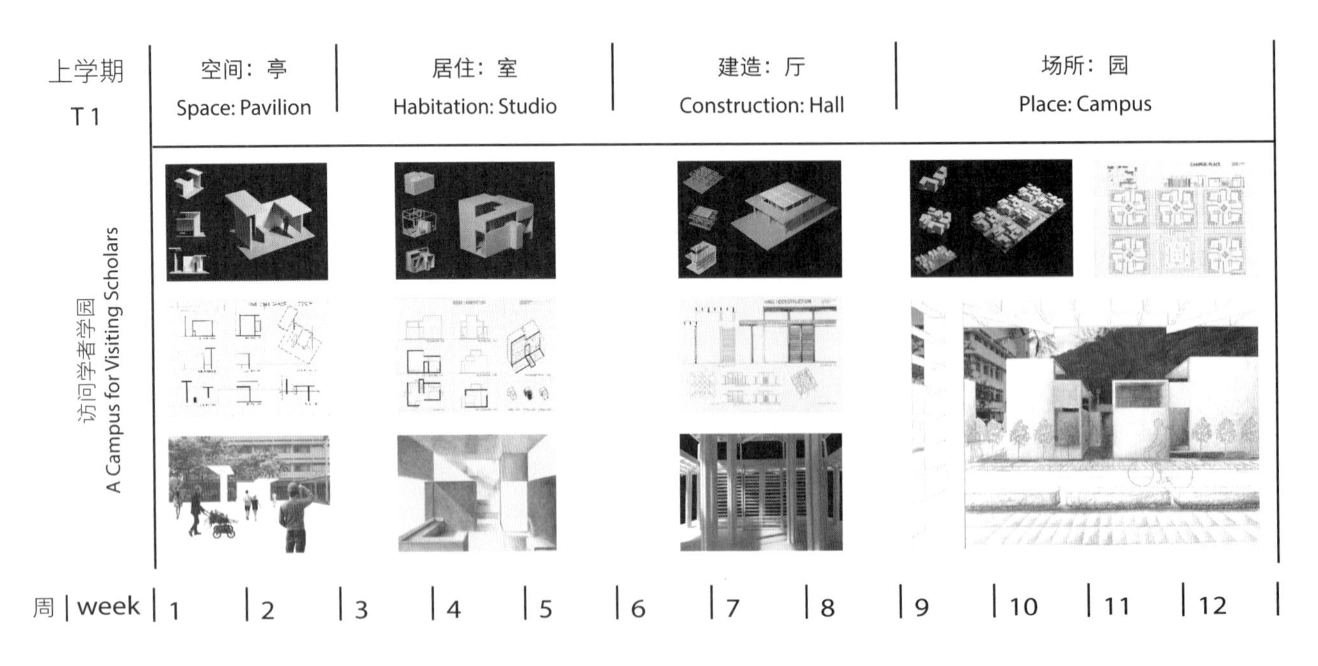

“建筑设计入门1”包括4个设计主题：展亭、居室（居住单元）、社区聚会厅和园区规划，这些主题与空间、居住、建造和城市主题相呼应。每项任务都涉及一系列处理特定问题并训练特定表现技能的练习。前3个任务分别使用平板、体块和杆件来形成单个空间容积，这些单元将在最后一个任务的场地中再组织。“建筑设计入门2”则在“建构工作室”课程基础上强化了“设计练习化”的教案编排。

延续了以前教程的方法，设计任务和表现方法之间存在对应关系。展亭的平面图、立面图和剖面图引入了正投影。带有背景和人物图形的展亭照片拼贴使学生能够在考虑光线和角度的情况下模拟实际视线高度，拍摄模型照片。居住单元内部的一点透视以单色（用石墨或铅笔）渲染空间氛围。构造研究引入了立面渲染的投影法则及材料模拟。以诺利图方式绘制的首层总平面图表达了室内空间与场地的关系。下学期设计除综合运用这些技法以外，还初涉了计算机辅助设计。

The U1 Studio comprises four thematic design tasks: pavilion, room (living unit), hall (community gathering place), and community, which echoes the themes: space, habitation, construction, and urbanization. Each task involves a series of exercises that deal with specific issues and train particular representation skills. The first three tasks respectively use slabs, space volumes, and sticks to form single-space volumes, which will be arranged on a site in the last task. On the basis of Tectonic Studio, U2 Studio strengthens the teaching plan arrangement of "design as a sequence of exercises".

As in previous programs, there is a correspondence between design tasks and representation methods. Orthographic projection is introduced to the drawing of plan, elevation, and section of the pavilion. A photo collage of the pavilion with background and human figures enables students to take a photo of the model at simulated eye level with consideration of light and angle. The one-point perspective of unit interior practices the rendering of atmosphere in monochrome (with graphite or pencil). The construction study introduces the techniques of shadow casting and material simulation in facade rendering. A Nolli plan is used to elaborate the interior space and its relation to the site. For U2 Studio, in addition to the comprehensive application of these techniques, the students are involved in computer-aided design.

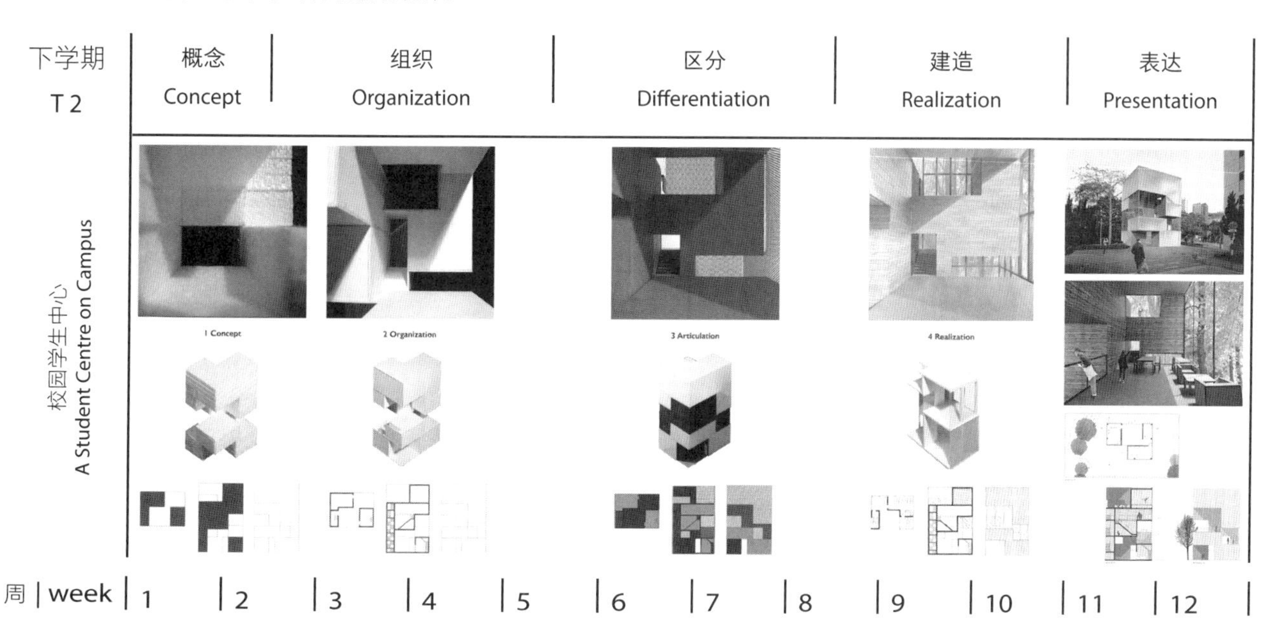

4 解读 | Interpretation

* 本文是对克莱默教学的解读和总结，由吴佳维的博士论文《建筑教学法的移植：苏黎世模式及其在中国的发展》（*The Transplantation of an Architectural Pedagogy: The Zurich Model and Its Developments in China*）中的一章改写而成。博士导师为顾大庆教授。原文分上、下篇发表于《建筑师》杂志 2018 年第 193 及 196 期。现合并收录于此，小节编号做了相应调整。

建筑师 The Architect June 2018/3 No. 193

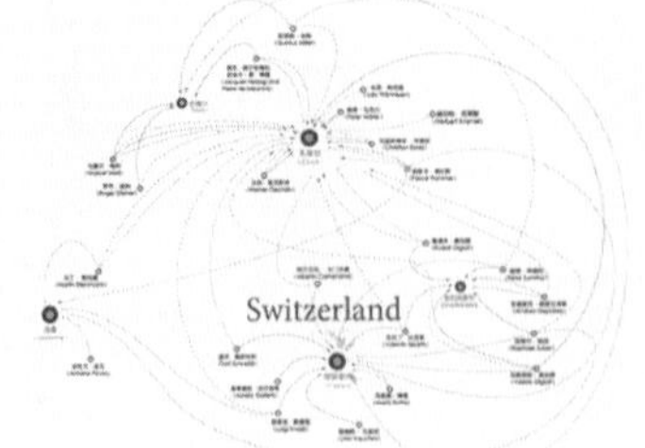

专辑：瑞士当代建筑
Special Issue: Contemporary Architecture in Switzerland

建筑师 The Architect December 2018/6 No. 196

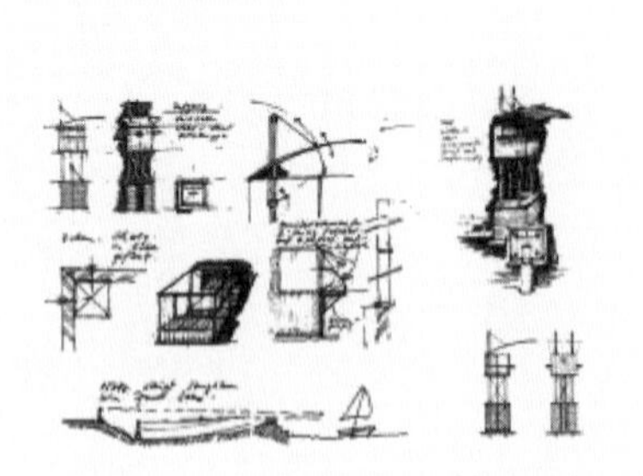

结构化设计教学之路：赫伯特·克莱默的“基础设计”教学*

The Road to Structured Design: Pedagogy of Herbert Kramel

上篇：一个教学模型的诞生

20世纪80年代中期，被公认为布扎建筑教育大本营的南京工学院（1988年后更名为“东南大学”）建筑系正在谋求适应当代的建筑教育模式。机缘巧合，一批青年教师在尝试教改的同时，陆续通过南工与苏黎世联邦理工学院（ETH）两所大学之间的“南京交流”互访计划接触到克莱默的“基础设计”教学。很快，他们发现这正是他们所寻求的建筑设计教学法。从1986年至2001年，先后来到ETH的大多数东南大学教师，如单踊、顾大庆、丁沃沃、赵辰、张雷、龚恺、冯金龙、吉国华、朱竞翔、张彤、鲍莉、胡滨等，都先后学习、工作于克莱默教席，并于归国后将这些心得与方法运用在设计教学改革中。随着其中一些教师任教于其他院校，“基础设计”的核心思想和方法进而很大程度上影响了南京大学和香港中文大学的设计教学，使“南京交流”的意义远远超出个体范畴，成为当代中国最具影响力的教学法移植案例。

那么，是什么特质使克莱默的“基础设计”区别于南工传统教学方法及当时盛行的包豪斯构成练习？首先，必须厘清塑成这一教学模型的思想，进而还原其应对彼时彼地的教学构建过程。接着，我们将借助实例来探讨“基础设计”的核心——教案设计——如何在结构化的框架之下随着时间的推移、随着设计者认识的更新而像有机体一般不断地演化，并谈及更多操作层面的细节。

1

一、思想与原理

1. 建筑是一门学问，建筑是一种职业

1978年，克莱默（图1）在一场与瑞士建筑师彼得·斯泰格（Peter Steiger）和达兴登（Justus Dahinden）的辩论中明确地表达了他的教育观点[1]。克莱默反对斯泰格主张的大学教育之意义在于帮助公民实现自我，而认为大学的首要义务是对国家与行业输出人才。他把建筑师教育比作外科医生的培训，学校必须保证扎实的专业素养。他也反对达兴登重想象力与个性培养，轻科学和技能训练的态度，认为艺术能力的发展是基于个体经历的，而学校必须搭建起基于理性原则的共性基础[2]。

克莱默总结出4种建筑教育的途径。其一，将建筑视为艺术，旨在开启学生的创造性。这种当今大多在英语国家所奉行的方法非常依赖于教师个体的偏好。第二种则是将建筑视为一门行业，有明确的惯例、需求及技术。教师通过课程及一系列举措向学生引介其未来的职业。第三种思路视建筑为一门学问，知识主体所定义的学习过程意味着逐层递增的复杂性，而非堆砌互不关联的知识包。最后一种思路则将大学视为一种研修生活，以信息搜集与研究为首，学习的过程非常个人化，相应的教学以指导为主。“如何”取代“什么”成为首要问题，借此磨砺思维，将研究的过程体系化[3]。

克莱默将第二及第三种角色赋予他的一年级教学，即视建筑为“一门学问”及“一种职业”，旨在传授“一门专业语言”。其他两种门径则退居次位。他把统一性置于个性之上，因为“只有当每一个学生在其一年级时应对了相同的问题，才能在整个学院建立共同点（common denominator）”[4]——共性之上才可能讨论个性。

克莱默的教育观与德语国家理工科大学的宗旨一致。这或多或少与他的求学经历相关。1936年生于维也纳，他先在维也纳技术学院（Technical College of Vienna，现名维也纳技术大学，TU Wien）学习建筑工程（1950-55），后求学于维也纳美院（Academy of Fine Arts of Vienna）建筑学院（1956-59）。技术学院奉行的建筑观务实而保守，一方面认为建造应实用、坚固、经济、适应长远需求；另一方面，在克莱默读书时，柯布西耶的作品被作为建筑进程的反例来谈论。技术学院要求通过修读多个相关课程来完成学业，而在不远处的维也纳美院，则沿用了布扎的大师图房模式。因为两所大学的设定不同，随着时间推移，美院更倾向于招收已具备基本建筑技能的学生，从而逐渐具有研究生院的性质，学生在教授的指导下进行3年的专题研究。1956年，克莱默在此遇到了他建筑道路上第一位重要的前辈，导师罗兰·莱纳（Roland Rainer）。

2. 功能、场地、建造决定建筑形式

莱纳比克莱默年长26岁，也毕业于维也纳理工学院。曾旅居荷兰及德国，在德国的不同院校担任过住宅、城市设计、土地使用规划和结构工程的教授。第二次世界大战后回到奥地利，1955年执掌维也纳美院建筑学院。1956年启动的维也纳体育馆（Wiener Stadthalle）工程成为他的代表作。1958年就职维也纳城市总规划师。莱纳的视野影响了克莱默。

20世纪50年代，莱纳开始对无名建筑学感兴趣，这与西比尔·莫霍利-纳吉

2

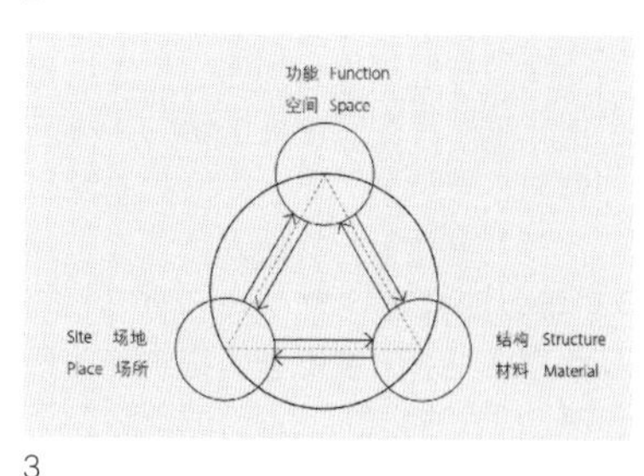

3

1. 赫伯特·克莱默教授 / 2. Breno 村部分调研成果 / 3. 建筑设计三要素

(Sibyl Moholy-Nagy)撰写《无名建筑中的自然风采》(*Native Genius in Anonymous Architecture*, 1957)及丹下健三展露对日本传统建筑的兴趣时间相当。莱纳曾游历土耳其、波斯及中国。作为他的学生和助手，克莱默为《诺伯根兰的无名建造》(*Anonymes Bauen Nordburgenland*, 1961)一书在各地拍摄照片。这触发了他对无名建筑学的兴趣。此外，他发现莱纳很注重书籍版式，有意识地统一了个人著作的开本，日后克莱默也形成了统一文本格式的习惯。

在其后的十多年间，一批关注无名建筑的书籍陆续出版，除丹下健三的《桂：日本建筑中的传统和创新》(*Katsura: Tradition and Creation in Japanese Architecture*, 1960)以外，还有本纳德·鲁道夫斯基的《没有建筑师的建筑：简明非正统建筑导论》(Bernard Rudofsky, *Architecture without Architects: A Short Introduction to Non-Pedigreed Architecture*, 1964)，阿摩斯·拉普卜特的《宅形与文化》(Amos Rapoport, *House Form and Culture*, 1969)，以及哈桑·法赛的《为穷人而建》(H. Fathy, *Architecture for the Poor*, 1973)等。20世纪60年代初，瑞士艺术与建筑理论家S·冯·慕斯(Stanislaus von Moos)把建筑分为"主要建筑"(architettua maggiore) 和"次要建筑"(architettura minore)，后者与英语世界中的"无名建筑"意义相近。

无名建筑学颂扬场所的价值，强调形式只有在文化、当地材料和建造技术的限制之下才得以发展。乡土建筑群通过共有的原型统一风貌。这些理论同时强调，建造的真实性不是建筑形式的阻力，而是生成力。受这些思潮影响，从20世纪70年代中期到20世纪90年代，与学院的研讨周结合，克莱默教席组织学生调研了瑞士南部提契诺区域的多个村庄，如马尔坎托内(Malcantone)、阿兰诺(Aranno)、皮亚诺(Madonna del Piano)等地。1974、1975年组织了在布列诺(Breno)的田野工作，学生记录了村落建筑的体量和屋顶平面、开放与私密空间、图一底关系、交通组织、当地建造技术及房屋类型，后续分析推演了主要建筑的历史演变(图2)。

克莱默得出了作用于建筑形式的三对要素：功能与空间(对建筑内部的人的活动、移动、尺度和氛围的考虑)、场地和场所(对外部环境和文脉的考虑)、建造和材料(对建筑技术和材料表达的考虑)[5](图3)。"三要素"对建筑设计具有普适性。在《无名建筑的自然风采》中，莫霍利-纳吉从三个方面回答了关于无名建筑"在哪里？""为什么？"及"如何做到？"的问题——场地与气候，形式与功能，材料与技艺[6]。同样可以在弗兰姆普敦(Kenneth Frampton)撰写的《建构文化研究》中提及的建筑环境三要素"地点、类型、建构"(the topos, the typos, and the tectonic)得到印证[7]。

3. 建筑是建造之艺术(The "art of building")

克莱默到ETH任教，实是机缘巧合。即将从维也纳美院毕业时，克莱默获得伊利诺伊大学的进修奖学金，但到美国不久，建筑研究生院的研修方式便让他觉得毫无意义，不到一年就返回欧洲。当时，维也纳的许多建筑学子对斯堪的纳维亚向往不已，克莱默也是其中一员。1961年，他本打算到北欧谋职，一位朋友建议他先到苏黎世看看。到达之后，朋友告诉他ETH建筑系的海因茨·罗纳(Heinz Ronner)教授这个学期正好缺一名兼职助教，他还可以同时在苏黎世的事务所实习。克莱默决定一试。没想

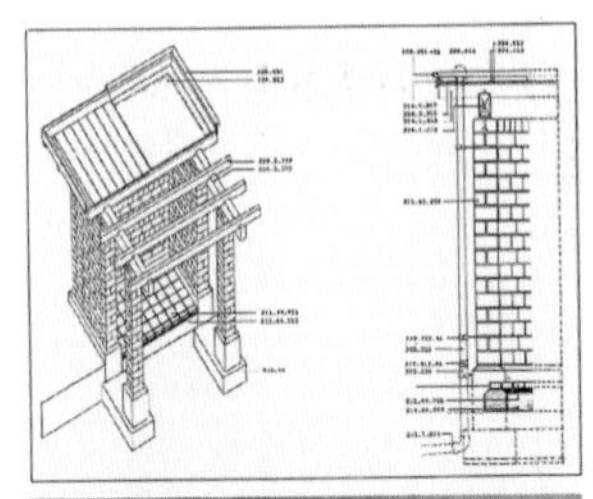

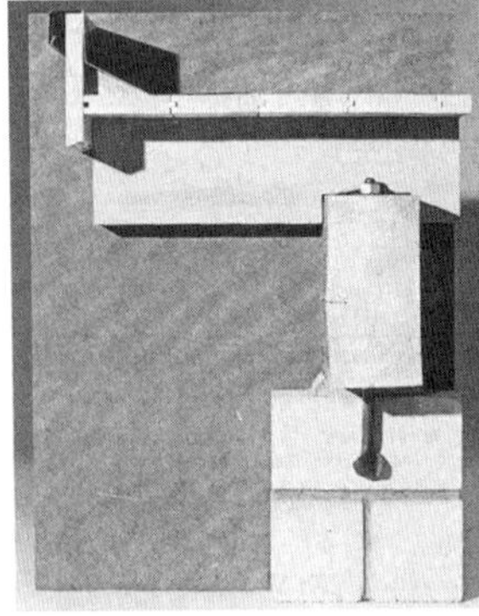

4

5

到，职业生涯便扎根在此。

当时，前“得州骑警”本纳德·霍斯利（Bernhard Hoesli）在该学院开展设计教学已逾两年，他对应理工学院的特点，结合在奥斯汀6年的教学经验和理论基础制定了新的课程结构，开发了新的教案。1961年正是新体系开始的第一年，罗纳的构造课是其中重要一环。作为助教的克莱默展现了化繁为简的教学才能。当时构造课的学时很少，必须高效地利用课堂练习时间。克莱默提议用立方体框架作为原型，讨论结构体系、承重与围合等相关问题——用一个可以不断重复叠加的单元来覆盖从节点到体系的知识跨度。罗纳非常认可，并采用了这个“使限制条件精确可控的教学模型”[8]。兼任助教期间，克莱默断断续续地任职于瑞士、奥地利及意大利的建筑事务所。1966年，克莱默再度前往美国，先后任教于佛吉尼亚理工学院（Virginia Polytechnic Institute, 1966-69）和肯塔基大学（University of Kentucky, 1969-70），直至1970年回ETH。4年的美国任教经历真正开启了克莱默的学术生涯，对其重视国际交流起了决定作用。

克莱默于1971年从罗纳手中接过构造课教席后，逐步改造了课程，将一个学年切分为4个具有独立设计题目的阶段，最终将全部设计内容统一在一块场地内。这成为1984学年克莱默同时担任设计课和构造课后将两门课进行整合的基础。他在构造课中把基础、墙身、洞口、屋面等呈现在一个轴测图里，使立方体模型具有人体尺度。学生根据轴测图中多个部件旁引出的数字编号在CRB（理性化建筑中心，Centre for the Rationalisation of Buildings，成立于1959年）的产品目录中找到真实数据，根据这些尺寸制作某节点的大尺寸拼贴剖面模型，由此强调了建造的标准化和工业化（图4）。

与罗纳一样，克莱默认为瑞士建筑的未来与工业化生产密不可分。康拉德·瓦克斯曼（Konrad Wachsmann）在他的著作《建筑的转折点》（*Wendepunkt im Bauen*, 1959. 英译本 *The Turning Point of Building: Structure and Design*, 1961）（图5）中指出，建筑工业化的影响在19世纪上半叶就开始显露；19世纪末，法国已经在钢筋混凝土和铝制构件生产上卓有成就，高层建筑则在芝加哥拔地而起，量产的芝加哥窗备受青睐。20世纪50年代幕墙的普及则标志了承重与非承重结构概念的分离。建筑工业化的时代已经到来，“社会如果无视这些新的可能性则将付出极大的代价；我们的任务是全面地识别、理解它们的潜力并驾驭之，它们是创造性活动的至上工具[9]。”在瑞士，大规模的预制工业则在20世纪60年代才兴起，逐渐在法语区出现能够生产大尺寸建筑构件的工厂。

“系统”深植于克莱默同期的许多瑞士建筑师的设计思维中——更重要的是构想出一个系统，然后才是适应环境的变体。例如当时的同事赫姆特·斯比格（Helmut Spieker）教授于20世纪60年代在德国研发了著名的玛堡建造系统（Marburg building system）。而罗纳也曾与建筑师雅各布·施灵（Jakob Schilling）、沃尔特·莫瑟（Walter Moser）一起研发了MRS模块木构系统，以127cmx127cm的木制模块满足不同的平面需求，先后建成约200套住宅。

工业化思维启发了装配部件（Kit-of-parts）的设计教学方法。1972学年的构造课，克莱默为引入“交接”问题，将练习设置为“由若干功能模块（楼梯、厨房、烟囱、卫生间）插入主体量构成居住单元”。而这个练习后来便成为“基础设计”第二个设计

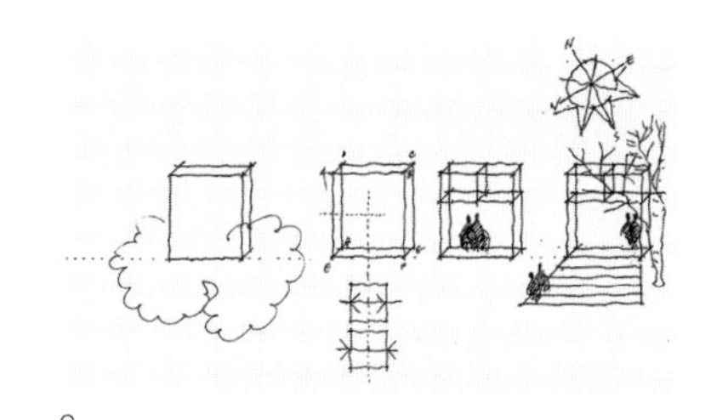

6

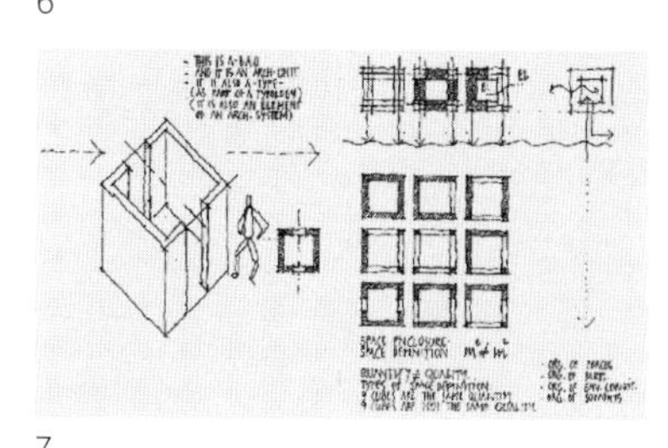

7

题目的蓝本。20世纪70年代，霍斯利的设计课中对模块的运用也明显增多。在教学中运用装配部件，使学生并非从零开始，而是在一个预设的形式系统中寻求各种可能性，这个系统本身就包含了最初级的形式概念。在此，设计不是一个推导的过程，而是直觉与批判性思考结合的试错—调节的过程——适应三要素所形成的条件与限制。其优点是众多形式变体中更容易发现和比较微差，并使庞大学生数目中的每个个体都能完成相当的设计深度。无疑，给定形式系统势必会损失一些学生自由发挥的空间，但正如克莱默认定的：共性之上才可能讨论个性，这是基础教学的使命。

立方体模型在1984年开始的“基础设计”教程中以“袖套”的形式作为总平面组织的工具，成为整个设计过程的开端。克莱默将它们所代表的边长3m或6m的立方体空间单元称为基本建筑单元——BAU（Basic Architectural Unit）（图6）。有趣的是，“Bau”在德语里正是建造的意思。BAU是一个抽象的空间容积，能够通过不同的诠释及单元组合适应多种活动。克莱默总结出9种空间类型——4个垂直界面的围合度不同，使得相同空间框架能够具有不同的空间品质（图7）。这一点类似于海杜克的“九宫格”问题，不同的是，克莱默空间基本类型的墙体并不游离于柱网之外，而暗示了不同的结构体系：框架结构、横墙体系或者实体式建造（Massivbau）。

其实，克莱默的基本空间类型本身在当时并非独创——类似图解已出现在岩本芳雄的户外空间研究和诺伯·舒尔茨的《建筑意向》（*Intentions in Architecture*, 1965）中，两者皆初版于1965年。然而，克莱默没有让空间原型止步在诠释性层面，而是把这种理解空间的方式与设计方法连结起来。借助空间分化的可能性，“基础设计”直接从三维操作切入形式和空间的讨论。至此，我们看到了结构主义建筑的影子。事实上，克莱默被阿尔多·凡·艾克（Aldo van Eyck）设计的阿姆斯特丹孤儿院所吸引。而它正是在克莱默提出将立方体作为构造课教学模型的前一年完工的——同一年，丹下健三的“东京湾规划1960”（Tokyo Bay Plan 1960）在建筑界造成不小震动——这是结构主义登场之时。克莱默认为，此前经典建筑都讲求恰到好处，多一分则繁，少一分则寡，然而孤儿院却是一个体系，增加或者减少一些单元并不影响其整体品质[10]。克莱默将它视为一种设计方法——建立严谨的设计逻辑并获得空间丰富性。

克莱默相信，BAU不但构成了建筑单体，往宏观尺度发展还可以成为应对建筑组群、集群，乃至城市尺度问题的基础。20世纪70年代，克莱默主持的工业化建筑研究深入分析了受结构主义建筑或“开放建筑”理念（Open Building）影响的多种荷兰房屋系统，如哈布瑞肯（N. John Habraken）领衔荷兰建筑师研究基金会（SAR）研发的，能够适应不同使用及变动的“支撑体”（support structure）。哈布瑞肯在《支撑结构：代替平民住房》（Supports: An Alternative to Mass Housing, 1961）中指出“支撑体不（仅仅）是建筑的骨骼，所有这些住宅的集合形成了城镇的骨骼；是一个鲜活而复杂的有机体[11]。”

回到场地组织所用的“袖套”模型，它们由卡纸带弯折而成，因为单方向穿透而具有空间指向性，从而，在用多个“袖套”组织总平面时，已经包含了结构方向和空间对位的初始考量。1985年，克莱默与他的教学团队接受委托设计非洲的摩洛哥罗校园综合体。团队用类似的方法规划场地。当时设计方与施工方一个在瑞士，一个在非

4. 构造课教学材料（上）和学生的拼贴模型，1983 / 5. 1959 年版《建筑的转折点》封面 / 6. 基本建筑单元（BAU）/ 7. 9 种空间类型

8

洲，工地上又缺乏有经验的工人，于是克莱默严格地以3m为结构模数规控建筑的定位，并适应基本构件——混凝土砌块尺寸25cmx25cmx50cm，墙、柱短边为25cm，钢构件截面为3cmx3cm——的模数[12]。后续沟通证明这一简明的控制法则行之有效，建成的校舍呈现一种节制清晰而不失变化的瑞士品质。

克莱默在构造教学中发现比例对引导设计的重要性。构造图必须对比例敏感，比例的大小决定其展露的细节，决定了思维层级。然而不同比例的图纸之间又有互承关系，例如完成1:200或1:100的图纸需要1:50或1:20的图纸提供信息，从而需要1:5图纸包含的构件信息。如此说来，节点也可以成为设计的起始，由技术客观性引发形式。这为“基础设计”中相互关联的设计练习在不同的比例中切换埋下伏笔。在这种互承关系中，介乎于“大”结构（体系）和“微”结构（节点）之间的比例显得尤为重要。克莱默偏爱1:20，因为其足够具体而又不会陷入太多细节，可以用它考虑建筑功能及其与场地的关系。

4. 设计是不断反馈的过程

当时霍斯利的设计教席和罗纳的构造教席办公室仅一墙之隔。因两位教授的密切合作，克莱默得以常常跟霍斯利交流。后者对克莱默的影响是不可替代的。通过霍斯利，他了解了“得州骑警”以及现代建筑的空间理论——这些从未在维也纳建筑圈谈论过的话题；在教学过程中耳濡目染，熟习了霍斯利的教学手段。事实上，克莱默认为是霍斯利将现代建筑带到了ETH。

为了“让现代建筑可教（teachable）”，霍斯利将设计教学视为培养建筑师工作方法的过程。他从建筑设计这一高度综合的赋形活动中识别出若干个工作阶段，从而把设计过程分解为一连串的分项问题，可以通过“准备练习”（Vorübung）逐一训练。同时，他自美国时期便认定高度结构化的课程是设计教学所必需。一年级基础课程（Grundkurs）第一学期的前半段由若干个时长各为一至两周的“准备练习”组成，例如1968/69学年度的练习顺序为空间定义、空间观察、使用功能分析、建筑名作分析、空间构成（复合的空间定义）、场地上的空间定义。这些练习具有明晰的给定条件、限制、操作方法；而后半段则是完整的建筑设计，将前面习得的知识和技能综合运用到一套步骤明晰的设计方法中。第二学期将这一套设计方法再次运用于一个以整个学期为跨度、空间复杂程度更高的设计任务。

克莱默视自己为霍斯利的下一代教师，但非一味秉承。克莱默观察到，学生完成场地、功能分析之后便常常陷入困境，苦于寻找一个“设计想法”。克莱默虽肯定练习作为专门化训练的效用，但质疑那些抽象的空间练习（图8）对于提高设计能力的直接意义，以及依靠分析寻找建筑形式的设计方法。此外，作为曾经的构造课教师，克莱默认为霍斯利的课程没有把结构作为积极的赋形因素。

那么，应当如何处理练习与设计（project）之间的关系？如何引导设计过程？这要从“系统论”说起。

发端于20世纪30年代的系统论在20世纪70年代初成为ETH及其他欧洲大学中的热门话题，科学家、工程师、建筑师都在谈论着各种“系统”和“结构”。系统论以卡

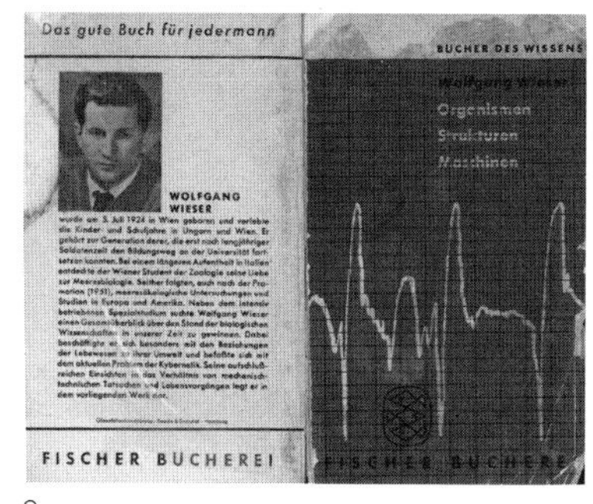

9

尔·路德维希·冯·贝塔郎非（Karl Ludwig von Bertalanffy）的“一般系统论”为代表。其基本假设是：所有系统，无论生物的还是人造的，都可被解析为相似的结构模式——元素及元素之间的联系。换言之，系统大于元素之和，强调连接机制的重要性。这一理论被用于自然与社会科学间的整合，不久又被引入设计方法运动（design method movement），致力于发现一套放之四海而皆准的设计法则，例如克里斯托弗·亚历山大的《形式合成纲要》（Christopher Alexander, *Notes on the Synthesis of Form*, 1964）。虽然设计方法运动随后就被批判与设计活动的本质相悖，但它的初衷与霍斯利追求的理性化的设计过程相符。在霍斯利列给一年级学生的参考书目里，沃夫冈·威瑟（Wolfgang Wieser）的《有机体，结构，机器：有机体的学说》（*Organismen, Strukturen, Maschinen: zu einer Lehre vom Organismus*, 1959）（图9）和赖特、勒·柯布西耶的作品集并列。霍斯利为教学创造的“使用功能系统”、“空间结构”及“空间系统”等术语就是受系统论的启发。罗纳则结合工业化思维将建筑视为构造的系统、过程的系统（建成-使用的时间性）和设计的系统（动态的设计过程）。

威瑟的书对克莱默更是影响至深——“此书对我而言是建筑师的圣经[13]。”但更吸引克莱默的是其中的“自调系统”（self-regulating systems）概念。从而，讽刺地，让他反思霍斯利的设计方法。自调系统的机制是指，通过不断的信息反馈调整在过程中所发生的偏差，直到最终达成目标。同理，设计过程也是一个不断试错—纠正的回路。克莱默虽然认同霍斯利把建筑形式视为“演变”而非“发明”的结果，却反对其教学中所引导的线性设计模式。他认为设计是互动的过程，应持续地在不同的问题和比例间转化以达至平衡。因此设计成果不意味着设计的终止，而仅仅代表某个平衡点，当加入新的需求或条件，又可以发展下去。设计教程必须凸显出这种反馈机制，交互关系不但发生在局部与整体之间，也发生在场地、使用需求和建造因素之间。

《有机体，结构，机器：有机体的学说》对克莱默的影响是在设计方法层面的，而在教育层面的另外一本关键著作是A·N·怀特黑德的《教育的目的》（Alfred North Whitehead, *The Aims of Education*, 1929）。怀特黑德在20世纪初提出一种新的教育模式。他认为现实不仅仅是互不相关的物体的总和，而是多个相互定义的“过程”（process）。因此，他反对向学生灌输大量相互孤立知识的传统教学法，提倡教授其中较为关键的概念，让学生借助这些概念联系起不同范畴的知识——为学生搭建知识的脚手架，帮助他们在未来形成自己的知识体系。

20世纪60年代杰罗姆·布鲁纳（Jerome S. Bruner）在关于构想学习及建构课程的经典著作《教育的进程》（*The Process of Education*, 1977）一书中响应了这一思路——主张在幼儿教育中，观念的结构（the structure of ideas）比单纯的事实记忆重要得多。他提出螺旋式的课程安排，每个科目间隔着重复出现，从基本到复杂，并鼓励学生在过程中自行发现新的联系。几乎同时，在建筑领域，瓦克斯曼（K. Wachsmann）认为建筑院校应该应对工业化的进程做结构上的改变。他提倡把通常耗时一整年的基础课程分成几段，分散在整个训练过程中，使学生时常有机会重温这些基本练习[14]。

这些思想随着20世纪70年代教学设计（instructional design）和系统论的兴盛而广

8. L形折板的空间限定，1963学年冬季学期，练习一 / 9.《有机体，结构，机器：有机体的学说》封底和封面为传播。它们成为克莱默关于结构化（structured）基础教学的思想来源。克莱默兼授一年级设计课的第二年，建筑系举办了当时若干个教席的教学成果展，后转化为一系列A3开本的图集。克莱默教席的展览内容呈现了从构造课到设计课程的转型过程，他将后续的这本专著命名为《教学作为教案》(Die Lehre Als Programm, 1986)，在其中写道："应明确的是课程设计的目的并不在于完成一个设计项目。它是一个框架，将一系列有针对性的练习有机地联系起来。因此，我们关注的不是设计的最终结果，而是设计过程；关注的不是学生给出的答案，而是在此过程中他们面对的问题[15]。"这可以视为他对结构化设计教学的展望与宣言。"基础设计"的"结构化"具有双重意味，不单是教程的结构化，也包含对教学行为的体系化组织。

二、教学模型

克莱默担任设计课与构造课意味着每周有两整天授课时间[16]。依照教席制度，克莱默作为主持教授，负责制定教案，主持讲座；课程教学分组进行，由助教讲解任务要求、演示绘图技法、与学生讨论设计等。这种等级化系统合理地配置了人力资源，保证面对大量学生时的教学质量。

20世纪80年代中期，一年级新生人数已经从20世纪70年代的约200人增加到300人，1989年更达350人。克莱默教席因兼任两门课而拥有20至25名助教，然而当中大多数是兼职，每周在教席的时间长短不一，所以每组由1至2名助教负责25名左右的学生。人数比决定了教学过程很难以一对一的深入讨论为基础。当然，克莱默在应对悬殊的师生比方面早有经验——1971年刚执教构造课时，面对200名学生的，只有他和5名助教。

克莱默还需应对助教们教育背景各异及流动性大的问题。虽然在构造课后期助教人数已增至15人，但其中只有5人留在"基础设计"团队。其他新加入的助教来自瑞士境内的不同院校以及来自中、美、德、日等国的访问学者。1984至1996年间，在克莱默教席连续工作5年以上的助教只有14名。因此可以说，在教学中不但学生是新手，助教往往也是"新人"。详细的教案设置成为增加教学的可操作度的必要条件。

克莱默将教学视为由（最核心的）教案、教师、学生和教学环境构成的整体。如本文第一部分所述，克莱默将建筑学系的教育使命放入行业及国家的范畴考虑。同样地，他把一年级的设计教学置于学校设定的教学框架中。20世纪90年代，建筑学院的教学计划分为三个阶段：一年的基础课程，两年的核心课程以及一年的文凭课程。基础课程旨在建立对设计和建造的基本认识，然后进行一次预文凭考试。接下来的核心课程补充和加强建筑形式发展的知识。毕业需要额外一年的建筑实习。在文凭课程结束后，学生要在10周之内完成毕业设计以获得文凭[17]。

克莱默为了实现他所设定的一年级教学目标：1.将空间理解为建筑现象以及功能/空间、场地/场所、结构/材料三者之间的关系；2.掌握建筑表现的常规技能；3.能够在设计过程中综合运用以上知识和技能；4.能够通过比较及与导师和同学的讨论分析评价设计结果——把教学过程总结为定义教案、策划教案、实施教案、过程督导、教学

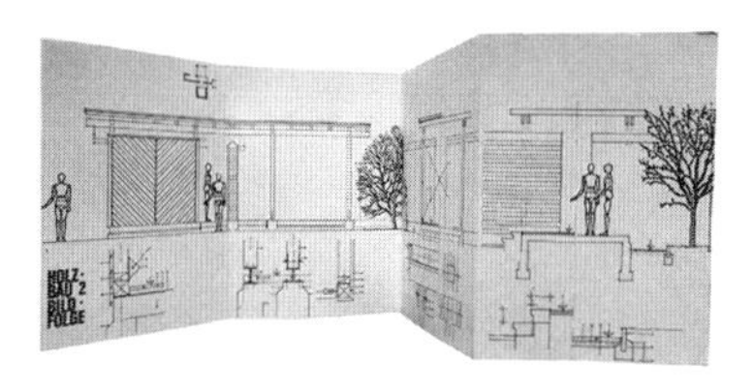

10

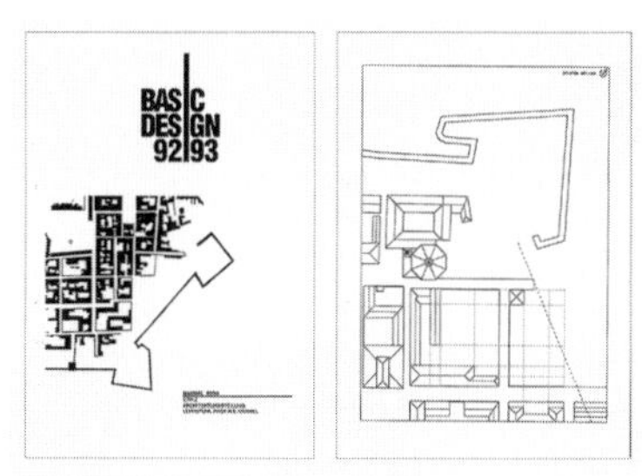

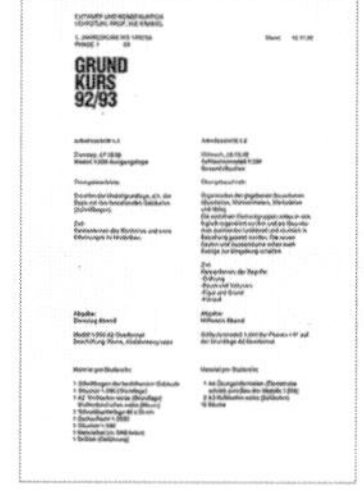

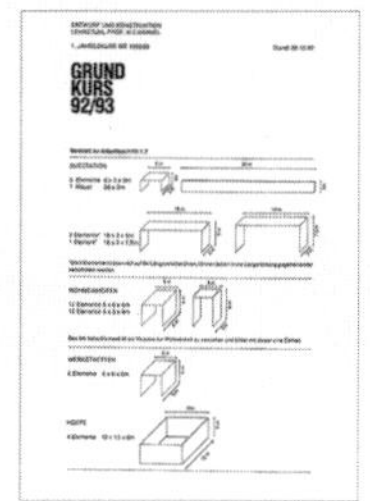

11

10. 克莱默构造课讲座所使用的"折页" / 11. "基础设计" 部分任务书，1992/93 学年

成果的评估及应用所构成的反馈系统，从而有效地传递知识、能力与技能、方法和媒介及形成经验的积累和领悟。教学过程运用了多种手段：用讲座传授一般知识，用课程作业实践方法，用研讨会探究重点问题，用参考读物扩展知识，用旅行和调研获取直观经验。

讲座从早上8点开始，约两小时。包含两大主题：周二早上由克莱默讲授现代主义建筑的范例和建筑设计原理，自1987年起，周三早上一般由讲师马库斯·吕舍（Markus Lüscher）讲授构造设计。

设计课讲座以建筑案例展开，但不是讲述现代主义建筑史，而是注重建筑案例中设计概念的发展，围绕18世纪的德国建筑、20世纪初的转折、社会住宅、与维也纳相关的建筑师如森佩尔(Gottfried Semper)、瓦格纳(Otto Wagner)、J·M·奥布里奇(Joseph Maria Olbrich)、卢斯(Adolf Loos)及约瑟夫·弗兰克(Josef Frank)等展开。介绍瑞士本土建筑时，克莱默把提契诺学派的建筑分为三类来讨论：功能主义——以塔米（Rino Tami）的作品为代表，文脉主义——以斯诺奇（Luigi Snozzi）作品为代表以及结构主义。

构造讲座的内容针对设计练习所需，所引用的建筑案例选取自各个时代。当吕舍挥舞臂膀在黑板上勾画剖面大样的时候，听课的学生们却不用着急抄图，因为教席已给每人派发一张折页形式的"构造纸带"（Konstruktionstreifen，下文称"折页"）（图10）。这是一条相当于两张A3纸短边拼接而成的构造"连环画"，上面印有经克莱默过滤掉多余细部的构造图，留空了全部的文字和尺寸标注，让学生一边听课一边根据理解填写。标准答案课后张贴在教室里，学生可以自我检验。

此外，20世纪90年代初，在克莱默的推动下[18]，系里成立了首个计算机教席，计算机数量也从20世纪80年代的几台增加到足以供3个设计大组的学生使用[19]。助教柏庭卫（Vito Bertin）作为当时计算机辅助设计（CAD）的专家向学生们授课。

讲座在十点前结束，助教们集中开会，其后再将克莱默布置的任务传达给自己的学生。设计练习基本上以天为单位完成。任务书列明了设计内容、时间安排、每一步骤的给定条件、成果及其版式要求，学期初的任务书包含整个学期所需用到的工具和材料（图11）。当时整个一年级的工作空间在HIL楼的一个长形大教室里，沿两个长边隔出各组的位置，中间通道上立展板供评图使用。练习基本上从午饭后开始，到傍晚就要集体评图。这种速战速决的方式有效地促使学生依据直觉即刻着手去做，而不是所谓的想清楚了再动手；教师能够及时给予反馈。

为了培养真正的专业工作习惯，防止学生为了最终评图才认真画分析图、绘制图纸、制作模型，克莱默参考了霍斯利"每日一课"（lesson of the day）的模式，详细规定每一个设计练习步骤的成果，而最终评图通常只需要把这些分步的成果按顺序组织起来。和一般先平面图纸再立体模型的呈现方法不同，与操作过程最直接相关的实体模型总是先于更为抽象的图纸和分析图解。与工科院校的传统一致，克莱默注重图纸的精确性超过美观，偏好轴测图多于透视图，评价学生作业时更看重其形式的逻辑发展。

在"基础设计"多年的教学积累中，克莱默对教学本身形成了体系化的思考，最初呈现在1994年他指导的，由助教卡罗·马克欧文（Carol McEowen）完成的设计教学

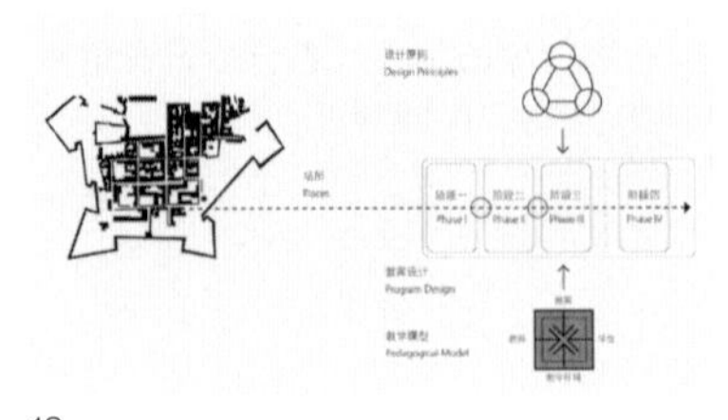

12

NDS（Nachdiplom，相当于硕士后文凭）毕业作品《构建一个操作基础》（*Postgraduate Program in Design Education 1993-1994: Building an Operational Base*）中（图12）。这本小册子结合1993学年的“基础设计”教学，从四个方面——教学模型（决定教学的条件因素），理论基础（对教学与建筑形式的理解），操作模型（教案结构）及参照环境（教案选取的场地）——全景式地描述了教学架构[20]。此后，对这个教学体系的总结归纳由后续修读该文凭的学生或助教逐步充实。像他的导师莱纳那样，克莱默统一了这一系列册子及教席制作的其他文本的规格——21cm（宽）x25cm——可以直接由A4纸裁切获得，并称之为“看上去大的最小开本[21]”。

下篇：教案的沿革与操作

回归到“基础设计”教学法最为核心的教案上，以“面”——历年教案的整体特征与沿革，“线”——一个典型学年的教案编排，“点”——一个设计问题自身的演变来呈现。

三、教案的沿革

1.“设计练习化”

从1984/85学年到1995/96学年，“基础设计”先后共有12个教案。它们延续了克莱默在构造课时期形成的4个阶段对应4个设计题目的模式，构成进阶式的设计训练。每个设计题目之下包含一系列练习，通过一连串可被单独检验的设计决策最终形成理性的工作方法。与霍斯利“先练习，后设计”的教案编排方法不同，克莱默将练习融入设计过程之中，或者说把设计的过程分解为一系列练习，即“设计练习化”[22]。从这一点来说，“基础设计”回归了工学院建筑教育的传统，即从学习的第一天开始，就着手于建筑设计本身。

练习重在训练学生在设计过程中灵活地运用模型、草图、图解及制图等媒介，借助不同的比例及媒介推进设计思维的深度。每一个题目内的练习顺序有时是可以调换的，但都实现从场地到细部构造的讨论，并且在前后题目之间形成空间复杂程度和结构构造难度的同步递增；学生得以不断回顾、演练之前所学的知识和技能。教案形成了类似脚手架的结构体。题目中变化的是内容，不变的是原理。前面题目的设计成果成为后题的参照物和限定条件。这种做法的一层考虑是为了避免学生养成在完成一次设计后把经验封存起来，以便所谓“重新开始”的不良习惯——这种习惯既不利于经验的积累，也不利于复合性设计思维的形成。

2. 4个题目

“基础设计”一个学年中前后4个题目的侧重点不同，其参考了克莱默此前构造教学多年积累所形成的设定。在构造课中，同样是为了使学生在运用过程中理解构造的基本原理，而非机械记忆构造知识，克莱默从1971年接手构造课开始，逐步将承重体系、材料交接、预制建造、不同材料的建造体系等构造问题与设计问题结合。例

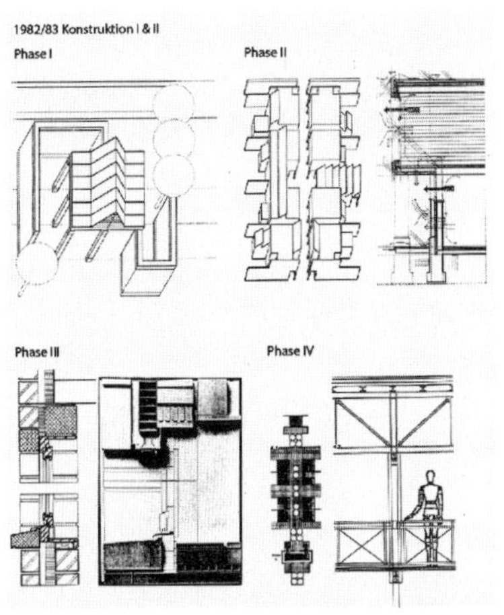

13

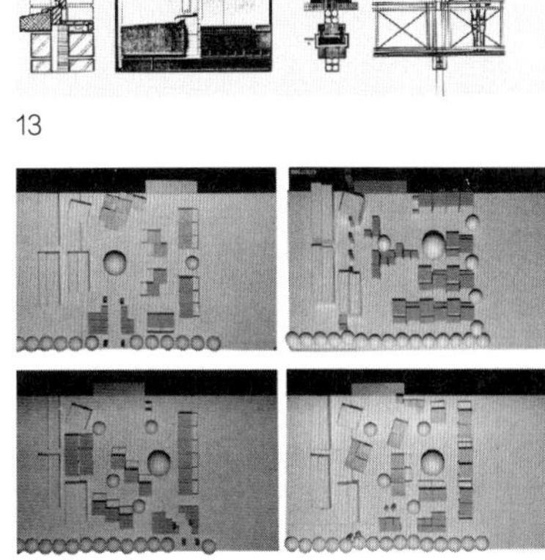

14

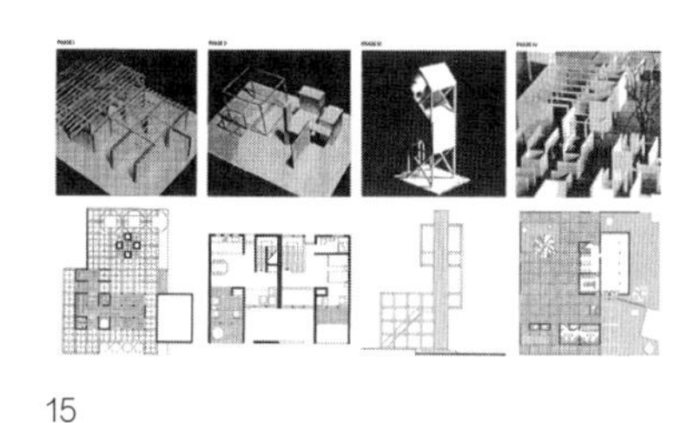

15

12. 教学结构图解（图左部为意大利古城撒比奥内塔平面图）/ 13. 1982/83 学年构造课的 4 个设计问题 / 14. 学生的总平面尝试，“基础设计”，1987/88 学年 / 15. “基础设计”的 4 个阶段，顾大庆试做，1987/88 学年

如，题目一是不设气候边界的半开敞木构候车亭或菜场售卖亭群组，题目二是有功能核及保温隔热要求的二层砖结构住宅，题目三深化题目二中的立面构造或与预制构件的运输和装配相关，题目四则涉及钢结构。在构造课后期，克莱默更有意识地将一学年中所涉建筑单体组织在同一块场地上，并使居住单元组团布置——“基础设计”教案的雏型诞生了（图13）。

3. 场地的选择

每年，设计主题在考虑建筑类别的差异性（如跨度、高度、氛围）、建筑类型为学生所熟悉以外，与当年所选的场地有关。场地曾先后选在学校所在地苏黎世、克莱默曾长时间居住的瑞士提契诺地区以及教席在研讨周组织调研过的意大利古城撒比奥内塔（Sabbioneta）。相应地，1987/88学年题为苏黎世湖畔的划艇俱乐部，1988/89学年为提契诺的小型社区，1990/91学年为古城文物修复研究所（见121-122页表）。选址周边的建筑尺度与设计内容相近有利于学生获得尺度感。如撒比奥内塔作为旧时地区性的贸易中心，古城结构自成一体，本身就成为空间观察的对象，为设计提供了许多空间线索和对位关系，而典型的意大利市场又为主题提供了一种建筑类型。

4. 教案的更新

为保持新鲜感，克莱默坚持“基础设计”教程从题目到练习的设置每年变化约1/4的内容。题目一主要解决场地组织的问题，引入基本空间概念和作图法；题目二“居住单元”是经典题目，相对不变；有一个复合空间的建筑题目，还有一个题目不确定，多年间一直在变。对教案的调整基于不同层面的考量。

首先，为教案的整体结构性进行调整。最初几年，题目一在设计开始之前设置了一个报纸拼贴练习，类似于霍斯利的“图一底”关系黑白拼贴，训练平面感知的视觉敏感度，但因其与后续教案的关联不大而取消了。20世纪90年代，经过一系列改变，前三个题目分别强调三大形式决定因素（场地、功能、建造）中的一项作为启动因素的模式愈渐清晰。必须说明的是，并不是一开始就有非常明确的教案整体结构的设想，而是在局部更新的过程中，每个设计问题产生了新的特点和关联，发现后，才促成了整体结构的明确化，这样看来，“基础设计”教案设计本身就是一个“反馈的过程”。

这些局部的更新，有根据学生表现作出的改变——如1987/88学年的木结构塔设计大大增加了学生对形式的自主性，但题目的限制还不足以抑制学生们过多的形式主义倾向，结果是发令塔的形式语言脱离于建筑群，甚至出现形式抄袭的现象，于是往后便取消了此题。有因应学生能力的变化作的调整——虽然每次面对“基础设计”都是建筑初学者，但在教学经验和成果积累到一定程度时，新一届的学生能够完成更高难度的题目，如20世纪80年代每个学年的第一个题目均维持在单层建筑，1990年代则一开始便挑战两层建筑，并加入钢结构空间桁架作为另一空间要素。随之，基本的空间组织模块也就不再是均一的3m边长的BAU（基本建筑单元），而包括代表了屋顶、坡道及功能单元的若干种模块。题目二的复杂度在20世纪90年代也有所提高，加入上人屋面的要求使其实际成为两层半的居住单元。有针对场地特点的设置，如一开始选

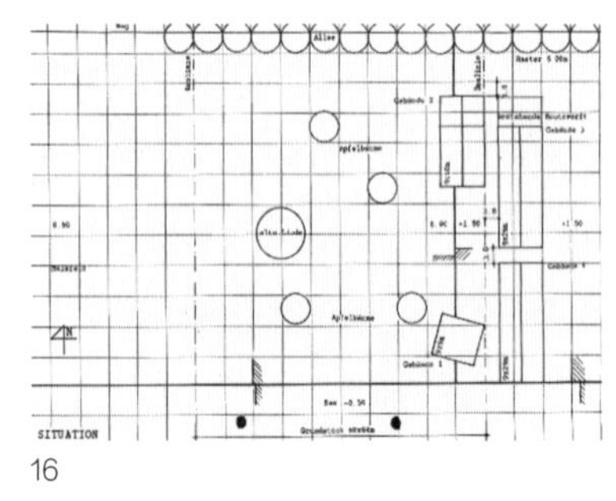

16

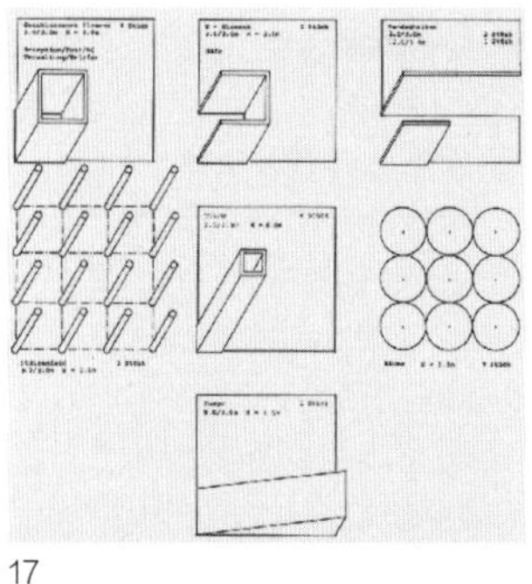
17

18

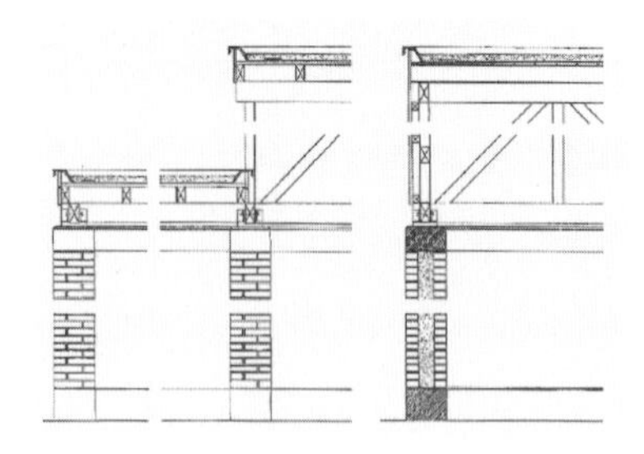
19

取的苏黎世湖畔并无太多文脉信息，而1988/89学年选址提契诺区村庄，题目三则改为外部过渡空间设计，1989/90学年开始均选址于古城撒比奥内塔，相较于早期在每一设计问题中都强调构造与立面表达之关系，后期任务设置中则有更多与城市文脉结合的考量。

最后，还有对一、二年级设计课衔接问题的考虑。早年，题目四同样具有清晰的步骤，但为了回应高年级教授关于“学生离开了严格的题目限定还能不能做设计？”的质疑，从1991学年开始改为开放性题目。撒比奥内塔提供了多样化的场地——城中心区、城墙外古塔的周边或城门交界地带等。学生自行选地、制定任务书、决定期末评图的表现形式。整一个月的过程中学生只需在周二和周三下午与助教自由讨论。

但总体而言，各个题目的时长和训练目标是相对稳定的。

题目一“场地与入口”一般长8周（包括评图），其目标是把空间作为一种建筑现象引入，以及它与人类活动的关系、场地与路径的概念，建立空间秩序及结构秩序的概念——“构思与建造是一体的”。设计内容首先是用BAU模块为本学年的整个建筑群做场地组织（图14），其二是设计场地上具有内外过渡空间性质的建筑，包括空间限定及基本的建造（墙、柱、梁、屋面等）。

题目二“居住单元”用时7周，关键性地引入了“空间域（带）”的概念。“建筑是生活的容器”这一理念通过反复涉及使用区、交通空间及活动空间的划分而强调。立面构成与其后方的室内空间之间的互动体现了建筑形式是空间组织的结果。在题目一承重结构的基础上，引入建筑围护的相关概念——保温隔热和防潮等。

题目三多年间变化最多，它只占据了第二学期开头的几周时间。一开始延续了构造课的方式，强调形式作为材料与构造的结果，突出工业化生产、预制与现场组装。如1985/86学年的主题是钢结构发令塔，构件尺寸及搭接的前后关系体现在1:5的局部模型上。1987/88学年转而更注重结构与表皮的整体关系，线性的木结构构件与部分围合的表皮形成了视觉对比。某些教案附加了案例分析练习。克莱默也曾在1990年前后把题目三作为题目二的延续（类似构造课时期），深化居住单元的构造设计。但当他意识到结构和功能的逻辑往往因显而易见而成为主导，深层的形式逻辑却容易被忽略时，就把重点从构造转移到形式和空间上来。从1991/92学年开始题目改为含有4个庭院的工作坊。

作为第一年学习的终结，题目四要激发学生综合运用所学知识和技能。主题通常是该建筑群的主建筑，如社区中心。其功能、空间序列及结构变化都较前3个题目复杂，同时应成为完善整个建筑群空间序列的重要空间节点。

1994/95学年，随着居住单元设计提前到题目一，题目二改为市场设计，“基础设计”的三大主题就变得明确：题目一关于建筑空间如何容纳活动；题目二关于如何处理大跨度和小跨度空间的关系；题目三关于如何利用庭院作为过渡空间以及建筑体量如何照应场地文脉。这同时产生了另一条发展线索——从单个建筑体量内的操作开始，接着是一群相似体量的组织，最后是若干变化体量与庭院的组合。然而，这也产生了一种远离空间讨论的倾向——恐有空间复杂性被体量复杂性替代之虞。

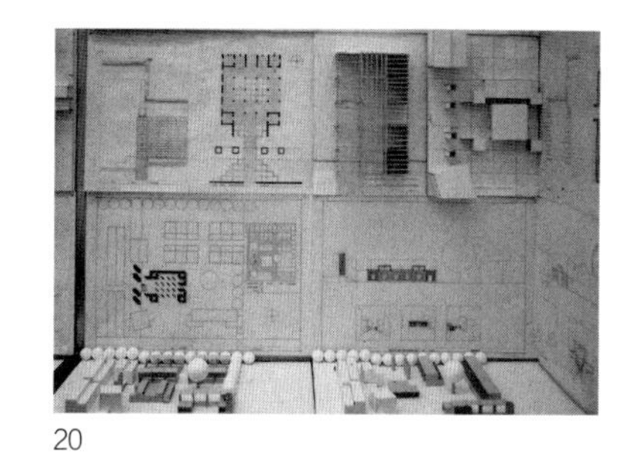

20

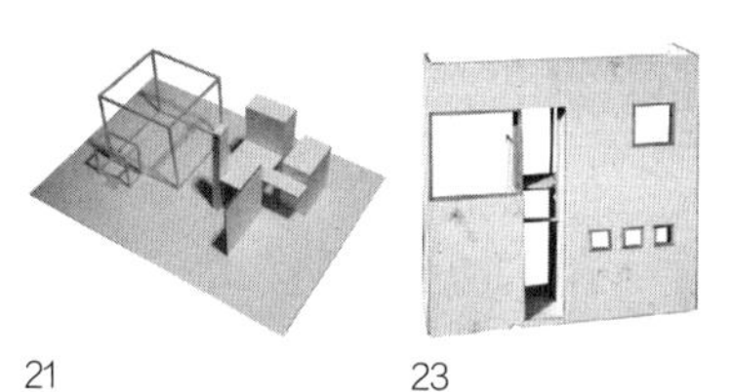

21　23

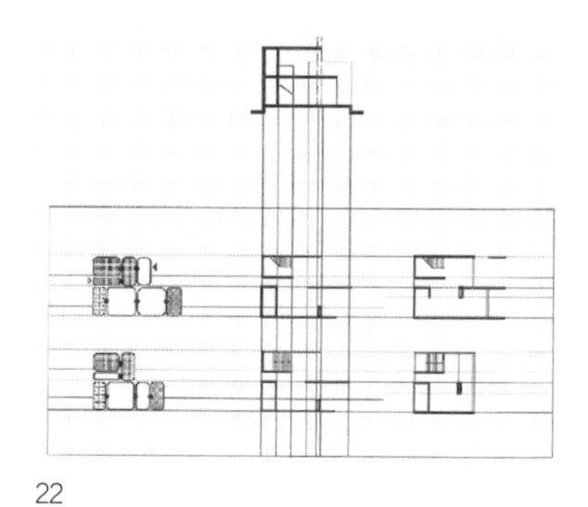

22

16. 场地与网格，“基础设计”，1987/88 学年 / 17. 题目一步骤 2，模块，1987/88 学年 / 18. 题目一步骤 6，屋顶结构模型，1:100，顾大庆试做，1987/88 学年 / 19. 题目一步骤 7，展亭剖面，1:10，顾大庆试做，1987/88 学年 / 20. 题目一评图展板，1987/88 学年 / 21. 题目二步骤 1，居住单元设计的预设组件，1:50，顾大庆试做，1987/88 学年 / 22. 题目二步骤 2，空间分析，1:100，顾大庆试做，1987/88 学年 / 23. 有空间进深的立面研究模型，1:20，顾大庆试做，1987/88 学年

四、一个典型教案

以下，以一个典型的“基础设计”教案为例，呈现“设计练习化”编织三条主线的方式——工具与思维之关系，单体与场地之关系，构思与建造之关系。

选取1987学年教案为例，因为它有较为完整的教学资料[23]并集中体现了早期教案的特点。主题是苏黎世湖畔的划艇俱乐部，4个题目分别为总平面与入口设计、住宅设计、发令塔及看台设计，划艇俱乐部设计（图15）。场地主体平坦，西侧有一处高差。根据模数3m调整场地尺寸为120m × 84m（1:200模型正好为A2图幅），红线内尺寸为60m × 84m（图16）。这是模块化操作的基础。

题目一：总平面组织与入口展亭设计[24]

首先，学生从教席已事先准备好的纸样上裁剪、弯折和粘贴快速地制作场地上现有的建筑并完成场地模型(1:200)，并根据设定用蓝色纸带折成若干“袖筒”模型。袖筒所具有的方向性和结构暗示使设计之初即要考虑与现有建筑的结构性对位。建筑群组织对于新生似乎是不可能的任务，然而，其本意就不是要产生一个成熟合理的方案，而是建立与教师讨论的基础，其结果日后还将调整。学生通过各种尝试来理解建筑和场地的图底关系，形成不同的功能区域，在入口、住宅群和公共建筑之间形成空间序列。

入口展亭设计在A3大的卡纸板上进行（1:100），对应于不同空间需求，具有不同围合度的无顶盖模块代替了此前均一的袖筒模型（图17）。平板与U形屋顶模块的加入进一步定义了下方的空间属性及整体体量。

在“基础设计”教案中，模型是思维工具，图纸则更多是一种记录。总平面图(1:200)在入口部分特别标示出墙柱的轴测投影，其余部分则只绘出轮廓线。而平面图(1:100)除表达了墙厚（40cm）、家具（空间的功能）及铺地之外，成为结构设计的底图——用三种颜色标示主要结构、次要结构及非承重结构的预设排布，U形屋顶所代表的区域转化为桁架。以同比例制作新的结构模型以检验稳定性并考虑屋架的光影效果。在实体模型上用纵向环绕模型的红线来呈现剖面的概念（图18）。学生一面对照模型，一面运用构造讲座中的木构屋顶细部、节点及砖石结构的知识绘制剖面图(1:10)（图19），构造课提供的“折页”起了关键作用。在此，最初的抽象元素转化为实在的建筑构件，“构思”和“建造”勾连起来了。

利用总平面图（1:500）区分软、硬质铺地，表达外部空间区域、绿化、动线等。根据这一反馈重新研究平面及构造节点。最后，从总平面模型上选择一个断面，设计展亭的立面，以考虑入口与其他建筑的形式关系。

阶段一评图时，学生只需把过程中完成的模型和图纸按一定的顺序钉在展板上。仅当最后的方案与前面有较大出入时，才需要重新制作前面的内容（图20）。后面的几个阶段由于正图要求更高，因而需要额外时间输出成果。

题目二：住宅设计

住宅设计包括单元设计以及它们在场地中的组织。用组件模块定义一个供两人居住的空间（1:50模型）（图21）。刚开始，学生往往会先将所有模块靠边角放置以让出中

24　25

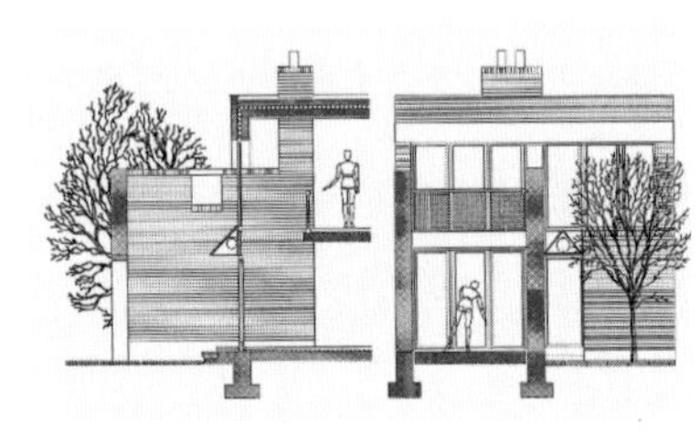

26

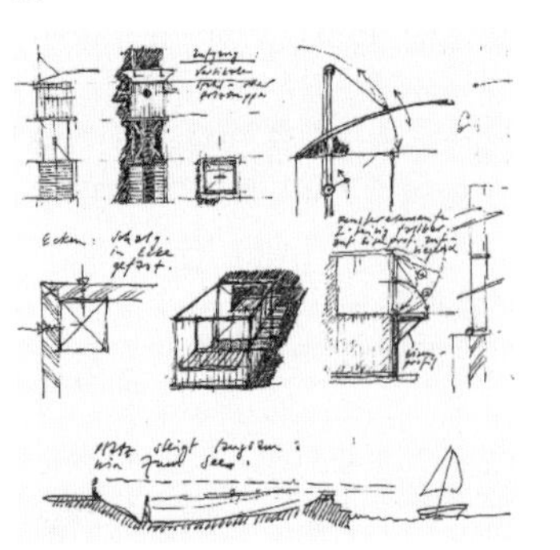

27

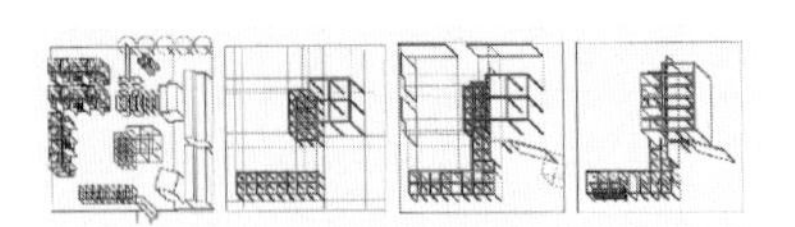

28

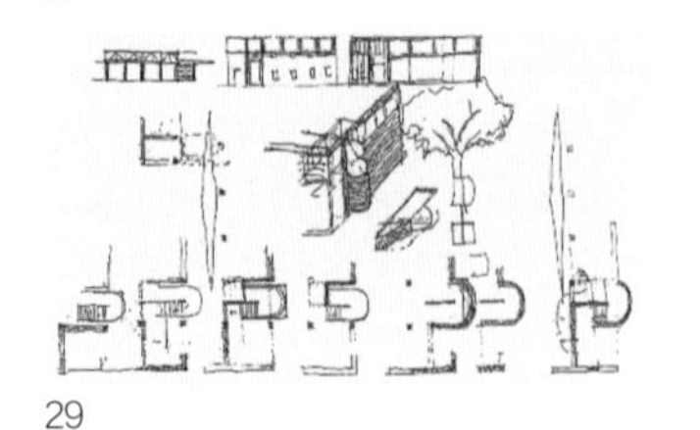

29

央的空间，但他们会发现这样并不能有效区分使用区域，反而令驻留空间受到交通空间的干扰；反之，如果敢于把某些体块（如烟囱）置于框架中央，其他体块再与之对位，便能够暗示出某些或宽或窄的带状空间，由此在框架内划分出若干区域，使空间具有层次。空间带同时反映在平面和剖面上。在移动功能块的过程中，学生的注意力从模块本身的形体表现力转移到它们所定义出的空间上，在发现空间秩序的同时产生了各自的设计概念。模块间的对位亦具有结构合理性。

随后，“空间分析图”（1:100）揭示抽象模型的空间结构，其实是对操作结果的检视，表明了模块对位与空间的因果关系（图22）。以模型组合尝试联排住宅的形式，行列两端如何收尾将影响立面设计。

接下来，设计就在立面—单元平面—场地—单元平面之间跳转。单元立面设计再次运用了模块，聚焦于立面构成及其后1.5m进深的空间（1:20），同时满足内部的使用需求、联排住宅的重复性和视觉比例（图23）。平面深化以一对镜像的单元为基础（1:50），墙厚统一为15cm，家具验证、强化了原本的空间朝向和比例（图24）。在场地组织中，单体回归到空间定义元素的角色上：围合、并排或形成组群。跳回到单元平面（1:50）时，不同部位墙体厚度的具体化引入了建筑外围护的讨论——38cm厚的多层保温外墙，15cm厚的内部承重墙和12cm厚的隔墙。在绘制平面图的过程中熟悉建筑制图规范，根据结构和构造要求调整墙、柱、开洞的位置，处理不同类型的墙体及窗口的交接[25]（图25）。围合体的构造进一步细化，沿着居住单元最有代表性的内部空间进行剖切，剖面图表达诸如单元入口和起居室等的空间关系，以及各个建筑部件的真实尺寸和层叠关系（图26）。

最后，总平面（1:200）表达了住宅建筑群和前面已经完成的入口亭子的首层平面，处理其附近的外部空间。

题目三：码头、发令塔及看台设计

通过一个小型木结构设计发掘结构体的塑形潜力。根据划艇训练比赛的功能需求，组织码头（6mx24m）、比赛发令塔（3mx3mx9m）及看台，考虑划艇运动涉及的系列行为和事件，如划艇的搬运、入水、登艇及入水方式，以及看台的视线等。以大比例模型（1:50）设计发令塔的结构体系，考虑其稳定性、木材的交接以及表皮（木板）和骨架的关系。由于较大的设计灵活性，除制作模型外，设计草图成为教学的一个重要内容（图27）。像题目一、二那样，设计反过来影响了场地的再组织。

题目四：划艇俱乐部设计

俱乐部包含赛艇存放、俱乐部及咖啡吧空间。作为学年最后一个设计，这个题目的难点在于必须通盘考量其周边已经呈现的文脉及其在整个场地中的角色，但操作的可能性也最多。设计的推进方式是泡泡图（bubble diagram）—体块模型（block model）—结构研究模型（structural model）—建筑模型（architectural model）（图28）。而相应的模型制作方式也从代表大空间（边长6m的立方体）和小空间（边长3m）的两种“门”型卡纸板，转化为代表混凝土结构构架和填充墙的厚纸板和木棒。在转变的过程中调整柱子的位置，异化均质的空间结构以强调空间尺度及引导的方向。经过一个学年的

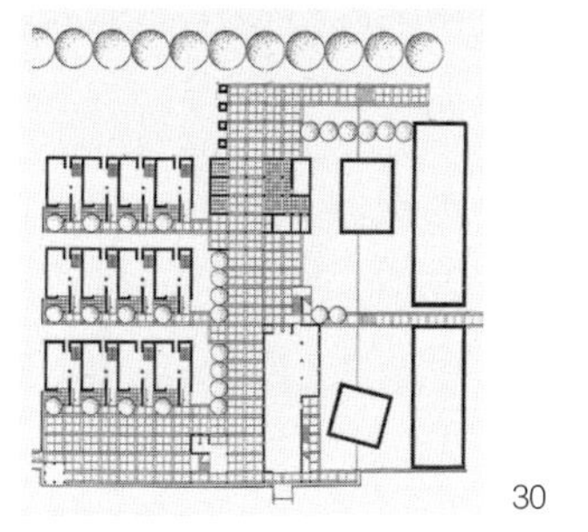

30

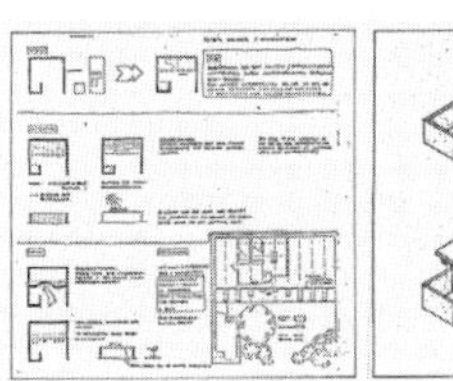
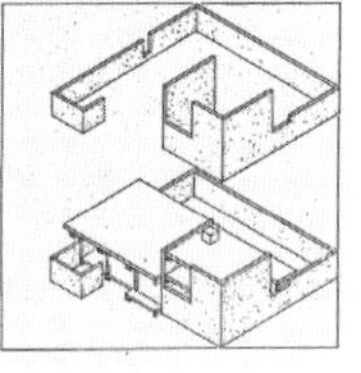

31

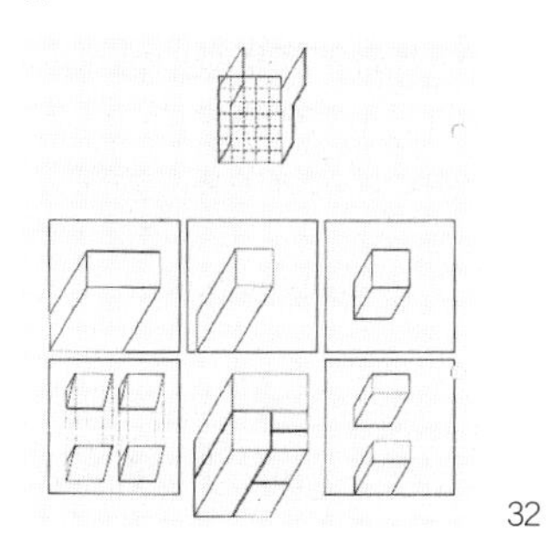

32

24. 题目二步骤 5，单层墙住宅平面，1:50，顾大庆试做，1987/88 学年 / 25. 题目二步骤 7，墙体厚度具体化的居住单元平面，1:50，顾大庆试做，1987/88 学年 / 26. 题目二步骤 8，居住单元剖面，1:20，顾大庆试做，1987/88 学年 / 27. 题目三步骤 2，瞭望塔。学生设计草图，1987/88 学年 / 28. 题目四，设计发展示意，顾大庆试做，1987/88 学年 / 29. 题目四，学生草图，1987/88 学年 / 30. 题目四，首层总平面，学生作业，1987/88 学年 /31. 题目二，1973/74 学年构造课 / 32. 题目二给定条件，1981/82 学年构造课

训练，有设计意识的学生已经能够自觉地推敲建筑元素之间（如楼梯间与外墙）的进退、连续或脱离的关系（图29）。最后回到场地，表达整个场地的首层平面，形成从入口到主建筑的空间序列（图30）。

五、“居住单元”设计问题的发展

作为“基础设计”中最经典，最稳定的一个设计问题，“居住单元”成功地运用模块奠定了形式语言的基础，并完成从构思到构造的转化（from conception to construction）。在历年的教案中，组件包括了一些基本的设定：一个主要的空间体量，1至2个附加的小体量，置入的元素有不同形态的功能块（烟囱、楼梯、厨房、浴室）、板片（作为二层楼板的水平板，作为隔墙的垂直板）。而要使这些模块在学生操作的过程中，激发前文所述的“空间带”意识，则取决于题目对部件从数量到尺寸的设定——模块的尺寸有共同的模数，其体积满足了最低限度的功能要求，而它们的体积之和仅占框架容积的不到1/3，使内部有很多组合的可能。

本篇开头提过，“基础设计”几个设计问题的雏形来自于构造课，那么，题目的设置是如何从以构造问题为先导转变为以形式逻辑为先导的呢？在多年间演化的推动力是什么？

构造课时期的设计问题二“居住单元”的主旨是在问题一“承重系统”的基础上讨论“非承重系统”及“部分与整体”的问题。早期以“给定（无结构布置的）住宅平面与院墙结合”为题温习承重系统之布置并考察室内外之构造处理（图31），后期以“一对山墙所限体量内的功能空间之组合”讨论各部位的材料交接问题（图32）——这一时期已经产生了联排住宅组合的设定。这些设定在“基础设计”初期有明显的投射——山墙与前院院墙的结合使得“单元”具有朝向性，与“袖筒”模型相应（图33）。而这时主体平面也不一定是正方形，虽然平面组织暗含了小尺寸的方形模数。

即便与构造课时期有易见的参照关系，“基础设计”问题二的模块设定却与前者有决定性的不同。本节开头所描述的那些可以彼此相离放置而又相互呼应的模块，产生了早前“功能空间堆积”所不具备的、能够被多重解读的“空间带”——这是从简单的功能—结构—构造问题向空间问题的转变。

从1985/86学年开始，为便于“袖筒”模型在方格网场地上排布总平面，居住单元的主体量也从此改为边长6m的正立方体。从第二年开始，主体量改为以空间框架表示，如此，与边长3m的附加体量、板片等分别对应了不同的空间特性：框架开放，功能块闭合，板片具有方向性。与原先山墙限定相比，框架限定在单元设计阶段，平面依然有向4个方向发展的可能（图34）。其后再根据单元组合或与地形结合的方式调整，如单体、坡地、庭院或联排（图35）。

这样以立方体为起点的小住宅是有现实参照的。正方形，是众多提契诺学派建筑师平面构形的起点。克莱默于1994年年底编纂的小册子《备忘录：1950年以来的提契诺建筑——一次回顾》（*Pro Memoria: Tessiner Architektur seit 1950- ein Rückblick*）专门收录了100个独户住宅案例。立方体是书中反复出现的母题，可见于马里奥·博塔（Mario

33

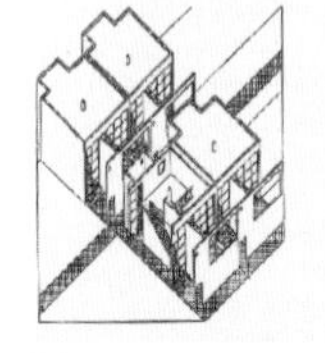

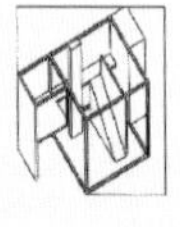

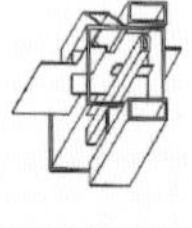

34

35

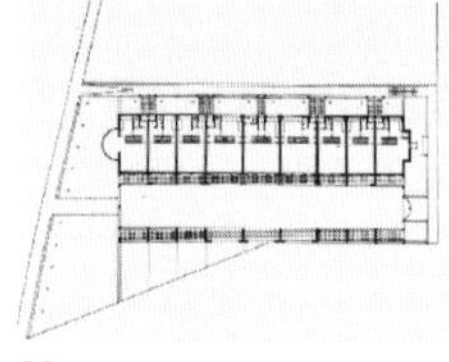

36

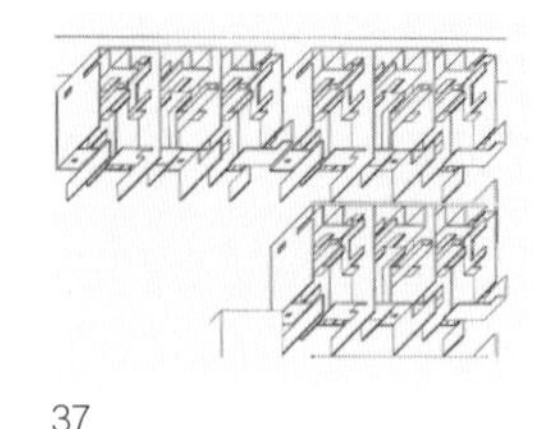

37　38

Botta）、法比欧 · 莱因哈特与布鲁诺 · 莱克林（Fabio Reinhard, Bruno Reichlin）、马里奥 · 坎比（Mario Campi）、利维奥 · 瓦契尼（Livio Vacchini）、路易吉 · 斯诺兹（Luigi Snozzi）等人的作品。克莱默有意识地向学生传递来自他们相对熟悉的文化圈的案例和对自身近期历史的认知，使其中蕴含的设计策略得到延续。如坎比与弗朗哥 · 贝西纳（Franco Pessina）设计的马萨尼奥（Massagno）联排住宅，在重复性单元排列的两个尽端，根据外部条件做了差异化的处理（图36）。这是克莱默在“基础设计”题目二居住单元群组织中不断重申“收尾处理”重要性的缘由（图37）。

1989/90学年在空间框架的前方加入了一个与主体同宽、高的“U”形构件，进一步增加了空间层（图38）。而这可以在伊凡诺 · 吉亚诺拉（Ivano Gianola）1970年设计的库尼亚斯科独户住宅（Einfamilienhaus in Cugnasco）找到高度吻合的原型（图39）（此例也被收录在《备忘录》中）；相似的还有设计师组合贝纳格、凯勒与夸利亚（Bernegger, Keller und Quaglia）于20世纪80年代初在卡斯拉诺（Caslano）建成的别墅（图40）。他们都属于新一代的提契诺建筑师。克莱默所采取的方法，不是放任学生自行寻找形式语言的参照对象，而是将这些语汇暗含在固有设定中，使学生在操作过程中体悟某种形式策略的妙处，并产生新的可能。1990/91学年开始，克莱默增高了这个“U”形空间带，使其成为单元顶层露台的遮阳并改变了立面比例（图41）。1993/94学年，又为单元增加了一层高的底座（图42），使其成为三层住宅，从而与当年的工作室建筑群体量相当。

1994/95学年，“居住单元设计”被调到阶段一，强调空间与功能的对应关系，但在形成空间的方法上出现了摇摆。在新的设定中，每一个空间模块皆对应于一个使用需求，紧靠彼此地置入套筒之中，而与主框架间仅剩一些边角空间（图43）。也就是说，“功能块与空间框架”的关系转化为“鞋盒中的多个空间模块”，这样，服务空间和被服务空间同化了，基本体量也就失去了它的空间意义，空间带和流动空间的概念大受影响。在最后几年，这个空间练习的设计方法趋同于“基础设计”的其他几个题目。

六、结论

与“得州骑警”一脉相承，“基础设计”的目的是让现代建筑“可教”，围绕空间、建造等建筑学核心问题寻求建筑形式的逻辑发展，使其与布扎传统的经验式教学以及包豪斯的抽象形式训练截然不同。在“基础设计”教学过程中，教师的角色是非个人化的，既非模仿的对象，也非简单的信息传递者，而是“认知活动教练”，利用针对性的练习把综合性的知识与技能分解为便于消化吸收的相互连贯衔接的模块，并在互动的逻辑讨论中及时给予学生反馈。通过系统化的教学组织和课程编排，相较于“先技法、后设计”的、耳濡目染式的师徒制教学，大大缩短了初学者掌握建筑设计工作方法的周期。教学过程的核心在于教案的设计，而非教师对学生个别设计的改良。在这种学习活动的终点，学生像运动员一样，可能超越其教练（老师）。设计过程以理性的方式推动形式发展，而不是在某个意象或形式模仿的基础上迎合功能需求。设计的结果是可预期、可评判的。反对极端的功能主义，主张形式概念的建立及实现

39

40

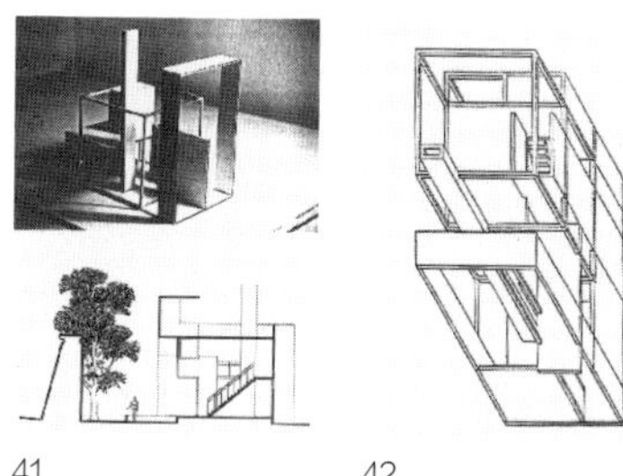

41 42

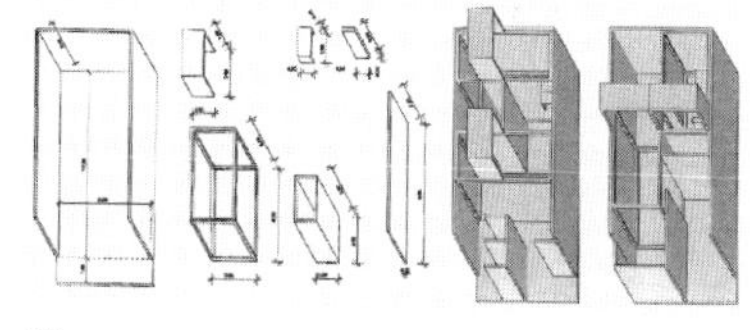

43

33. 题目二，1984/85 学年构造课 / 34. 左起 1985/86，1986/87，1987/88 学年的“居住单元”设计 / 35. 居住单元根据外部条件进行的变体，1987/88 学年 / 36. 马萨尼奥联排住宅，卢加诺，1985 / 37. “住宅设计”，学生作业，1988/89 学年 / 38. “居住单元设计”，学生作业，1989/90 学年 / 39. 库尼亚斯科独户住宅，提契诺区，1970 / 40. 卡斯拉诺别墅，1982 / 41. “居住单元设计”，学生作业，1991/92 学年 / 42. “居住单元设计”，学生作业，“基础设计”，1993/94 学年 / 43. “居住单元设计”，给定模块与学生作业，1995/96 学年

才是建筑设计的核心。直接着手于三维空间操作，而非始于平面图解。在获得空间形式复杂度的同时，保持了建造的清晰性以及与环境的连续性。

如何切分问题则很有技巧，克莱默的方式是基于设计过程中的阶段性问题，而非分项的技能训练，唯此才能使学习者自如地在不同情境下运用相应的知识与技能，最终掌握一套思维方式。对于建筑设计的入门教学，虽然知识、规范和技法的习得很重要，但更重要的是建筑观的树立及工作方法的培养，这才是最根本的专业素养。同时，题目的设置要求教案的设计者自身具备良好的形式感和观察力。

“基础设计”探寻理性化赋形的过程吸收了当时建筑界、科学界的多种观念、理论和思潮。正如“得州骑警”将现代主义建筑大师的作品作为参照，克莱默则继续引入了荷兰结构主义和提契诺学派建筑，在学生作品中能明显读到后者的形式语言；工业化系统思维所启发的装配部件操作方式规限了形式语言，克莱默以此回应了20世纪80年代他在美国目睹的正在兴起的后现代主义符号化倾向——这些是“基础设计”所包含的时间性因素。另一方面，正如“得州骑警”的贡献不在于开启了以格式塔原理阅读空间的方式，而在于将其变为教学中可以操作的设计工具那样，克莱默的贡献也不在于构建某种完备的建筑学或教育学理论，而在于将它们融汇于行之有效的设计训练模式之中——而这是超越时间的。

这种模式出现在后继的不同教学环境中。1996年克莱默前助教马克·安吉利尔（Marc Angélil）接手后的一年级设计教学虽然在教案内容上大相径庭，却延续了结构化的教程编排。虽然克莱默本人从未在华做系统性的讲学，但参与“南京交流”的东南大学教师们却把这套方法带回中国。在南京，丁沃沃、赵辰和张雷于20世纪90年代在东南大学二年级的设计课试验，赵辰、吉国华等在2000年左右进行的三年级课程设计改革以及张雷、吉国华在南京大学建筑研究所的研究生设计教学均显露了类似的结构模式。顾大庆与柏庭卫在香港中文大学以严谨的结构化方式组织建筑设计入门教学，从集装箱建筑主题（1994-2001）、建构工作室（2001-2009）到近年以装配部件操作为基础的上学期设计课程（U1 Studio），以空间定义要素形态操作启动的下学期设计课程（U2 Studio），均延续了相同的教学法原则。

可见，“基础设计”的教学组织和教案结构的现实意义比具体的练习设置更为重大。教案的结构是骨骼，它提供了通过系统化的学习方法掌握作为一门学问和一种技能的建筑设计的途径，而具体的内容则是可以不断更新的皮肉。当今的问题不仅是如何使现代建筑可教，而更具体地表现为如何使当代建筑可教，如何解释不断涌现的新的建筑现象，如何从中提炼形式策略和设计方法，如何使学生应对不同的文脉和需求，灵活运用这些策略和方法。而我们依然可以借助“基础设计”这副骨骼，以结构化、系统化的方式回应这些问题。

注释

[1] Justus Dahinden, Herbert Kramel, Peter Steiger. "Architektur im Gespräch: zur Ausbildung" [J]Bauen + Wohnen: Internationale Zeitschrift, 1978, 32(1):5-8.

[2] Herbert Kramel. Die Lehre als Programm: Grundkurs 1985 =Teaching as Program: Basic Design 1985[M]. Zürich: ETH Zürich, Institut für Geschichte und Theorie der Architektur gta, 1985:11.

[3] Herbert Kramel, Tracy Quoidbach. A Structural Approach to Basic Architectural Design: The Zurich Model 1985-95[M]. Zürich: ETH Zürich, Architekturabteilung, Lehrstuhl Kramel, 1997: 3.

[4] Herbert Kramel. Die Lehre als Programm: Grundkurs 1985 =Teaching as Program: Basic Design 1985[M]. Zürich: ETH Zürich, Institut für Geschichte und Theorie der Architektur gta, 1985:15.

[5] 相比之下，霍斯利对建筑的理解则更为复杂，他认为建筑是存在于具体的场所与时间中的"使用—空间—建造—形式"的整体系统。

[6] Sibyl Moholy-Nagy. Native Genius in Anonymous Architecture[M]. New York: Horizon Press, 1957: 31.

[7] Kenneth Frampton, John Cava. Studies in Tectonic Culture: The Poetics of Construction in Nineteenth and Twentieth Century Architecture[M]. Cambridge, Mass.: MIT Press, 1995: 2.

[8] Ulrich, Pfammatter."Die Lehrtätigkeit an der ETH Zürich 1961-1991"[A]. Heinz Ronner[C]. Zürich: ETH Zürich, Architekturabteilung, 1991:7.

[9] Konrad Wachsmann. The Turning Point of Building: Structure and Design[M]. Thomas E. Burton 译为英文. New York: Reinhold Pub. Corp, 1961: 135.

[10] 吴佳维对克莱默教授的采访，2014 年 12 月 2 日，苏黎世。

[11] Nicolaas John Habraken. Supports: An Alternative to Mass Housing[M]. London: Architectural Press, 1972: 69.

[12] 后来在实践中发现，因为墙体本身已占据 25cm 的厚度，墙中距 3m 所剩余的空间不便于放置家具，因此最后一个非洲项目（Mafia project）的结构模数改为 3.25m，为室内留出 3m 的净距离。

[13] 吴佳维对克莱默教授的采访，2015 年 2 月 23 日，苏黎世。

[14] Konrad Wachsmann. The Turning Point of Building: Structure and Design[M]. Thomas E. Burton 译为英文. New York: Reinhold Pub. Corp, 1961:203.

[15] Herbert Kramel. Die Lehre als Programm: Grundkurs 1985 =Teaching as Program: Basic Design 1985[M]. Zürich: ETH Zürich, Institut für Geschichte und Theorie der Architektur gta, 1985:15.

[16] 1986 学年及之前，一年级设计与构造课设在每周一和周三；1987 学年开始改为周二和周三。一学年分为 18 周的冬季学期（从 10 月 1 日至二月，中间包含圣诞假期，实际授课 15 周），及 13 周的夏季学期（从 4 月 1 日至 7 月），每个学期有一周是学院的公共研讨周。

[17] 瑞士大学学制原使用学时计法，并且没有本科 - 硕士阶段的划分。1999 年瑞士与欧洲其他 28 国签订《博洛尼亚宣言》后，开始致力于建立以两阶段模式为基础的高等教育体系及学分体系。2004 年开始，ETH 的学制及学时由此经历了一段转变时期。从 2007 学年至今，学制改为 3 个学年的本科课程及 1.5 个学年（3 个学期）的硕士课程。学生要获得硕士学位，必须在此基础上完成毕业设计以及 12 个月的实习（其中有 6 个月需在修读硕士课程之前完成）。

[18] 作为 20 世纪 60、70 年代设计方法运动的分支，在 20 世纪 80 年代整合了之前两个十年中产生的独立的技术和理论成果而逐步成为一个可辨识的领域。克莱默很早就预判到这种新兴设计工具的

前景，虽然自己不会用，却极力推进计算机的普及应用，并于1987至1993年担任欧洲计算机辅助设计教育协会ECAADE（Association for Education in Computer Aided Architectural Design in Europe）的主席。

[19] 通常以2至3名学生为一组使用一台电脑。

[20] Carol McEowen. Postgraduate program in Design Education 1993-1994: Building an Operational Base[D]. Zürich: ETH Zürich, Architekturabteilung, Lehrstuhl Kramel, 1994:2

[21] 原话是"...the smallest size that looks large." 吴佳维对克莱默教授的采访，2014年12月2日，苏黎世。

[22] "设计练习化"称法为顾大庆总结。

[23] 任务书、学生作业及顾大庆对试做过程的详细记录。本文展示的顾大庆试做的部分图纸为其在当时学期结束后重新以计算机绘制。

[24] 20世纪80年代，ETH一学年分为冬季和夏季两个学期，前者较长，从10月至次年2月，中间包括圣诞假期，有效学时为15周；夏季学期从4月下旬至7月，为13周；每学期的第6周为学院公共的研讨考察周。设计课和构造课分别在每周二和周三，"基础设计"因合并了这两门课，所以在连续的两天内进行，通常早上讲座、下午练习，每天完成一个步骤，而每个小组在这两天由不同的助教指导设计。

[25] 值得注意的是，单元外墙的额外厚度是在原来单层墙平面上向外扩展的，因此并未改变内部空间的尺寸，却使单元面宽大于6m，虽然这对单元立面的影响不大，但尺寸的累积会在联排住宅的总面宽中变得明显，然而，这种变化最终并没有体现在模数化的总平面上，而且首层总平面图依然以单层墙表示——这种理想化的处理方式体现了学习过程的工具性。

图片来源

1. 2015年4月摄于克莱默家中 / 2. 调研成果集，1974 / 3. 笔者重绘 / 4. Herbert Kramel. Der Konstruktionsunterricht an der Abteilung für Architektur der ETHürich[J]. Schweizer Ingenieur und Architekt, 1983,101(47):1113. / 5. 来自网络 https://www.amazon.de / 6, 7 克莱默手绘 / 8. GTA档案馆 / 9. 克莱默藏书 / 10, 11, 14-26, 28, 30, 35, 40 顾大庆提供 / 12. Carol McEowen. Postgraduate program in Design Education 1993-1994: Building an Operational Base[D]. Zürich: ETHürich, Architekturabteilung, Lehrstuhl Kramel, 1994 / 13，31，32: 克莱默构造教席教学资料 / 27, 29 David Bushnell, Herbert Kramel. Grundkurs 87: chronologische Darstellung der vier Arbeitsphasen des Grundkurses[M]. Zürich: ETHürich, Architekturabteilung, Lehrstuhl Kramel, 1988. / 33. Herbert Kramel. Die Lehre als Programm: Grundkurs 1985 =Teaching as Program: Basic Design 1985[M]. Zürich: ETH Zürich, Institut für Geschichte und Theorie der Architektur gta, 1985:27. / 34. 整理自"基础设计"教学年鉴：Grunkurs 85, Grunkurs 86, Grunkurs 87 / 36. Mario Campi, Franco Pessina, Werner Seligmann, J. Silvetti, E. Hueber, & K. Frampton. Mario Campi-Franco Pessina, Architects [M]. New York: Rizzoli, 1987. / 37, 38 ETH建筑学系年鉴 / 39. Ivano Gianola. Einfamilienhous in Cugnasco TI: Architekt Ivano Gianola, Riva San Vitale TI[J]. Das Werk, 1971(9):605. / 41, 42 克莱默"基础设计"教席教学资料 / 43. 吉国华 . Structure, Organization and Form in Basic Architectural Design[D]. Zurich: ETHürich, Architekturabteilung, Lehrstuhl Kramel, 1997.

附录 | Appendix

致谢 Acknowledgement

本项研究得到中国香港特别行政区研究资助局（项目编号：CUHK443812，2013-14）及香港中文大学2014-15年度全球卓越研究奖学金的资助。

著者曾就“基础设计”、建筑设计入门教学及“南京交流”的历史走访了相关院校机构，受访者细致的口述历史无疑对重构这一段教学发展史起了不可替代的作用。其中一些还提供了珍贵的历史档案。特此致谢：香港中文大学柏庭卫及朱竞翔教授，东南大学黄伟康、单踊、郑炘、龚恺、张彤、鲍莉教授，南京大学鲍家声、丁沃沃、赵辰、张雷、吉国华、傅筱教授，瑞士建筑师马库斯·吕舍、贝蒂·纽曼女士，皮娅·西蒙丁格博士。

感谢苏黎世联邦理工学院建筑系建筑与建造过程教席的沙赫·曼兹教授，迈克尔·艾登本兹博士和伊冯·慕斯曼女士，建筑历史及理论档案馆费里尼·瓦格纳女士；洛桑联邦理工学院皮埃尔·弗雷教授、现代建筑档案馆（ACM）伊莲娜·波亚尼奇女士和乔艾乐·诺因施万德·费尔女士；南京大学唐莲博士及刘铨博士；欧姿兰·阿颜博士，黄旭升博士，建筑师汪弢，顾田，以及比亚翠斯·詹氏-克莱默女士以各种方式对本研究的帮助。

感谢陈乐根据课程资料重绘了第31和36页上的构造图。感谢徐亮博士对版式设计的建议。

The work described in this book was cosponsored by the Research Grants Council of the Hong Kong SAR, China (Project no. CUHK443812, 2013-14) and the Global Scholarship Program for Research Excellence of The Chinese University of Hong Kong for 2014-15.

The author visited related universities and institutions for the topic on Basic Design Program, the introductory pedagogy of architectural design and the history of Nanjing Exchange. The detailed recounts by the interviewees undoubtedly make up an irreplaceable part of the reconstruction of the history of pedagogic development. In addition, some of them provided valuable documents. We are grateful to: Prof. Vito Bertin and Prof. Zhu Jingxiang from Chinese University of Hong Kong; Prof. Huang Weikang, Prof. Shan Yong, Prof. Zheng Xin, Prof. Gong Kai, Prof. Zhang Tong, Prof. Bao Li from Southeast University; Prof. Bao Jiasheng, Prof. Ding Wowo, Prof. Zhao Chen, Prof. Zhang Lei, Prof. Ji Guohua, Prof. Fu Xiao from Nanjing University; Swiss architects Markus Lüscher, Betty Neumann, and Dr. Pia Simmendinger.

We wish to thank Prof. Sacha Menz, Dr. Michael Eidenbenz, and Yvonne Moosmann from the teaching chair of Architecture and Building Process at D-Arch, ETH; Filine Wagner of the GTA Archive; Prof. Pierre Frey from EPFL; Jelena Bojanic and Joëlle Neuenschwander Feihl from Archives de la construction modern (ACM) in EPFL; Dr. Tang Lian and Dr. Liu Quan from Nanjing University; Dr. Ozlem Ayan and Dr. Huang Xusheng, architects Wang Tao and Gu Tian, and Beatrice Tschanz-Kramel for their helps.

We would also like to express our gratitude to Chen Le for redrawing the construction details on page 31 and 36. We thank Dr. Xu Liang for his advice on the overall layout.

参考书目 Bibliography

Angélil, Marc, Herbert E. Kramel, and Eva Debrunner. *Sabbioneta: eine Teilstudie im Rahmen der Untersuchungen zum Thema: das Ländliche und das Urbane*. Ueberarbeitete Fassung SS 94 / verantwortliche Mitarb. für die Ueberarbeitung 94: Eva Debrunner. Vol. 2/84 Ed.1994. Werkstattberichte / ETH Architektur-Abteilung. Zürich: ETH Zürich, Architekturabteilung, Lehrstuhl Kramel, 1994.

Bao, Li 鲍莉. *Nanjing Jiaoliu* 南京交流 =*Nanjing Exchange*, 2016.

——. *Structural Transformation: The Dar Es Salaam Run: Experiments in Design Education*. Vol. Teil 1, 160/1. [Grundkurs Basic Design / Design Methodology]. Zurich: ETH Zürich, Architekturabteilung, Lehrstuhl Kramel, 2000.

——. *The Tongli Experiment: Function and Meaning in a Chinese Town*. Vol. Teil 1, 160/2, 8/98; 11/98; 5/99. Experiments in Design Education. Zurich: ETH Zürich, Architekturabteilung, Lehrstuhl Kramel, 1999.

Blaser, Werner, and Dennis Quibell Stephenson. *Architecture 70/80 in Switzerland*. Basel: Birkhäuser, 1981.

Bushnell, David, and Herbert E. Kramel. *Grundkurs 87: chronologische Darstellung der vier Arbeitsphasen des Grundkurses 1988*. Vol. 21/88. Werkstattberichte. Zürich: ETH Zürich, Architekturabteilung, Lehrstuhl Kramel, 19.

Caragonne, Alexander. *The Texas Rangers: Notes from an Architectural Underground*. Cambridge, Mass.: MIT Press, 1995.

Coray, Monika, Peter Eberhard, and Ueli Schäfer. "Entwicklungsland China: Gespräch Mit Zwei China-Reisenden = China as a Developing Country: Conversation with Two Travellers in China." Bauen + Wohnen = Building + Home: Internationale Zeitschrift, 501073-1, 30, 1976, 7-8, 307, 1976.

Ding, Wowo, Javier Rimolo, and ETH Zürich. Abteilung für Architektur. "Postgraduate Program in Architectural Design Education 1994-1996." ETH Zürich, Architekturabteilung, Lehrstuhl Kramel, 1996.

Dongnan daxue jianzhu xueyuan 东南大学建筑学院, Gong, Kai 龚恺, Ding, Wowo 丁沃沃, and Bao, Li 鲍莉. *Dongnan daxue jianzhu xueyuan jianzhuxi ernianji shej jiaoxue yanjiu* 东南大学建筑学院建筑系二年级设计教学研究, 02: 空间的操作. 第1版. Dong nan da xue jian zhu xue yuan 80 zhou nian yuan qing xi lie cong shu. Beijing: Zhongguo jianzhu gongye chubanshe, 北京：中国建筑工业出版社，2007.

Dongnan daxue jianzhu xueyuan 东南大学建筑学院, Gong, Kai 龚恺, Gu, Daqing 顾大庆, and Shan, Yong 单踊. *Dongnan daxue jianzhu xueyuan jianzhuxi yinianji sheji jiaoxue yanjiu* 东南大学建筑学院建筑系二年级设计教学研究, 01: 设计的启蒙. 第1版. Dong nan da xue jian zhu xue yuan 80 zhou nian yuan qing xi lie cong shu. Beijing: Zhongguo jianzhu gongye chubanshe, 北京：中国建筑工业出版社，2007.

Dongnan daxue jianzhu xueyuan 东南大学建筑学院, Gong, Kai 龚恺, Zhao, Chen 赵辰, and Li, Biao 李飚. *Dongnan daxue jianzhu xueyuan jianzhuxi yinianji sheji jiaoxue yanjiu* 东南大学建筑学院建筑系二年级设计教学研究, 03: 专题 · 阶段 · 整体. 第1版. Dong nan da xue jian zhu xue yuan 80 zhou nian yuan qing xi lie cong shu. Beijing: Zhongguo jianzhu gongye chubanshe, 北京：中国建筑工业出版社，2007.

ETH Zürich. Abteilung für Architektur, and ETH Zürich Jahresausstellung der Abteilung für Architektur. *Entwurf ... an der Abteilung für Architektur, 86/87 = zum Stand der Dinge 2*. Vol. 86/87 = zum Stand der Dinge 2. Entwurf ... an der Abteilung für Architektur, 1987.

ETH Zürich. Departement Architektur, and ETH Zürich. Abteilung für Architektur. *Jahrbuch = Yearbook*. Zürich: gta Verlag, 1996.

ETH Zürich. Lehrstuhl Architektur und Konstruktion Professor Herbert Kramel. *Experiments in Design Education*. Zürich: ETH Zürich, Architekturabteilung, Lehrstuhl Kramel, 1981.

Goechnahts, Ernst, and Herbert E. Kramel. *Grundkurs 86: chronologische Darstellung der vier Arbeitsphasen des Grundkurses 1986*. Vol. 20/87. Werkstattberichte. Zürich: ETH Zürich, Architekturabteilung, Lehrstuhl Kramel, 19.

Graves, Charles, and Yijin Wen. *Basic I-IV: Überprüfung Des Einsatzes Des PC's Im Grundlagenunterricht Aufgrund Der Übungen Im Studienjahr 1985/86*. CMA Centre for Microcomputing in Architecture 3. Lehrstuhl Prof. H.E.Kramel, 1986.

Gu, Daqing. "Introductory Education in Architectural Design: The Design Studio: Its Formation and Pedagogy." ETH Zürich, Architekturabteilung, 1994.

——. "The Nanjing Exchange 1987-1988: Documentation of Basic Design Course," 1988.

Gu, Daqing 顾大庆. "Jianzhu Sheji Jichu Jiaoxue Gaige Zonghe Yanjiu Baogao 建筑设计基础改革综合研究报告." Dongnan daxue jianzhuxi, 1991.

Gu, Daqing 顾大庆, and Bertin, Vito 柏庭卫. *Introduction to Architectural Design* 建筑设计入门. 第1版. Beijing: Zhongguo jianzhu gongye chubanshe, 北京：中国建筑工业出版社, 2010.

——. *Space, Tectonics and Design*= 空间，建构与设计. 第1版. Beijing: Zhongguo jianzhu gongye chubanshe, 北京：中国建筑工业出版社, 2011.

Gu, Daqing 顾大庆, and Jianzhu sheji jichu jiaoxue xiaozu 建筑设计基础教学小组. "Jianzhu Sheji Jichu Jiaocheng 建筑设计基础教程". Dongnan daxue jianzhuxi, 东南大学建筑系, 1991.

Gubler, Fredi, Günther Zöller, and ETH Zürich.Lehrstuhl Architektur und Konstruktion Professor Herbert Kramel. *Vernacular Architecture = Anonyme Architektur*. Erweiterter Katalog. Vol. 6. Experiments in design education. Zürich: ETH Zürich, Architekturabteilung, 1982.

Gut, Roland, Kenneth Kao, Peter Steiner, and ETH Zürich. Abteilung für Architektur. *Konstruktion I & II: Dokumentation der Bildfolgen (Unterrichtshilfen im Konstruktionsunterricht) an der Abteilung für Architektur*. Vol. 16/85. Werkstattberichte. Zürich: ETH Zürich, Architekturabteilung, Lehrstuhl Kramel, 19.

Hueber, Eduard, and ETH Zürich. Abteilung für Architektur. *Architektur & Konstruktion: Dokumentation des Konstruktionsunterrichtes an der Abteilung für Architektur in den Jahren 82-83*. Vol. 11/85. Werkstattberichte. Zürich: ETH Zürich, Architekturabteilung, Lehrstuhl Kramel, 19.

Hueber, Eduard, Markus Lüscher, and ETH Zürich. Abteilung für Architektur. *Baukurs - Baustelleneinsatz - Werkseminar: Dokumentation der in den Jahren 82-85, im Rahmen der Abteilung für Architektur durchgeführten Baukurse und Werkseminare*. Vol. 9/85. Werkstattberichte. Zürich: ETH Zürich, Architekturabteilung, Lehrstuhl Kramel, 1985.

Ji, Guohua, and ETH Zürich. Abteilung für Architektur. *Structure, Organization and Form in Basic Architectural Design*. Vol. 114, 7/97. Nachdiplomstudium Architektur, ETH Zürich, Diplomarbeiten. Zurich: ETH Zürich, Architekturabteilung, Lehrstuhl Kramel, 1997.

Koeberle, Martina, and ETH Zürich. Abteilung für Architektur. "Postgraduate Program in Architectural Design Education 1996-1998." ETH Zürich, Architekturabteilung, 1998.

Kölliker, Paolo, Herbert Sägesser, ETH Zürich. Abteilung für Architektur, and Madonna del Piano Zürich). Madonna del Piano: Zusammenfassung der Studie "Madonna der [i.e. del] Piano": Katalog zur Ausstellung. Vol. 8/85. Werkstattberichte. Zürich: ETH Zürich, Architekturabteilung, Lehrstuhl Kramel, 1985.

Kramel, Herbert E. *A Structural Approach to Basic Architectural Design: The Zurich Model 1985-95*. Edited by Tracy Quoidbach. Vol. 6/97. Experiments in Design Education. ETH Zürich, Architekturabteilung, Lehrstuhl Kramel, 1997.

——. "Der Konstruktionsunterricht an Der Abteilung Für Architektur Der ETH Zürich." *Schweizer Ingenieur Und Architekt*, 162359-X, 1420-3367, 101, 1983, 47, 1111, 1983.

——. *EXCH 1984-1999: Nanjing Exchange 1984-1999*. Edited by Zina Gasser. Zurich: ETH Zürich, Architekturabteilung, Lehrstuhl Kramel, 1999.

——. *Grundkurs 85: chronologische Darstellung der vier Arbeitsphasen des Grundkurses 1985 = Chronological description of the four phases of the basic design courses, 1985 Basic Design Program*. Vol. 19/86. Werkstattberichte. Zürich: ETH Zürich, Architekturabteilung, Lehrstuhl Kramel, 19.

——. "Konstruktion und Systemdenken (Dargestellt am Beispiel der anonymen Architektur)." ETH-Bibliothek Prod, 1973.

——. *Nanjing Exchange 1984-2004*, 2011.

Kramel, Herbert E., Wowo Ding, and Lei Zhang. *Structural Transformations: The Second Year at Southeast University, Nanjing*. Edited by Guohua Ji. Experiments in Design Education. ETH Zürich, Architekturabteilung, Lehrstuhl Kramel, 1998.

Kramel, Herbert E., Entwurfsunterricht an der Architekturabteilung (3, and Zürich). *Die Lehre als Programm: Grundkurs 1985 = Basic Design 1985*. Vol. Band 3. Entwurfsunterricht an der Architekturabteilung. Zürich: ETH Zürich, Institut für Geschichte und Theorie der Architektur gta, 1985.

Leinweber, Detlef, ETH Zürich. Abteilung für Architektur, and ETH Zürich Jahresausstellung der Abteilung für Architektur. *Entwurf ... an der Abteilung für Architektur, 87/88*. Vol. 87/88. Entwurf ... an der Abteilung für Architektur. Zürich: ETH Zürich, 1988.

Li, Pang. "NDS Design Education -02/2003." ETH Zürich, 2003.

——. "On Architecture Design & Design Process: A Case Study in Kindergarten Design." ETH Zürich, 2003.

——. "On Form and Space: Search and Research in Basic Design." ETH Zürich, 2003.

——. "Tectonic Explorations: A Joint Project between Chinese University of Hongkong and ETH." ETH Zürich, 2003.

Lüscher, Markus, and Oya Atalay Franck. *Grundkurs Architektur und Konstruktion ETH Zürich 96/97*. S.l.: sn, 1997.

McEowen, Carol, and ETH Zürich. Abteilung für Architektur. "Postgraduate Program in Design Education 1993-1994: Building an Operational Base." ETH Zürich, Architekturabteilung, Lehrstuhl Kramel, 1994.

Moholy-Nagy, Sibyl. *Native Genius in Anonymous Architecture*. New York: Horizon Press, 1957.

Müller, Doris, ETH Zürich. Abteilung für Architektur, and ETH Zürich Jahresausstellung der Abteilung für Architektur. *Entwurf ... an der Architekturabteilung, 85/86*. Vol. 85/86. Zürich: ETH Zürich, Architekturabteilung, 1986.

Pan, Guxi 潘谷西, and Shan, Yong 单踊. Dongnan daxue jianzhuxi chengli qishi zhounian jinian zhuanji 东南大学建筑系成立七十周年纪念专集 Memorial symposium for 70 anniversary of the Architectural Department of Southeast University. 第1版. Beijing: Zhongguo jianzhu gongye chubanshe, 中国建筑工业出版社：北京, 1997.

Peters, Tom F. "Report of an Ongoing Experiment: Case Studies in Construction as Examples of Theoretical Approaches to Teaching Technology in Architecture." *Journal of Architectural Education* 39, no. 4 (1986): 11–21. doi:10.2307/1424791.

Zhang, Lei. "Appendices 01: Test-Set 1." ETH Zürich, Architekturabteilung, Lehrstuhl Kramel, 1993.

——. "Appendices 02: Basic Design 93 : A Documentation of the Teaching Procedure." ETH Zürich, Architekturabteilung, Lehrstuhl Kramel, 1993.

——. "Appendices 03: Flight to Prague: An Introduction to Basic Architectural Design." ETH Zürich, Architekturabteilung, Lehrstuhl Kramel, 1993.

——. "Building an Operational Base: (The BAU-Model): 1 / 2 / 3 / 4." ETH Zürich, Architekturabteilung, Lehrstuhl Kramel, 1993.

Zhang, Lei, and ETH Zürich. Abteilung für Architektur. "NDS Postgraduate Program in Design Dducation 1993-1994." ETH Zürich, Architekturabteilung, Lehrstuhl Kramel, 1993.

Zhang, Lei, and Zhaojin Ying. *Basic Design and Design Basics: Teaching Basic Design in Architecture 1992-94*. Edited by Tracy Quoidbach. Vol. 5/97. Experiments in Design Education. ETH Zürich, Architekturabteilung, Lehrstuhl Kramel, 1997.

Zhang, Tong. *The Tongli Experiment II: Function and Meaning in a Chinese Town*. Vol. Teil 2, 160/2, 8/98 ; 11/98 ; 5/99. Experiments in Design Education. Zurich: ETH Zürich, Architekturabteilung, Lehrstuhl Kramel, 1999.

作者介绍

赫伯特·克莱默　瑞士苏黎世联邦理工学院荣休教授。1955年毕业于维也纳技术学院（现名维也纳技术大学）。1959年于维也纳美院建筑学院获硕士学位。1962年起任教于苏黎世联邦理工学院建筑系。1966-70年旅美任教。1976年起担任苏黎世联邦理工学院建筑与建筑技术讲席教授。1976-78年、1990-94年任建筑系系主任。1978-80年任欧洲建筑教育协会主席。于2001年荣休。

顾大庆　东南大学建筑国际化示范学院教授。1982年于南京工学院（今南京东南大学）建筑系获学士学位，1985年获硕士学位。1994年于瑞士苏黎世联邦理工学院获博士学位。1994-19年任教于香港中文大学建筑学院，现为荣休教授。

吴佳维　南京大学建筑与城市规划学院副研究员。2008年于华南理工大学获建筑学学士学位。2011年于同济大学获建筑学硕士学位。2014-15年瑞士苏黎世联邦理工学院访问学者。2017年于香港中文大学获哲学博士学位。2017-20年香港中文大学建筑学院高级研究助理及博士后研究员。

图书在版编目（CIP）数据

基础设计 = Basic Design · Design Basics. 设计基础 : 汉英对照 / (瑞士) 赫伯特 · 克莱默, 顾大庆, 吴佳维著. -- 北京 : 中国建筑工业出版社, 2020.5（2024.8 重印）
ISBN 978-7-112-25004-2

Ⅰ. ①基… Ⅱ. ①赫… ②顾… ③吴… Ⅲ. ①建筑设计－汉、英 Ⅳ. ①TU2

中国版本图书馆CIP数据核字(2020)第051855号

策划：顾大庆
责任编辑：滕云飞　徐　纺
版面设计：吴佳维
责任校对：李美娜

基础设计 · 设计基础
Basic Design · Design Basics
[瑞士] 赫伯特 · 克莱默　顾大庆　吴佳维　著
*
中国建筑工业出版社出版、发行（北京海淀三里河路9号）
各地新华书店、建筑书店经销
北京中科印刷有限公司印刷
*
开本：889毫米×1194毫米　1/20　印张：14⅕　插页：1　字数：295千字
2020年11月第一版　2024年8月第五次印刷
定价：58.00元
ISBN 978-7-112-25004-2
（35760）

柏庭卫 · 顾大庆 · 胡佩玲 | 中国建筑工业出版社

Vernacular Contained
香港集装箱建筑

Vito Bertin · Gu Daqing · Woo Pui Leng | China Architecture & Building Press

顾大庆 · 柏庭卫 | 中国建筑工业出版社

Introduction to Architectural Design
建筑设计入门

Gu Daqing · Vito Bertin | China Architecture & Building Press

顾大庆 · 柏庭卫 著 | 中国建筑工业出版社

Space, Tectonics and Design
空间、建构与设计

Gu Daqing · Vito Bertin | China Architecture & Building Press

柏庭卫 著 | 中国建筑工业出版社

leverworks: one principle, many forms
杠作: 一个原理、多种形式

Vito Bertin | China Architecture & Building Press

[美] 布鲁斯·朗曼 徐亮 著 | 中国建筑工业出版社

Abstract Composition and Spatial Form
抽象构成与空间形式

Bruce Lonnman Xu Liang | China Architecture & Building Press

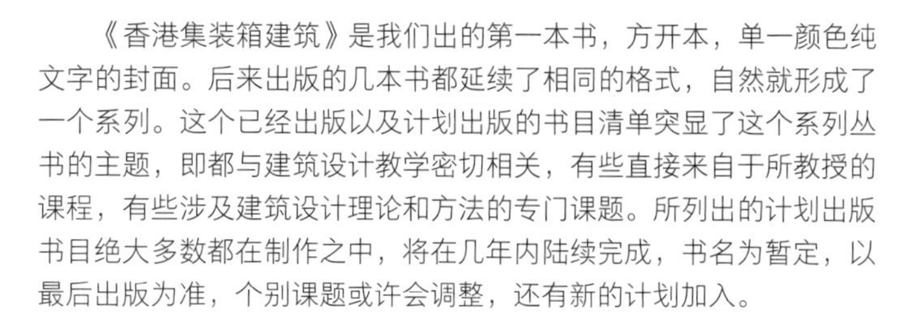

《香港集装箱建筑》是我们出的第一本书，方开本，单一颜色纯文字的封面。后来出版的几本书都延续了相同的格式，自然就形成了一个系列。这个已经出版以及计划出版的书目清单突显了这个系列丛书的主题，即都与建筑设计教学密切相关，有些直接来自于所教授的课程，有些涉及建筑设计理论和方法的专门课题。所列出的计划出版书目绝大多数都在制作之中，将在几年内陆续完成，书名为暂定，以最后出版为准，个别课题或许会调整，还有新的计划加入。

已出版：

香港集装箱建筑 柏庭卫 顾大庆 胡佩玲 2004

建筑设计入门 柏庭卫 顾大庆 2009

空间、建构与设计 柏庭卫 顾大庆 2011

杠作：一个原理、多种形式 柏庭卫 2012

基础设计 · 设计基础 赫伯特 · 克莱默 顾大庆 吴佳维 2020

抽象构成与空间形式 布鲁斯 · 朗曼 徐亮 2020

待出版：

建筑设计入门之基本问题 顾大庆

建筑设计入门之基本方法 顾大庆

观察：日常建筑学 顾大庆

体验素描与设计素描 顾大庆

设计与视知觉（新版）顾大庆

建筑分析 柏庭卫

剖碎与透明 顾大庆

勒杜：九宫格立方体 顾大庆 陈乐

空间：从绘画到建筑 顾大庆

空间：从概念到建筑 顾大庆

模型之空间操作 顾大庆 柏庭卫 朱竞翔

教学法：练习、设计与教案 顾大庆

中国建筑教育的转折点 顾大庆

香港现代建筑 顾大庆 柏庭卫 韩曼